MW01252981

Business Transformations in the Era of Digitalization

Karim Mezghani
Al Imam Mohammad Ibn Saud Islamic University, Saudi Arabia & University of Sfax, Tunisia

Wassim Aloulou
Al Imam Mohammad Ibn Saud Islamic University, Saudi Arabia & University of Sfax, Tunisia

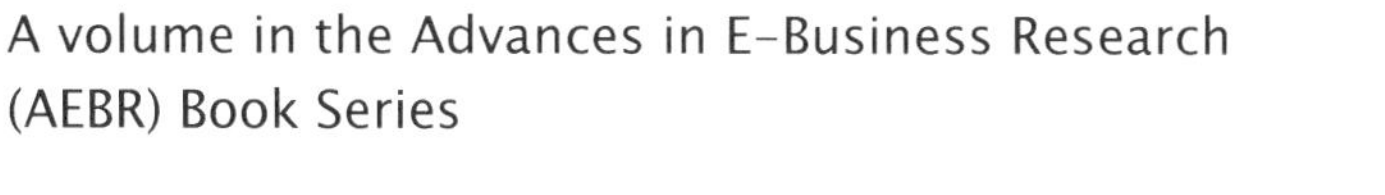

A volume in the Advances in E-Business Research (AEBR) Book Series

Published in the United States of America by
IGI Global
Business Science Reference (an imprint of IGI Global)
701 E. Chocolate Avenue
Hershey PA, USA 17033
Tel: 717-533-8845
Fax: 717-533-8661
E-mail: cust@igi-global.com
Web site: http://www.igi-global.com

Library of Congress Cataloging-in-Publication Data

Names: Mezghani, Karim, 1979- editor. | Aloulou, Wassim, 1972- editor.
Title: Business transformations in the era of digitalization / Karim Mezghani and Wassim Aloulou, editors.
Description: Hershey, PA : Business Science Reference, 2019.
Identifiers: LCCN 2018023867| ISBN 9781522572626 (hardcover) | ISBN 9781522572633 (ebook)
Subjects: LCSH: Information technology--Economic aspects. | Technological innovations--Economic aspects.
Classification: LCC HC79.I55 H3333 2019 | DDC 658.4/06072--dc23 LC record available at https://lccn.loc.gov/2018023867

This book is published in the IGI Global book series Advances in E-Business Research (AEBR) (ISSN: 1935-2700; eISSN: 1935-2719)

British Cataloguing in Publication Data
A Cataloguing in Publication record for this book is available from the British Library.

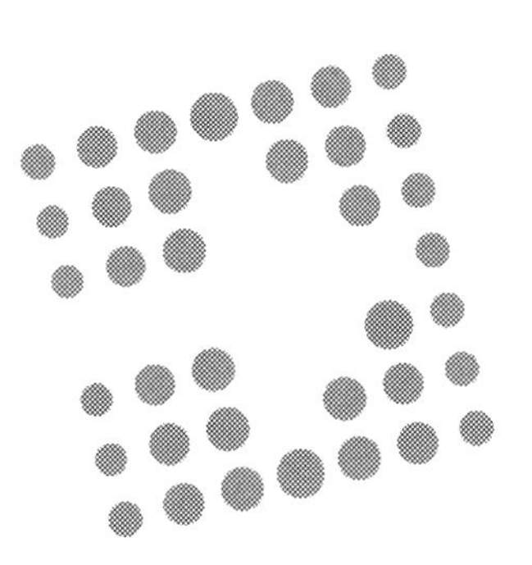

Advances in E-Business Research (AEBR) Book Series

In Lee
Western Illinois University, USA

ISSN:1935-2700
EISSN:1935-2719

Mission

Technology has played a vital role in the emergence of e-business and its applications incorporate strategies. These processes have aided in the use of electronic transactions via telecommunications networks for collaborating with business partners, buying and selling of goods and services, and customer service. Research in this field continues to develop into a wide range of topics, including marketing, psychology, information systems, accounting, economics, and computer science.

The **Advances in E-Business Research (AEBR) Book Series** provides multidisciplinary references for researchers and practitioners in this area. Instructors, researchers, and professionals interested in the most up-to-date research on the concepts, issues, applications, and trends in the e-business field will find this collection, or individual books, extremely useful. This collection contains the highest quality academic books that advance understanding of e-business and addresses the challenges faced by researchers and practitioners.

Coverage

- B2B e-marketplaces
- E-Business Management
- Interorganizational information systems
- Electronic Supply Chain Management
- E-procurement methods
- Virtual Collaboration
- Economics of e-business
- Social Network
- Trends in e-business models and technologies
- E-marketing

IGI Global is currently accepting manuscripts for publication within this series. To submit a proposal for a volume in this series, please contact our Acquisition Editors at Acquisitions@igi-global.com or visit: http://www.igi-global.com/publish/.

The Advances in E-Business Research (AEBR) Book Series (ISSN 1935-2700) is published by IGI Global, 701 E. Chocolate Avenue, Hershey, PA 17033-1240, USA, www.igi-global.com. This series is composed of titles available for purchase individually; each title is edited to be contextually exclusive from any other title within the series. For pricing and ordering information please visit http://www.igi-global.com/book-series/advances-business-research/37144. Postmaster: Send all address changes to above address.

Titles in this Series

For a list of additional titles in this series, please visit: www.igi-global.com/book-series

E-Manufacturing and E-Service Strategies in Contemporary Organizations
Norman Gwangwava (Botswana International University of Science and Technology, Botswana) and Michael Mutingi (Namibia University of Science and Technology, Namibia)
Business Science Reference • copyright 2018 • 366pp • H/C (ISBN: 9781522536284) • US $205.00 (our price)

Multi-Sided Platforms (MSPs) and Sharing Strategies in the Digital Economy Emerging Research and Opportunities
Sergey Yablonsky (St. Petersburg University, Russia)
Business Science Reference • copyright 2018 • 192pp • H/C (ISBN: 9781522554578) • US $165.00 (our price)

Crowdfunding and Sustainable Urban Development in Emerging Economics
Umar G. Benna (Ahmadu Bello University, Nigeria) and Abubakar U. Benna (Durham University, UK)
Business Science Reference • copyright 2018 • 343pp • H/C (ISBN: 9781522539520) • US $205.00 (our price)

Improving E-Commerce Web Applications Through Business Intelligence Techniques
G. Sreedhar (Rashtriya Sanskrit Vidyapeetha (Deemed University), India)
Business Science Reference • copyright 2018 • 363pp • H/C (ISBN: 9781522536468) • US $225.00 (our price)

Optimizing Current Practices in E-Services and Mobile Applications
Mehdi Khosrow-Pour, D.B.A. (Information Resources Management Association, USA)
Business Science Reference • copyright 2018 • 366pp • H/C (ISBN: 9781522550266) • US $210.00 (our price)

Mobile Platforms, Design, and Apps for Social Commerce
Jean-Éric Pelet (ESCE International Business School, Paris, France)
Business Science Reference • copyright 2017 • 411pp • H/C (ISBN: 9781522524694) • US $215.00 (our price)

The Internet of Things in the Modern Business Environment
In Lee (Western Illinois University, USA)
Business Science Reference • copyright 2017 • 340pp • H/C (ISBN: 9781522521044) • US $205.00 (our price)

Key Challenges and Opportunities in Web Entrepreneurship
Alexandru Capatina (Dunarea de Jos University of Galati, Romania) and Elisa Rancati (University of Milan – Bicocca, Italy & Beta Consulting, Italy)
Business Science Reference • copyright 2017 • 291pp • H/C (ISBN: 9781522524663) • US $205.00 (our price)

701 East Chocolate Avenue, Hershey, PA 17033, USA
Tel: 717-533-8845 x100 • Fax: 717-533-8661
E-Mail: cust@igi-global.com • www.igi-global.com

Editorial Advisory Board

Table of Contents

Section 4
Digitalization Experiences and Applications

Detailed Table of Contents

This chapter attempts to study the intentions to use cloud-based CRM applications through a combination between a Technology Acceptance Model (TAM) and a Theory of Planned Behavior (TPB). To test the different links identified in the research model, a research questionnaire was prepared and sent to marketing managers within Saudi SMEs in Saudi Arabia. A total of 41 useful questionnaires were collected. The authors opted to the structural equation modeling (SEM) using the Partial Least Squares (PLS) to analyze data. The tests are prepared with XLstat software since it integrates both factor analysis and PLS modules. Among the main statistical analyses, the authors conclude that the TPB-TAM is suitable to study cloud CRM. From a managerial perspective, the authors expect that cloud CRM is perceived with good impression and that this new technology should be implemented strongly and gradually in SMEs to improve the quality of services provided to customers and organizations.

This chapter critiques trends in enterprise resource planning (ERP) in respect to contemporary multi-organizational enterprise strategy in order to identify under-researched areas. It is based on the premise that multi-organization strategies and information systems span more than one legal company entity and are becoming increasingly important as digital Internet based systems become more prolific, and outsourcing and collaboration between companies becomes more widespread. This chapter presents a critique of literature covering theoretical, methodological and relational aspects of enterprise resource planning systems and multi-organizational enterprise strategy. The critique gives a unique perspective and highlights four major gaps in current research and points towards a trend which is referred to in this chapter as 'enterprization.' This research could help organizations make more effective use of their information and operations systems strategies when used across more than one company. It should interest researchers, teachers, IS developers and managers.

there are only few studies on digital enterprises studies emanating from the developing and emerging countries in the Middle-East and Africa continents. Based on the evidence from the literature, this paper provides an overview of digital entrepreneurship, identifies key variables that determine a successful digital entrepreneurship and then provides a conceptual model to guide the understanding of a successful digital entrepreneurship development within the context of developing and emerging economies. The paper offered some implications for digital entrepreneurs, policy makers and some other people in the business of digital entrepreneurship in Saudi Arabia.

On November 8, 2016, Government of India declared demonetization of all Rs. 500/- and Rs.1000/- currency notes towards a cashless society and create a digital India. The point of sale (PoS) and prepaid instruments are the most popular systems currently installed by merchants and service providers for receiving payments from customers. The primary focus of the study is to understand the adaptability, affordability, acceptability, and sustainability of the payments system as seen from the point of view of small merchants. A total of 221 responses were collected in Chennai. Results show that cash remains the most preferred mode for business. It is required for the working capital, payment of employee remuneration, wages, and others. With regards to the use of payment systems such as POS and prepaid instruments, awareness needs to be created of the benefits in having non-cash transactions. Improving credit worthiness and eligibility to receive loans from banks is one such benefit which would convince the merchants. However, too many systems could confuse the merchants and customers.

Digital transformation is imperative for gaining and sustaining a firm's competitive advantage. Hence, understanding the dynamics of technology evolution becomes salient for both scholars and practitioners. This chapter aims to provide a complementary perspective to the field of innovation by mapping and visualizing the patterns of digital transformation at the industry level with a particular focus on the role of technology convergence. The authors tracked 20 years of the technology of the U.S. communications industry in order to investigate how digital transformation has shaped the industry technological structure, which are the technological gaps and potential future technology trends. The results show a deep transformation of the industry with many interconnections between technology domains and a high degree of overlap between technology areas.

This chapter aims to provide a better tool for implementing the marketing technique known as influencer marketing in the tourism industry. To do so, a results-oriented influencer marketing manual for the tourism industry has been created. Despite the success of influencer marketing, the few previous studies in this field do not include verified measures to ensure its effectiveness. For this reason, the approach that is presented here could be crucial to support these marketing activities. As this topic is new and often little-understood, the data compiled was based on the case study methodology. This chapter proposes the following phases: (1) campaign planning, (2) search for influencers, (3) evaluating the best profiles, (4) contacting influencers, (5) proposing a project, (6) execution, and (7) analyzing the results. This work could help companies considering influencers as a new communication channel to successfully run their campaigns.

This chapter aims to discuss how the rapid evolution of digital technologies is creating opportunities for new agricultural business models. First, it provides an overview of what the authors consider to be part of the digitalization in agriculture. Then it addresses the emergence of a community of practice based upon the data exchange and interconnections across the agricultural sector. New business opportunities are presented first through an overview of emerging start-ups, then discussing how the inventor farmer profile could create opportunities for new business models through the appropriation of technologies, eventually highlighting the limits of some classic farm business models. Finally, the chapter presents an example of farmer-centered open innovation based on the internet of things and discusses the related business model. The conclusion provides some perspectives on the use of agricultural digitalization to increase the share kept by farmers in the value chain of agricultural productions.

Foreword

More than 4000 years ago, somewhere in Mesopotamia, a king, named Hammurabi, had decreed that all the information on all the transactions taking place in his kingdom should be transcribed in a ledger in stones, animal skins, or animal bones. The required information was the date of the transaction, the object of the transaction, the name of the buyer, the name of the seller, and the value of the exchange. Without doubt, Hammurabi had invented the first information system of humanity. An information system working without technology !

It is also to him that we owe what is now known as the Hammurabi Code of Commerce, which, among others such as that of Ur-Nammu or that of Eshnunna, were the first Codes of Law, having inspired religions and constitutions. Now available in European museums, theses Codes of Law adopted the first casuistic structures in the form of [if crime then punishment]. The programming of the first computers, appearing 19 centuries later is, to this day, based on these several millennia old casuistic structures. And so is the programming of the first AI inference engines.

An era of frenetic automation followed in which information technologies (IT) have transformed the way we live, work and play. Humans endeavored to automate anything that was repetitive and structured, just as Hammurabi began to do.

That was the automation era: do with computers what we have always been doing. Until came the day when it was discovered that IT could do more than just assist us in what we already do by helping us to do it better (cheaper, faster, etc.). Technology had also the capability to help us to do better things, to transform our tasks, mostly continual and bureaucratic, into more thoughtful and meaningful processes. That was the informating era. IT started to help us not only to reinvent organizations and the work in organizations, but also to relate to each other and communicate instantly with anyone and across great distances.

Humans quickly discovered that from mere innovation, IT has become both a source and an enabler of innovation. At the same time, IT almost surreptitiously evolved into information and communication technologies (ICT), and the convergence between computing and telecommunications that was so much touted in the 1980s and 1990s, became common place. Today information and communication are governed by computers and phones, jostling for our apps, competing for our attention. Once limited to microprocessors, Moore's law has become prevalent and applicable to everything we use to live, work and play.

Today, words such as Smartphones, Smartwatches, Smart Offices, and Smart Cities have become as common as AI, Blockchain, Big Data, Digital Twins, 4D printing (soon 5D ?), IoT, Machine Learning, Deep Learning, VR/AR, Quantum Computing, etc.

So, from being an instrument of change and improvement, ICT, have become disruption agents. ICT for business moved to a whole business of IT. And words such as digitization (simply converting paper to numeric) and digitalization (transforming organizations through improving business processes) popped up in the literature. Quite an odyssey indeed in 4000 years !

We will not finish without closing the loop.

Isn't it true that Hammurabi's ledger has become a distributed ledger ? Blockchain's central concept is nothing other than the distribution on a huge P2P network of what the king was talking about; "that all the information on all the transactions taking place in his kingdom should be transcribed". If the kingdom is the business, then the ledger is an ERP. If the kingdom is the world, then the ledger is a Blockchain, which, in addition, is immutable! If it is particles, then it is quantum entanglement. In all cases it is the generalized and almost instantaneous replication of information that is widely and instantly shared.

Isn't it true that smart contracts, another Blockchain technology central concept, is nothing other than the generalized sharing of a Code (in the legal sense) in the form of a code (in the computing sense). The Code of Law thus became law of code, with code governing not only expert systems and robots, but also search engines, chatbots, and social networks, and all the apps that we use on a daily basis.

It is clear that it is in a world in continuous development, governed by the technologies that are mentioned here, that this book was conceived. With 15 chapters divided into four sections, this book examines the digitalization of companies and positions itself as a punctual witness to some of its successes in each of its individual contributions, focusing on such ICTs as ERP, 3D Printing, Big Data, all along the digital transformation vector. Both the digitization and digitalization will be addressed through both information and communication technologies activating through both their improvement and disruption roles.

Thus, although not exhaustive, this book tackles the arduous task of describing companies' response to change, using technologies that are themselves constantly changing.

Mohamed Louadi
University of Tunis, Tunisia

Preface

The development of mobile devices, cloud solutions, social media use within firms and the extension of Internet applications to the "physical" through Internet of Things (IoT) have favored the link between "business" and "digital". As a result, large volumes of high velocity, complex and variable data (Big Data) are more and more stored, specific tools for data analysis are more and more needed leading to more digital transformations.

Besides transformation of IT deployment, the adoption of such disruptive tools and technologies imposes additional challenges to firms that are pushed "to create new business models that consider and leverage the increased digitization" (Gebhart et al., 2016; Khare et al., 2017). Moreover, Schallmo and Williams (2018), and earlier Gens (2016), expect that every (growing) enterprise will become "a "digital native" in the way its executives and employees think and how they operate".

To highlight this interrelation between "business" and "digital", this book focuses on examining the concept of digitalization which is related, according to Gartner (2017), to the use of digital technologies in order to improve businesses, create new revenues and value-producing opportunities. Thus, digitalization allows organizations to change their business model and value chains with the use of disruptive technologies and digital data in the digital economy (Skilton, 2016).

Such business transformations should be understood and mastered by providing practical tools and frameworks to assist managers in dealing with digitalization challenges. In this book, a special focus will be on transitional economies facing big concerns and challenges linked to unexpected events and additional global pressures. Digitalization and digital transformation can be, at the time, sources of opportunities and risks for such economies when dealing with globalization challenges.

This book aims to:

- Identify latest trends, business opportunities and challenges linked to digitalization.
- Define concepts in relation with business transformations caused by digitalization.
- Share useful experiences and best practices to deal with the new technological changes.
- Present new theoretical and empirical frameworks related to the topics.
- Identify the effects of new digital transformations at the organizational, economic, social, societal and environmental levels.

Regarding its contributions, this book is divided into four interrelated sections.

The first section is titled "Disruptive Technologies in the Era of Digitalization". Three chapters (1-3) are included in this section dedicated to the study of disruptive technologies use in different contexts.

In the first chapter, the author examines how disruptive innovation could be a powerful tool for broadening and developing new markets and providing new product functionality, which, in turn, may disrupt existing markets, platforms and ecosystems. He focuses on 3D printing technology and explores the fact that such disruptive technology is having an impact on the strategic scope of manufacturing activities in terms of product scope, market scope, geographical scope and competence scope.

The second chapter is dedicated to the study of the feasibility, suitability, and sustainability of the blockchain from the viewpoint of experts in the field. Since blockchain is considered as an emergent technology, the author opts for a qualitative approach by conducting interviews with stakeholders of the blockchain. Among results, the author notices the lack of communicational maturity in the blockchain community which can prevent the ultimate achievement of the distributed dimension of the blockchain mainly during periods of crisis. This chapter focuses on the social dimension leading to more debate on blockchain concerns.

In the third chapter, the authors attempt to study the intentions to use cloud-based CRM applications within Saudi SMEs through a combination between the Technology Acceptance Model (TAM) and the Theory of Planned Behavior (TPB). To do so, the authors developed a research model and then a questionnaire sent to marketing managers. The collected data were analyzed with the Partial Least Squares (PLS) technique. Among the main statistical analyses, the authors conclude that the TPB-TAM is suitable to study cloud CRM. From a managerial perspective, the authors expect that cloud CRM is perceived with good impression and that this new technology should be implemented strongly and gradually in SMEs to improve the quality of services provided to customers and organizations.

Section 2 is dedicated to the presentation of "New Strategies Under Digital Transformation". Four chapters (4-7) are integrated in this section.

In the fourth chapter, the authors attempt to perform a critical analysis of trends in enterprise resource planning (ERP) in respect to contemporary multi-organizational enterprise strategy in order to identify under-researched areas. The authors based their study on the premise that multi-organization strategies and information systems span more than one legal company entity and are becoming increasingly important as digital Internet-based systems become more prolific, and outsourcing and collaboration strategies become more widespread. This chapter presents a critique of literature covering theoretical, methodological and relational aspects of enterprise resource planning systems and multi-organizational enterprise strategy. The critique gives a unique perspective and highlights major gaps in current research and points towards a trend which is referred to 'enterprization of multi-organizational enterprises'. This concept tries to explain the fits between different types of ERP systems and different types of multi-organizational enterprises. Through this concept, the authors attempt to catalyze inter-dependent operations management and information systems management research; which, in turn, should help practitioners in the 'Internet of Things' (or Industry 4.0) era connect interdependent IS and operation management development strategy.

In line with the increasing digitalization of firms, the fifth chapter proposes an integrated view of big data management through a conceptualization of big data value chain. Based on a literature review, the authors attempt to provide a holistic approach to big data management that begins with the identification of business problems explaining the need to carry out big data projects, and ends with the value creation by generating and leveraging deep customer insights. From the beginning to the end of the value chain, three interrelated processes are positioned which are related to big data management process, big data architecture and business process reengineering. The suggested value chain focuses on the interrelations between strategic, organizational and technological elements in order to extract value from big data.

As data is more and more considered as a source of value for decision making, the proposed conceptualization of big data value chain should help managers in their quest to take advantage and gain valuable insights from huge sets of data.

The sixth chapter deals also with the strategic perspective of nascent technologies under digitalization by discussing new insights of business-IT alignment under digital transformations. In this chapter, the author attempts to illustrate the main shifts in the business-IT alignment framework in such context. In fact, the author guesses that digitalization supposes permanent transformation of competitive advantages and critical success factors. Thus, through an analytical literature review, the author proposes a novel framework to help managers re-align their strategies in the era of digitalization.

Regarding the managerial issues linked to big data, the seventh chapter provides an overview of big data analytics with a focus on their use in the supply chain field and underlines their potential role in the supply chain transformation. Through a deep literature review, the author attempts to provide an understanding of the business value created from different types of big data analytics use within the supply chain. The author uses SCOR (Supply Chain Operations Reference) as a tool to study processes concerned by big data-enabled supply chain transformation. He concludes that big data analytics can play a critical role in supply chain management on strategic, tactical and operational levels by improving decision-making, helping in the creation of new products and services, enabling to discover needs and customization, and assisting the achievement of operational excellence.

Section 3 is titled "Entrepreneurship and Innovation in the Digitalization Era". This section contains four chapters (8-11) from several proposals that were submitted by researchers from different countries.

In the eighth chapter, the author provides an overview of digital entrepreneurship, identifies key variables that determine a successful digital entrepreneurship. Then he proposes a conceptual model to guide the understanding of a successful digital entrepreneurship development within the context of developing and emerging economies.

The ninth chapter highlights the challenges of demonetization of the economy of India by studying the adaptability, affordability, acceptability and sustainability of the payments system currently installed by merchants and service providers (such as Point of Sale PoS and Prepaid instruments). For achieving their goal, the authors surveyed 221 small merchants and analyzed their point of view about such payments system. Their study showed the needs of awareness about such system in having non-cash transactions.

The tenth chapter aims to provide a complementary perspective to the field of innovation by mapping and visualizing the patterns of digital transformation at the industry level with a particular focus on the role of technology convergence. For that, the authors tracked 20 years of the technology of the U.S. communications industry in order to investigate how digital transformation has shaped the industry technological structure in terms of technological gaps and potential future technology trends. The main results of their study showed a deep transformation of the industry with many interconnections between technology domains and a high degree of overlap between technology areas.

The eleventh chapter is dedicated to the understanding of digital entrepreneurship and innovation in the digitalization era. Based on key concepts such as digitalization, entrepreneurship and innovation, this chapter contributes to the literature of digital entrepreneurship and innovation by adopting an ecosystem approach. Then, this chapter provides an overview of the digital entrepreneurship and innovation ecosystem and its main components. Within this new philosophy of digital entrepreneuring, the chapter presents new trendy phenomena as precursors and enablers to boost digital entrepreneurial ventures; and certain uncharted territories that need to be explored. In the end, the chapter advances new directions for future research in digital entrepreneurship and innovation. It concludes with the idea of democratization gained for entrepreneurship, innovation and digitalization in this era.

Section 4 is titled "Digitalization Experiences and Applications". This section contains also four chapters (12-15) from different disciplines of management and sectors of activities.

The twelfth chapter investigates the effectiveness of different types of display advertising by means of an eye-tracking study combined with a pre- and a post-test questionnaire with the purpose of collecting quantitative and qualitative data concerning ad visibility and interaction. In their chapter, the authors aimed to look into whether banner blindness still applies regardless of the type of display ad used, whether the visual pattern remains F-shaped, the effect of placing ads below the fold, how effective trick banners are and which types of ads are annoying to users.

The thirteenth chapter explores the question of the influence of the use of mobile technologies by the customers on recruitment in the banking sector and on the number of the branch banking networks. Particularly, the author seeks to investigate the effects of the use of mobile technologies by customers on recruitments and the number of branches in the banking sector in Tunisia. For bringing answers to this question, she analyzed the annual reports of the last seven years published by Tunisia's Professional Association of Banks and Financial Institutions and the findings were shown to be contrary to those observed in foreign countries concerning the reduction of the number of branches and recruitments following the digital transformation in the banking sector in Tunisia.

The fourteenth chapter aims to provide a better tool for implementing the marketing technique known as influencer marketing in the tourism industry. To do so, a results-oriented influencer marketing manual for the tourism industry has been created. Despite the success of influencer marketing, the few previous studies in this field do not include verified measures to ensure its effectiveness. For this reason, the approach that it is presented here could be crucial to support these marketing activities. As this topic is new and often little-understood, the data compiled was based on the case study methodology. This manual proposes the following phases: (1) Campaign planning; (2) Search for influencers; (3) Evaluating the best profiles; (4) Contacting influencers; (5) Proposing a project; (6) Execution; and (7) Analyzing the results. This work could help companies considering influencers as a new communication channel to successfully run their campaigns.

The fifteenth chapter aims to discuss how the rapid evolution of digital technologies is creating opportunities for new agricultural business models. First, it provides an overview of what the authors consider to be part of the digitalization in agriculture. Then, it addresses the emergence of a community of practice based upon the data exchange and interconnections across the agricultural sector. New business opportunities are created for the new business models through the appropriation of technologies. Finally, the chapter presents an example of farmer-centered open innovation based on the internet of things, then, discusses the related business model. The conclusion provides some perspectives on the use of agricultural digitalization to increase the share kept by farmers in the value chain of agricultural productions.

We hope this book can help readers to better understand the different types of business transformations in the era of digitalization and the tools used to achieve successfully these transformations and thus can inspire other researchers to explore new related issues.

Karim Mezghani
Al Imam Mohammad Ibn Saud Islamic University, Saudi Arabia & University of Sfax, Tunisia

Wassim Aloulou
Al Imam Mohammad Ibn Saud Islamic University, Saudi Arabia & University of Sfax, Tunisia

REFERENCES

Gartner. (2017). *Digitalization*. Retrieved from https://www.gartner.com/it-glossary/digitalization

Gebhart, M., Giessler, P., & Abeck, S. (2016). Challenges of the Digital Transformation in Software Engineering. *ICSEA*, *2016*, 149.

Gens, F. (2016). *IDC FutureScape: worldwide IT industry 2017 predictions*. Framingham, UK: International Data Corporation.

Khare, A., Stewart, B., & Schatz, R. (2017). *Phantom Ex Machina*. Springer International Publishing. doi:10.1007/978-3-319-44468-0

Schallmo, D. R. A., & Williams, C. A. (2018). *Digital Transformation Now! Guiding the Successful Digitalization of Your Business Model*. Cham: Springer Briefs in Business.

Skilton, M. (2016). *Building the digital enterprise: a guide to constructing monetization models using digital technologies*. Springer.

Acknowledgment

The editors would like to acknowledge authors and reviewers for their engagement in the process of developing this challenging book, through collective sense-making efforts aiming to explore the different ways of transforming businesses in the era of digitalization.

Their sincere gratitude goes also to Maria Rohde (Assistant Development Editor at IGI Global) and Ms. Jan Travers (Director of Intellectual Property and Contracts at IGI Global) for their ongoing support.

The editors wish to thank in advance the future CXO and other business transformers, who will read this book and find certain aspects related to digitalization shared by the authors of the chapters included in this book.

Karim Mezghani
Al Imam Mohammad Ibn Saudi Islamic University, Saudi Arabia & University of Sfax, Tunisia

Wassim Aloulou
Al Imam Mohammad Ibn Saudi Islamic University, Saudi Arabia & University of Sfax, Tunisia

Section 1
Disruptive Technologies in the Era of Digitalization

Chapter 1
3D Printing as a Case of Disruptive Technology:
Market Leverage and Strategic Risks for Traditional Manufacturing

Vincent Sabourin
University of Quebec at Montreal, Canada

ABSTRACT

Disruptive innovation is a powerful tool for broadening and developing new markets and providing new product functionality, which, in turn, may disrupt existing markets, platforms and ecosystems. From the case of 3D printing technology, we find a clear example of disruptive technology. This chapter examines how disruptive the technology of 3D printing on the strategic scope of manufacturing activities in terms of product scope, market scope, geographical scope, and competence scope.

INTRODUCTION

In an environment where innovation drives a fundamental objective in developing a competitive position of developed and heavily industrialized economies, disruptive technologies offer a strategic option of profoundly transforming the manner and process in which goods are manufactured (Seidle, 2015). Therefore, disruptive technology is far associated with successful technology also known as "technological success", which emerges to take the place of already dominating technologies and innovations in the 3D printing market (Capdevila, 2015).

Chris Anderson observes the current revolution and dynamic emergence of 3D Printing technologies as a part of post-modern industrial revolution. In this regard, developed economies are making substantial investment in 3D printing sector (Leila and Beaudry, 2015). According to Cutting et al., (2015), there is an upsurge in the perception among business practitioners, leading scholars and industry influencers that these technologies will create a new phase of technology consumption and embracement, that could be linked to an emergence of next generation entrepreneurs, exciting business opportunities and as well as new business models altogether.

DOI: 10.4018/978-1-5225-7262-6.ch001

For instance, in the United States investments plan of 2013, President Barack Obama, cited, that America and its leading innovation companies should invest in the creation of 3D printing center to revive the innovation front and create more job opportunities (McKinsey & Company, 2013). On this front, European economies have proactively moved to forge an entry into the sector of 3D Printing technologies. Commencing January 2013, the project "*European Amaze of Spatial Agency*", was developed to shift the focus on the creation of mobility within 3D printing for all industrial, collapsible segments in the aerospace sector and other commercial divisions of advanced composite materials (Smith, 2015).

According to Christen & Raynor (2016), "...disruption has a paralyzing effect on industry leaders. With resource allocation processes designed and perfected to support sustaining innovations, managers are constitutionally unable to respond to the emerging trends and challenges in the disruptive process..." This research finds that although these threats and trends are considered to be inferior when measured against value propositions on which sustaining innovation has been focused, these disruptive technologies have other hidden functional attributes such as being less expensive, simpler and more convenient. Thus, from the value of disruptive innovation theory, managers should be to adopt disruptive technologies to improve on the sustaining innovation process of their companies and in the long run sustain their competitive position in the market.

LITERATURE REVIEW

Exploring 3D Disruptive Technologies

This chapter highlights the fact that the 3D disruptive technologies have proved to have different meaning according to authors (Yu and Hang, 2011). In a more general perspective, research on the concept revealed certain criticisms in particular as to the definition of the disruptive character of the technologies identified by Christensen (Rao & Cull, 2006). According to these authors, one of the criticisms on 3D printing concept results from the fact that the analyses which research has, are essentially ex-ante and that the robustness of the concept of 3D disruptive technologies remain weak to study and forecast disruption. The fragile attribute of the predictability of 3D printing concept regarding disruption ability highlights the necessity for studying this phenomenon in a finer and detailed way to take into account the variety of strategic impact of disruptive technologies with a more structured categorization. In line with this, the researcher studies Adner et al., (2002) proposed methodology to better understand the disruptive process when disruptive technologies appear in the market. This is also similar to the case of Rafii and kampas (2002) who propose a six phase model, allowing for the explanation of various parties to understand the disruptive character of disruptive technologies.

Disruptive Technology

The concept of disruptive technologies comes from Bower and Christensens whose works were published in Harvard Business Review and revealed how the "disruptive technologies" are potentially destructive to already established companies (Bower & Christensen,2015).

According to Bower and Christensens (2015), one of the most consistent patterns in business is failure of leading companies to stay at the top of their industries when technologies or markets change. This has caused other companies which have previously been considered less profitable to exploit ways and technologies that are now creating a disruptive wave to the current leading technological sectors (Krech et al., 2015). As argued by Amshoff et al. (2015), disruptive innovation is a powerful means of broadening and developing new markets and providing new functionalities, which, in turn, may disrupt existing market linkages and structures. According to Bower & Christensen (2015), disruptive technologies are technologies that provide different values from mainstream technologies and are initially inferior to mainstream technologies along the dimensions of performance that are most important to mainstream customers. Bower & Christensen (2015) introduce the important aspects of changing performance with time, plots the trajectories of product performance provided by firms and demanded by customers of different technologies and market segments, and shows that technological disruptions occur when these trajectories intersect. According to recent studies, managing the disruptive nature of technologies such as 3D printing will require proper indulgence of the strategic scopes of business (Amshoff et al., 2015).

The Context and Business Environment of 3D Printing Innovation

Practically, Grynol (2015) finds that, within the North American manufacturing segment, global competitiveness has become intense and growth of the industry has experienced a consistent decline in a couple of years. According to Grynol (2015), the US, once considered as the global manufacturing hub, currently supports fewer manufacturing companies than it did previously. The challenge here is that for economic growth to occur, there must be innovation, and innovation cannot therefore occur in the absence of technologically-complemented manufacturing capabilities and potential. This hesitation is limiting numerous countries in terms of GDP forecasts (Edmond, 2015). With this challenge, it's imperative for manufacturing companies and firms across diverse industry segments, to initiate 3D printing capacities with the intention of ensuring sustainable innovation (VanderPanne et al., 2015).

In many economies, traditional forms of manufacturing processes, still occupy an important place in their business scope. Quite invariably, it becomes challenging to match the available scale of economies through traditional manufacturing once products have been fully developed and commercialized (Smith, 2015; Krech et al., 2015). Originally, due to their foundational defects, traditional manufacturing process were created to help in the instant prototyping framework, however, currently the 3D printing process has now become more often adopted for use in the manufacturing process.

The traditional limitation for 3D printing technologies emerged from the fact that materials failed to created room for massive use of 3D printing technology which was suitable to aid in the production of prototypes (Smith, 2014, 2015). Though 3D printing technologies have been in use since 1980s, it is only as from 2010, that a majority of business segments felt their impacts (Business Week, 2012). Beginning 2013, the potential for small-scale production and for value-addition capability begun to be implemented (Le Monde, 2012).

3D printing sector has witnessed sustained double-digit growth in recent years, and it is within reality to project that the sector will be worth more than $7.5 billion by 2020 (McKinsey & Company, 2013; Whalen, & Akaka 2015). There are clearly great opportunities for the embracement of this platform in key economic sectors such as medical devices, medical implants, aerospace, power generation, and automotive manufacturing (Green, 2015).

Majority of companies have already evaluated this technology or have begun making use of it on a limited scale. Additionally, 3D printing innovations could lower material use, energy, and water by removing waste, in addition to with all other harmful process drivers, thus having a positive effect on sustainability (Cozmei & Caloian, 2012).

Due to their nature in digital space, 3D printing innovations are proactively being made to be part of Internet, which helps users to directly engage in the design process, and enables customer co-creation and personalization (Pearce, 2015). The embracement of 3D printing is anticipated to stimulate the growth of alternative supply-chain management and business models as well as approaches by preventing the need for capital intensive investments such as tooling, availing working capital within the supply pipelines, and minimizing business risk in new product development and innovation (Bessen et al., 2012).

3D printing which is also termed as additive manufacturing is currently considered as one of the fastest growing innovations with the ability to generate massive profits in market with a forcefully disruptive power (Hollister et al., 2015). After a status review and improvements lasting close to three decades in reworking the technology, 3D printing is swiftly moving from being a developmental prototype with high rapid wave within the industry to graduating to mainstream procedure in the manufacturing process, to be adopted by consumers and industry players in an equal measure. Focusing on the global landscape of entrepreneurship, innovation and technological scopes, developments with interest on 3D printing have achieved a sustainable momentum leading to a pool of breakthrough news and creating high anticipations concerning 3D's capability in revenue generation (Bland & Conner, 2015).

The Competitive Dynamic of companies and the Global Competition

As we explained it previously, the 3D technologies were anciently limited to the manufacture of prototypes. However, beginning 2013, the opportunities of small-scale commercialization and for value-addition sections made their entry into 3D printing sector (Bhardwaj et al., 2015). This disruptive technological breakthrough was made achievable by companies as Shapeways and Ponoko who developed less expensive and more affordable 3D printing technologies enabling an aggressive distribution and marketing models for 3D printing technological products (Bland & Conner, 2015). With a reduction in the manufacturing costs of 3D printing technologies which made the technology within affordable reach, more manufacturers in similar line can now fabricate short terms in technology production (Olander & Hurmelinna-Laukkanen, 2015).

The market for 3D printing technology is the fastest growing sector, though its industry still lags behind with only two companies having sizeable market share and growth spectrum: 3D Systems Corporation and Stratasys Inc. with 15% and 3.4% of market share respectively. We note that in the next few years, there shall be approximately between 8 and 12, 3D printing manufacturers controlling the market share and directly influencing the market structuring in several key business sectors (Capdevila, 2015; Seidle, 2015).

Example of players includes: Stratasys Inc. with $750 million turnover, 3D Systems Corporations, $81m, ExOne $43.9 m, Arcam $39 m and these companies have already reached the critical sizes enabling them to strategically position within the 3D printing global market.

During the next few years, the level of entry into this market will become more expensive and difficult for traditional manufacturers. The structuring of 3D printing market will also have strategic implications on the whole chain, because technology players are expected to bring a restructuring and a revolution of business models in a majority of industries; especially those operating within the aeronautics and health sectors. According to Forrester (2013), there has been an explosion in the technology of 3D printing; the number grew from 155,000 to 23000 units between 2008 and 2013.

Figure 1. Examples of 3D companies
Source: (Wohlers Report, 2015)

	Market share
3D Systems Corporation	15.2%
Stratasys Inc.	3.4 %
Shape ways	Less than 1.0%
Applied Rapid Technologies (ART)	Less than 1.0%
eMachineShop.com	Less than 1.0%
Ponoko	Less than 1 %

Figure 2. Illustrates this situation
Source: Forrester (2013)

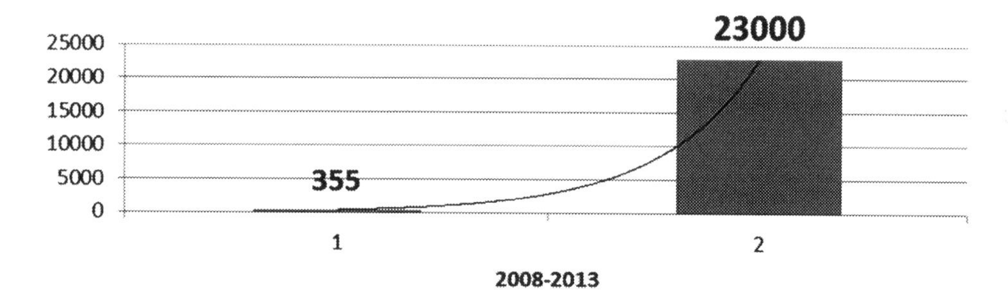

From this contextual illustration, we note that a majority of industrialized countries have thriving companies in this domain. So, we find among the classification of 25 large players in the sector of 3D printing technology, companies coming from more than ten different countries.

Research Objectives

The review of literature and background information has shown that very limited research has been carried out to categorize and compare the disruptive nature of 3D printing technologies and their impact on the scope of businesses. Thus, the main objective of this chapter was to explore and categorize how the disruptive nature of 3D printing technologies is having an impact on the strategic scope of manufacturing activities in terms of product scope, market scope, geographical scope and competence scope. As defined by Cohen & Lefebvre (2005), categorization is a process in which ideas and objects are recognized, differentiated, and understood, usually for some specific purpose. A research effort of categorization is fundamental in language, prediction, inference, decision making and in all kinds of variable interactions within a research (Frey & Saake, 2011).

More specifically objectives were the following ones:

1. To study the impact the strategic scope of 3D printing as a disruptive technology
2. To assess the strategic risks of 3D printing technology on traditional manufacturing processes.

METHODOLOGY

This research design framework is based on action-response research design (Chen et al., 1991; Chen et al., 1992, 1992). This Chapter's analysis focuses on the strategic disruptive innovations responses adopted by key competitors in the industry to challenge the incumbents, especially with traditional manufacturing models. This study relied on a similar research approach that other scholars in previous analysis had conducted on competitive disruptive responses (Chen et al., 1992; Smith et al., 1991; Tang & Shapira, 2011).

STRATEGIC SCOPE AND DEVELOPMENT OF CONCEPTUAL FRAMEWORK

Strategic Scope

The strategic scope of a business entails fundamentally the markets and products. Strategic scope seeks as well to differentiate what is or is not part of a business (Burk, 2013). The design of a firm's strategic scope is informed by four fundamental choices which serve to define the firm's nature of engagement. These innate elements include: product scope, market scope, geographical scope and scope of competencies.

- Product scope refers to a range of products and services offered by a firm/company (Fisher and Henkel, 2012).
- Market scope refers to the number of markets and segments (often channels of distribution).
- Geographical scope resonates to the number of geographical markets served.
- Finally, scope of competencies has a focus on the range of skills possessed by a company, and activities within the confines of a firm or the decision to subcontract part of firm's activities from external partners (Hang et al., 2011; Cohendet and Caloian, 2012).

From this evaluation, we found that 3D printing technology has an impact on a speck of strategic scope of 3D printing business.

Development of Conceptual Framework

In the course of developing this conceptual framework, the author relied on the previous work of Abell's (1980) conceptual model as well as from the work of Allaire and Firsirotu (1993) on the generic categorization scheme; which is built on the premise that a business can be defined by the scope of its offerings (product scope), its degree of capability and skill differentiation (scope of competence), and the extent to which its offerings are able to reach diverse customers (market scope).

Abell's matrix is a three dimensional tool most often referred to as the three dimensional model of business definition. The model is used to analyse the scope of business's operation. This may include areas such as technologies and products a business offers in a market or the audience that it targets. A detailed analysis of the business's current activities can help create future strategies that will help the business stay tuned to the changes that may occur within the market.

The three dimensions of a business are the customer groups (who will be served by the business), customer needs (what are the customer needs that will be met) and technology or distinctive competencies (how are these needs are going to be met). A major point of importance in this matrix is to focus on understanding the customer rather than the industry and its products and services. Through these three dimensions, this tool helps define a business by its competitive scope (narrow or broad) and the extent of competitive differentiation of its products/services (Abell, 1980).

According to Martin (2015), the major point of significance in this matrix is to focus on understanding the customer rather than the industry and its products and services. Through these three dimensions, this framework helps define a business by its competitive scope (narrow or broad) and the extent of competitive differentiation of its products/services.

Data and Sample Collection

To collect information on this analysis' cases the author followed five steps:

Step 1: The author collected information and surveyed 25 large 3D printing disruptive technologies for their disruptive character from a technological and economic perspective. The following figure 4 illustrates in a preliminary manner of these 25 selected 3D printing companies. The data collection was completed with a survey of reports and articles on the topic of disruptive innovation with a focus on 3D printing (Wohlers Report, 2015).

Step 2: All these reports were supported with economic data (generally quantitative) and technological data (generally descriptive) describing their disruptive nature. Wohler's report of 2015 has produced an integrative and very complete report and also a full set of detailed reports surveying and assessing specific 3D printing companies for specific disruptive technologies. Since the objective of this research was not to provide an assessment the technological and economic impact rather than the extent of strategic scope, the author listed these 3D printing companies.

Figure 3. Strategic scope of the business

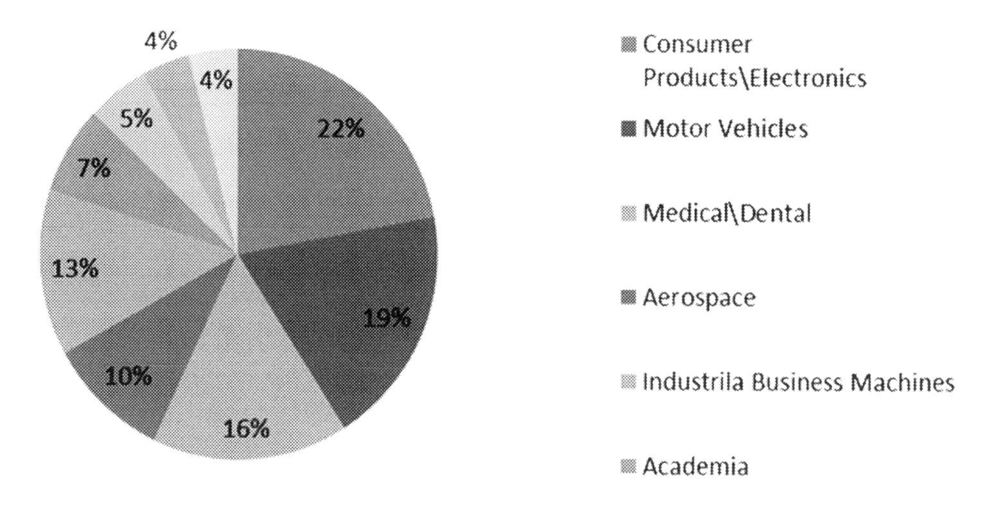

**For a more accurate representation see the electronic version.*

Step 3: The author also provided a geographical categorization by nationality using the dominant drivers according to the description of the technological and economic impact of the literature. Several of the twenty-five disruptive innovations combined several aspects and were at the same time product but also market driven. Using Abel's matrix, the researcher selected the dominant factor among the three factors of the model.

Step 4: The disruptive extent of 3D printing disruptive technologies was assessed from both technological perspective and economic perspective. From a technological perspective, the disruptive character of 3D printing was assessed by the radical nature of the substitution provided by the technology. When the innovation could be accommodated and easily assimilated by competitors, it was not considered as being disruptive. When it has to be assimilated and not easily integrated (determined by the ease of substitution) and therefore required a transformation of the strategic scope of the firm, it was considered as being "disruptive."

Step 5: The selection was completed using a technological and an economic perspective. From a technological perspective the author assessed the disruptive character of these innovations in terms of functionalities (products), reach and competencies based on the sources from engineering, computer science and software and life science.

Data Selection

The author provided a categorization using dominant drivers according to the description of the economic and the technological impact provided by the conceptual framework. The selection was completed using both a technological and an economic perspective. From a technological perspective the author assessed the disruptive character of these technologies in terms of functionalities (products), distribution and competencies based on the sources from engineering, computer science and software and life science. For an economic perspective, the author relied on economic impact assessment. For the final selection, the author only kept 3D printing technologies that combined these three strategic scopes.

Figure 4 illustrates the survey of disruptive technologies (by alphabetic order).

Figure 4. Large companies in 3D printing
Source: (Wohlers Report, 2015)

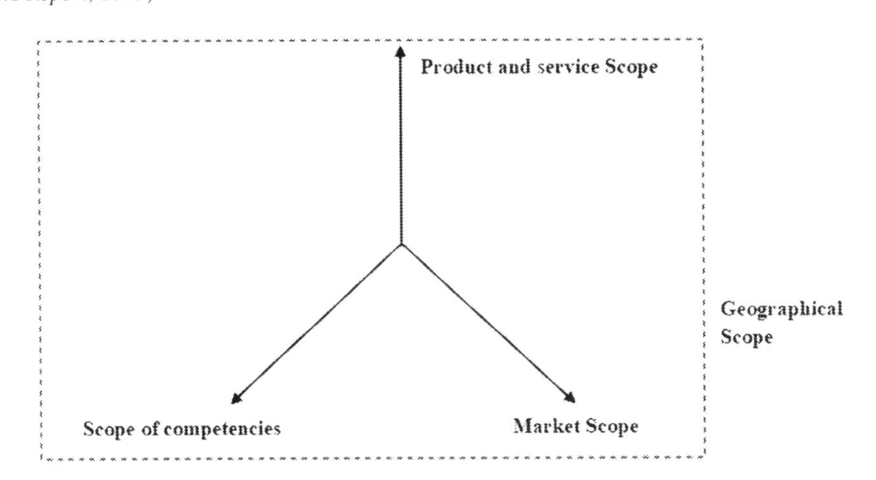

DATA ANALYSIS

The researcher systematically analysed each component of the conceptual framework; for each innovation using the economic literature on the subjects. For each innovation, the author identified a predominant driver among the product scope, the market and the scope of competencies with specific effects on the b1usiness of 3D printing companies. A summary of the same was based on:

1. In this systematic analysis, product scope was defined as having an impact on the substitution of product or services functionalities for 3D printing disruptive
2. Market scope was defined as an impact on the substitution of channels of distribution or the segmentation of the market, such as the long tail effect on the 3D disruptive technologies.
3. Scope of competencies was identified as having an impact on substitution of the core competencies for the 3D disruptive technologies.

Finally, the author developed brief analysis of findings, discussion and strategic implications to build on the systematic review of literature.

RESEARCH FINDINGS

This section analyses the literature in terms of opportunities and strategic risks to traditional manufacturing companies. The leverage and strategic risks in 3D printing are analysed and assessed under key fundamentals of product scope, market scope, geographical and competence scope.

Allen (2015) asserts that embracing 3D printing is a new medium for technnovation, as such an economy that adopts this technology will achieve a sustainable economic growth due to sustained revenues from additive manufacturing. The successive adoption and monetization of 3D printing technology will depend on how effectively and efficiently companies use strategic scopes: product scope, market scope, geographical scope and competence scope

Impact on the Product Scope of the Business

Product scope refers to the number of products and services offered by a firm/ company (Ke and Pan, 2014). The term "product" designates a function vis-à-vis the client. Product scope is a key factor in the company's strategy. Product scope varies greatly from one firm to another due to the effects of mass customization and product personalization. According to Grynol (2015), mass customization and personalization of re-defined product range leads to proliferation of products. This has a risk of destroying the existing product scopes as well as economies of scope. On the same note, a market specialist might offer a single product, while a generalist will offer a complete range of products. To compete, a company has to select its clients, across the dominant stages of the platform. The first popular usage of the term platform seems to have been in the context of new product development and incremental innovation around reusable components or technologies. Baldwin and Woodward (2009) refer to product platforms as internal product development in which a company, either working by itself or with suppliers, can build a family of related products or sets of new features by deploying reusable components. Makinen et al., (2014), for instance, describe how "product platforms" can meet the needs of different customers simply by modifying, adding, or subtracting different features.

Bechthold et al., (2015) cite that in order to benefit from the applications and opportunities of 3D printing technologies moving forward, companies in virtually every industry must be fast, flexible and capable to understand the implications that 3D printing will have on the nature of their businesses. Failing to do so will lead to a potential loss of market share, due to increased competition from new companies that create market changing products, and disruptive technologies. And competition won't stop there due to diverse product scopes by key players (Gyrnol, 2015; Bechthold et al., 2015).

Since 3D printing technology is easier to use and more price friendly, companies are finding different uses for the technology. Many are even starting their production from scratch as firms centered entirely on 3D printing technology. Researchers Choi et al., (2015) argue that the 3D printing space was valued at $2.5 billion in 2013, and projects the worldwide revenue in the space to grow to $19.2 billion by 2019. In Prototyping, Grynol (2015) argues that as a result of high utilisation of product scope, single items can be produced inexpensively without incurring the molding and tooling costs of traditional manufacturing.

Lower Manufacturing is a critical business opportunity for 3D printing vendors, especially when it comes to prototyping. Companies, small and large that have new products or ideas headed for mass production benefit immensely from 3D printing owing to the fact that 3D printing is a fast, inexpensive way to build prototypes. It is also easy to make changes to a prototype in 3D modelling software. If there is a small or large custom change that needs to be made in a prototype, it can be fixed quicker, cheaper and easier with a 3D printer. Many major companies, including Nike, already use 3D printers for prototypes, so the market already exists and is only going to grow, to accommodate the diverse product extensions in the industry. To the extent that there is a high market scope signifies that the 3D product scope is on the verge of take-off.

According to Edmond (2015) and Grynol (2015) since 3D printers are an affordable way to make custom items, and some of these machines are able to print items out of all sorts of materials, including metal, many companies are ordering custom parts from 3D printing technology vendors. These small companies have been taking orders from customers who walk in with designs or ideas for products and models.

Additionally, 3D scanning technology has caught up to with 3D printing, so users can scan any product and create multiple copies of these products through 3D printers. The striking aspect about additive manufacturing technology is a wide range of possible applications, reaching from industrial manufacturing, medical manufacturing and 3D bio printing with living cells to 3D home printing with desktop applications.

The different areas, do however, differ in terms of maturity (Piller, Weller, & Kleer, 2015; Stroh, 2015). As Gartner (2014) reports a technology's life cycle stage indicates excerpt, 3D printing in the industrial area is already an established technique, whereas 3D printing technology in the consumer area or 3D bio printing technology is at earlier stages of development. Thus, the future of additive manufacturing technology is still uncertain and dependent on the specific area of technological application.

Whether or not a specific 3D application is technically possible or economically feasible is largely dependent on its production volume, part size, complexity, and material cost. 3D printing technology is most widely used in applications with low production volumes, small part sizes, and having complex designs. According to Fischer and Henkel (2012), 3D printing technology is cost effective with plastic injection molding on production runs of 50 to 5,000 units. According to Lu (2012), 3D printing technology is a competitive technology, whose plastic injection molds on production runs of 1,000 items. In the future, some experts (McKinsey & Company, 2013; Royal Academy of Engineering, 2013) feel that the range of efficient production may be further increased as raw material costs drop in price; this would occur as more companies adopt 3D printing technology to produce finished goods and as final consumers begin to purchase 3D printers. 3D printing technology is also especially suitable for manufacturing parts that are small in size, such as ball bearings (Lu, 2012). Advances in 3D printing technology have enabled this technology to be employed across a broader range of applications (Smith, 2011).

Impact on the Market Scope of Business

In order to fully exploit the strategic scope of business success in 3D printing technology, key players must carefully select their fields of market scope due to the fast-paced digital distribution platforms. However, an emergence of a local integrated channel of supply with local delivery being made via real time Web delivered the next day. The competitive edge of this technology is being capitated by the disintegration of traditional technology. Disintegration of traditional market scope channel is caused by large vertical integration of market generalists relying on price discrimination of distribution channels to control the market. These choices apply to markets segments and channels of distribution. Market scope has to do with price discrimination and the specific approaches to promote and sell products in channels of distribution (Tekic & Kukolj, 2013). Market scope is related to segmentation strategies and is used when the firm serves several markets or segments (Smith, 2015).

The sector of 3D printing technology manufacturing expected to experience a robust growth of more than 10% during the next 5 years. Wohlers Associates consider that, by 2015, the sale of products and services of additive manufacturing will reach 3.7 billion US dollars at the world level, and that, by 2019, it will exceed $ 6.5 trillion.

According to Piller, Weller, & Kleer (2015), products can be customized for a single purpose or created in small and economical production runs. Many 3-D printing technology companies and experts argue that they have tapped into a market they did not originally intend to go after. According to magazine Business Week, the income of the manufacturers of equipment in 3D printing technology is estimated at 800 million dollars with a profit margin of 96 million profits. This technology experienced a 10.3% growth in market scope on average since 2013 and should show a growth of 12.3% of 2013-2018. We find in North America, 112 active companies operating in the 3D printing. The 3D technology has the following impacts, which define the market scope: Defence, aeronautics and health. For the defence and aeronautics sectors, the advantages are the reduction of weight and the consolidation of numerous elementary parts and operations of assembly in a single complex element. The manufacturers of engines of aircrafts plan to use metallic components of additive manufacturing in the next generation of jet engines. In aeronautics sector, large companies of Quebec are involved in the production of parts leading to a decrease in the weight of planes and as well as to an increase in the efficiency of energy consumption via the realized products.

For the Healthcare sector, the 3D printing technology has greater stake in biomedical engineering with the production of organs such as ears and live tissues. For instance, the Chinese developed a specialization in the printing of human organs through biological printing using living cells. The medical and dental segments adopted the additive manufacturing because of its capacity to produce unique parts for a patient, without resorting to any muscle, matrix or tool whether it is. An example can be cited from the hearing aids which marry the shape of the ear, the invisible parts, the crowns and the dental bridges. When new technologies emerge in the market, it is quite clear that there is a broad market scope to be covered in terms of product reach. In this regard, the decline of industries is never rapid, nor is it immediate; implying that new radical changes have to be made if the already existing market players have to remain relevant in the wave of changing technology drive market platforms. Several other sectors such as that of automobile and transport, cinematographic industry, and jewelry already use 3D printing technology in the consumption of parts technically difficult to realize.

In the areas of customizable 3D printed implants, as well as growing number of non-load bearing prosthetic devices, 3D printing technology is enabling treatment of patients with non-life-threatening injuries or conditions where previously treatments weren't viable (Leila and Beaudry, 2015; Lu, 2012). In summary, we can argue that whether its hearing aid shells, spinal implants, standard hip cups, or surgical guides, a key reason 3D printing technology has so much market potential to revolutionize medical treatment is because of opportunities for scale production aligned with markets where high volumes of parts are required. Before businesses can capitalize on the 3D emerging opportunities for innovation, they need a framework for market success; one which covers emerging diverse technological product scopes. Therefore, establishing a 3D printing market scope driven is only possible through value creation and product technovation.

Impact on Geographical Scope of Business

Geographical scope designates the number of geographical markets served. There is a need for 3D printing market players to develop their strategic reshoring and enhancement of multi-domestic geographical market based on cultural, climatic differences (lower cost to adapt to local tastes). In terms of 3D printing, strategic reshoring and enhancement of multi-markets that favor global products and brands should standardize all of a company's products for all of their global markets. The strategic approach adopted by 3D printing players should reduce a firm's overall costs by spreading investments over larger market scopes both at the local and international level.

Though economies of scale in the production and marketing of 3D printing technologies can be achieved through supplying global markets, global geographical players have experienced value destruction for standardized product traditionally sold in global markets. This is due to the fact that customer needs and interests worldwide are becoming more homogeneous.

Valverde, Chakravorti, & Fernandez (2015) state that in a more relevant trend, markets are often defined with respect to different categories of users; sometimes residential and business users fall within the same relevant market, sometimes they fall within two separate markets.

However, it is clear that first, some businesses have very different needs than residential users, and second that business users may not form a uniform category given their different nature and size (Hildebrand, 2015). In principle, market analysis and definition process should be conducted separately for every individual geographic market (Weyl, & White, 2014). Liebl (2015) cite that such cluster market definition helps to identify impending strategic risks in the markets that could hamper the competitiveness of any given geographical market scope.

In 3D printing technological platforms relating to disruptive technologies, regulatory and competition authorities in most economies have traditionally defined relevant markets as national in geographical scope (Staykova & Damsgaard, 2015). According to Ware (2013), this is not surprising given that geographical scope is often defined by the licensing regime: if the license is national, then so likely is the market. However, as the number, coverage and market share of alternative networks and alternative platform players' increases, competitive conditions may no longer be homogenous across the country. This, in turn, as an argument goes, should be reflected in competition analysis adopted by multisided players (Hildebrand, 2015).

Impact on Business Scope of Competencies

The scope of competencies has to do with the range of skills possessed by a company, and the decisions to complete an activity within boundaries of the company or to subcontract part of the company's activities. With respect to 3D printing scope of competencies, researchers have asserted that there is a need to create value chain with the market so as to have a sustained monetization path. The value capitalization and monetization of multiple scope economics coming from new strategic skills in customized should be based on complex additive manufacturing plan and digitization of product manufacturing process. Destruction of strategic competencies should be based on economies of scale of large and fully run plants and radically lower minimum plant efficiency scale.

Seidle (2015) states that strategic scope capitation in 3D printing technology can also be realized at the competence level. It results from accumulated knowledge and specialized skills, and constitutes an intangible asset, useful in a number of activity sectors. Scope of competence ensues from the utilization of an intangible asset for instance, an accumulated know-how of the company (Burk, 2013). The strategy involves using the firm's internal competences systematically, when these competencies have an economic potential for external transactions (Krech et al., 2015).

The main advantage of 3D printing technology is that a low number of goods can be produced at an inexpensive cost, as compared to traditional manufacturing, which typically requires higher volumes to at least cover costs (Seidle, 2015). In 3D printing, overhead required to invest in an inventory, and warehouse it, is reduced since items can be printed as needed. Traditional manufacturing methods typically require larger volumes of inventory to be produced and warehoused at one time (Turban et al., 2015). As more and more individual consumers gain the ability to engineer and produce their own goods, and 3D printing technology becomes a more efficient and cost-effective way to produce goods, there will be an opportunity for individuals to create new 3D printing technologies, disrupt industries, and potentially generate new sources of wealth (Geradin et al., 2011). As long as the technology is accessible, new businesses will continue to emerge.

Though 3D printing technology has leveraged on process improvement by offering strategically cheap but high-quality technology, its data management is a key potential risk: issues associated with data management are related to the need for substantial memory storage capacity, and not the manufacturing technology itself (Huang, 2015). In this sense, "rather than advancements in the machines themselves, software developments are what will 'drive the industry forward'" (Royal Academy of Engineering, 2013). It might be worth looking for insights from the development of electronic design automation (EDA) industry, which could be quite useful in predicting some of the future trends in the evolution of 3D printing software design tools (Macmillan et al., 2000). In this regard, companies failing to invest in 3D printing technologies may find it difficult to manage their data appropriately posing risk of data theft or any unplanned breakdown (Huang, 2015).

DISCUSSION

In this section, the author discusses the impact on the strategic scope of 3D printing technology. Strategic scope of the firm refers to the breadth of activities a business engages in. Scope of 3D printing encompasses mainly products and markets. The scope also seeks to delineate what is or is not involved in 3D printing ventures (Lu, 2012; Krech et al., 2015). The design of strategic scope of a firm is determined by four fundamental choices which serve to establish a company's field of operations. These fundamental factors are: product scope, market scope, geographical scope and scope of competencies. Product scope refers to the range of products and services offered. Market scope refers to the number of markets and segments served often channels of distribution (Vander Panne et al., 2015). Geographical scope designates the number of geographical markets served. Finally, the scope of competencies has to do with the range of skills possessed by the company, and the decision completes an activity within the boundaries of the firm or to subcontract part of the firm's activities (Fischer and Henkel, 2012).

On the basis of this research, first, we found that 3-D printing technology has an impact on a speck of strategic scope of a business (Bessen et al., 2012). The first impact of 3D printing technology is on project scope defined as the range of products and services offered. In this case, this research paper finds that the main leverage for traditional manufacturing comes from mass customization and personalization of redefined product ranges leading two forms of product proliferation (Burk, 2013; Krech et al., 2015). As for product scope, the main strategic risk to traditional manufacturing comes from the destruction of existing product scope and the value of economies of scope coming from product diversification on which is based tradition a traditional manufacturing to control the market offering. Secondly, it also has a strong impact as to do with the market Scope.

Second, there is an impact of market scope defined by channels of distribution and segmentation that businesses go through while offering products and services in the market (Leila and Beaudry, 2015). Market scope is related to relevant segmentation strategies and is used when a firm serves several market niches or segments. In terms of leverage for traditional manufacturing, the emergence of local integrity channels of supply via local delivery being made in delivering in real time by representing the main leverage. The strong customization combined with a real time analytics of the delivery process will lead to our re-definition of market scope of traditional manufacturing (Cozmei & Caloian, 2011; Krech et al., 2015).

On the other hand, 3D printing technologies with a significant impact represent a strategic risk for traditional manufacturing in terms of market Scope. This risk may lead to the disintegration of traditional market scope channels mainly, large vertically integrated market for market generalization list relying real time information on price discrimination by existing distribution channels to control the market (Leila & Beaudry, 2015).

Third, another impact of 3D printing technology on strategic scope has to do with geographical scope. Geographical scope could be defined as the rate of business function such as transportation delivery and servicing the market. Geographic scope also refers to the specific cultural and climatic conditions of an existing local regional market (Hang et al., 2011; Leila and Beaudry, 2015). The main leverage for traditional manufacturing comes from offshoring of business abroad. It is also associated with emergence of multi-domestic geographical market bays on cultural diversity and climatic differences of markets. In other terms, 3D printing technology lowers the cost to adapt to local standards, facilitating the emergence of multi domestic strategy. On the flip side, 3D printing technology leads to strategic risk for traditional manufacturing with a risk of value destruction for standardized product that are traditionally manufactured and sold in globally standardized market reaching mainly homogeneous markets or segments.

Finally, 3D printing technology has an impact on the scope of competencies of a business. The scope of competencies refers to the proportion of activities managed internally or outsourced externally (Pohlmann and Opitz, 2013). When a company manages all its activities internally without relying on suppliers or distributors, it is said to have a large scope of competencies. On the other hand, it represents leverage for traditional manufacturing since it facilitates value captation and monetization coming from multiple scope economies of mass customization; leading to development of new strategic scales, and the customization of products and services into small batches. It also leads to new skills regarding digitization of products and services and well as new skills in complex additive formula for 3D printing.

On the other hand, it represents a strategic scope to traditional manufacturing since it is it leads to destruction of 3D printing competencies mainly based on economy of scale to companies with large plant runs. 3D printing technology also lowers the minimum efficiency scales and therefore destroys the competencies associated with traditional large run scale-based competencies.

Strategic Recommendations and Implications

Our research though theoretical in nature has established that 3D printing technology is set to reconfigure diverse ecosystems. However, key question here is: what do companies need to own to be unique in terms of their 3D strategic scope? 3D printing technology which falls within the confines of additive manufacturing is a concept that allows companies to strategically acquire 3D printers that can produce many products. From these analyses, managers in 3D printing technology need a strong sense of involvement in their company's role in the world to make key decisions about which resource they will

Figure 5. Presentation of 3D impacts, leverage and strategic risks

Impact of the strategic scope on 3D Printing	Leverage for traditional manufacturing	Strategic risks for traditional manufacturing
Product scope	Mass customization and personalization of re-defined product range leading to product proliferation.	Destruction of existing product scope and economies of scope.
Market scope (segmentation and distribution channels	Emergence of a local integrated channel of supply with local delivery being made via real time Web delivered the next day.	Disintegration of traditional market scope channel for large vertically integrated of market generalists relying on price discrimination of distribution channels to control the market.
Geographical scope	Strategic reshoring and enhancement of multidomestic geographical market based on cultural, climatic differences (lower cost to adapt to local tastes).	Value destruction for standardized product traditionally sold in global markets.
Scope of competencies	Value captation and monetization of multiple scope economics coming from new strategic skills in custom and small batch product made based on complex additive formula and digitization of product manufacturing.	Destruction of strategic competencies based on economies of scale of large of full run plants and radically lowers minimum plant efficiency scale.

invest in—or divest themselves of. This is due to the fact that as more companies become freed from the numerous logistical requirements, strategic risks and challenges of traditional manufacturing, managers have to assess the value of their capabilities and other resources and how those elements and resources complement or compete with the capabilities of others.

In summary, given all the potential solutions and effectiveness of highly integrated 3D printing technology, business process re-engineering and management is set to become the most vital capability around the globe. Some of the leading companies that have excelled in this area have built out proprietary coordination systems and intellectual property (IP) to secure competitive advantage. Other emerging players have moved with speed to adopt and help to reshape standard packages and platforms created by leading global software companies.

And with these divergent competitive edges, new hybrid models will emerge which can be leverage up on. However, given the contrasting cost advantages between traditional manufacturing and 3D printing technology, each 3D printing company will have its place in market in the future. Practitioners have noted with keen interest that decentralization will most certainly occur only in scenarios where 3D printing products are not too complex and don't require any serious completion and/or assembly.

DIRECTIONS FOR RESEARCH

This chapter's analysis has provided reliable insights and opened a perspective for additional research to better categorize 3D printing management innovations according the benefits derived from the use of 3D printing technologies. However, additional work will be required to further structure this research perspective. Based on the concept of 3D printing technology and its application in industrial and healthcare sectors, this chapter concludes diverse research on the concepts discussed might facilitate better the categorization of disruptive 3D printing technologies and lead to a better study into this complex, economic and social phenomenon using a market/ industry perspective.

CONCLUSION

Assessing beyond the company-level analysis, key scholars, experts and practitioners of 3D printing technology have searched for motives of disruption from the context and environment of a company, because companies are often locked into commitments reflected in the anticipations of investors and analysts, public promises and goals, and existing relationships with financiers and suppliers (Hang et al., 2011; Fischer and Henkel, 2012). For instance, individual technologies cannot be commercialized unless aligned with other related technologies that have been developed prior to or together with their introduction. Thus, relationships with technological suppliers and partners are of great significance to the commercialization of disruptive products (Cohendet and Pennin, 2011; Bessen et al., 2012). In addition, early research has established that effective companies with substantial technological, legal or social uncertainty tend to undertake reorientations or quantum innovations that include disruptive innovation such as 3D printing (Layne-Farrar & Schmidt, 2010). Successful product design requires regular review and inputs from diverse origins. Fast collaboration with engineering, marketing and quality assurance which can empower designers' data integrity and security is paramount for 3D printing technology to functional effectively in a competitive environment (Olander and Hurmelinna-Laukkanen, 2015). When evaluating 3D printing systems, players should consider facility requirements and expertise needed to run the system; accuracy, thoroughness and diverse skills and proficiencies that will suffice proper accumulation of competent skills necessary to turnaround 3D printing technology (Bessen et al., 2012).

REFERENCES

Allen, K. R. (2015). *Launching new ventures: an entrepreneurial approach*. Cengage Learning.

Amshoff, B., Dülme, C., Echterfeld, J., & Gausemeier, J. (2015). Business Model Patterns for Disruptive Technologies. *International Journal of Innovation Management*, *19*(3). doi:10.1142/S1363919615400022

Bechthold, L., Fischer, V., Hainzlmaier, A., Hugenroth, D., Ivanova, L., Kroth, K., & Sitzmann, V. (2015). 3D printing: A qualitative assessment of applications, recent trends and the technology's future potential. *Studien zum deutschen Innovations system*.

Bessen, J., Ford, J., & Meurer, M. M. (2012). The private and social costs of patent trolls: Do nonpracticing entities benefit society by facilitating markets for technology? *Regulation*, *34*(4), 26–35.

Bhardwaj, G., Agrawal, A., & Tyagi, R. (2015). Combination therapies or standalone interventions? Innovation Options for pharmaceutical firms fighting cancer. *International Journal of Innovation Management Vol.*, *19*(03), 1540003. doi:10.1142/S1363919615400034

Bland, S., & Conner, B. (2015). Mapping out the additive manufacturing landscape. *Metal Powder Report*, *70*(3), 115–119. doi:10.1016/j.mprp.2014.12.052

Bower, J. L., & Christensen, C. M. (2015). Disruptive technologies: Catching the wave. *Harvard Business Review*, *12*, 350–365.

Burk, D. L. (2013). Patent reform in the US: Lessons learned. *Regulation*, *35*(4), 20–25.

Capdevila, I. (2015). Co-working spaces and the localized dynamics of innovation in Barcelona. *International Journal of Innovation Management*, *19*(03), 1540004. doi:10.1142/S1363919615400046

Choi, J. Y., Das, S., Theodore, N. D., Kim, I., Honsberg, C., Choi, H. W., & Alford, T. L. (2015). Advances in 2D/3D printing of functional nanomaterials and their applications. *ECS Journal of Solid State Science and Technology*, *4*(4), 3001–P3009. doi:10.1149/2.0011504jss

Cohendet, P., & Pénin, J. (2011). Patents to exclude vs. include: Rethinking the management of intellectual property rights in a knowledge-based economy. *Technology Innovation Management Review*, *1*(3), 12–17. doi:10.22215/timreview/502

Cozmei, C., & Caloian, F. (2012). Additive Manufacturing Flickering at the Beginning of Existence. *Procedia Economics and Finance*, *3*, 457–462. doi:10.1016/S2212-5671(12)00180-3

Cutting, S. T., Meitzen, M. E., Wagner, B. P., Backley, C. W., Crum, C. L., & Switzky, B. (2015). Implications of 3D printing for the United States Postal Service. In Postal and Delivery Innovation in the Digital Economy (pp. 43-54). Springer International Publishing.

Edmond, R. (2015). Five business opportunities surrounding 3-D printing. *The Channel Co*. Retrieved from http://www.itbestofbreed.com/slide-shows/five-business-opportunities-surrounding-3-d-printing/page/0/1

Fischer, T., & Henkel, J. (2012). Patent Trolls on Markets for Technology– An Empirical Analysis of NPEs' Patent Acquisitions. *Research Policy*, *41*(9), 1519–1533. doi:10.1016/j.respol.2012.05.002

Gasparin, M., Micheli, R., & Campana, M. (2015). Competing with networks: a case study on the 3D printing. In *Proceeding 1st International Competitiveness Management conference, EIASM*, Copenhagen

Geradin, D., Layne-Farrar, A., & Padilla, A. J. (2011). Elves or trolls? The role of nonpracticing patent owners in the innovation economy. *Industrial and Corporate Change*, *21*(1), 73–94. doi:10.1093/icc/dtr031

Hang, C. C., Chen, J., & Yu, D. (2011). An Assessment Framework for Disruptive Innovation. *Foresight*, *13*(5), 4–13. doi:10.1108/14636681111170185

Hollister, S. J., Flanagan, C. L., Zopf, D. A., Morrison, R. J., Nasser, H., Patel, J. J., ... Green, G. E. (2015). Design control for clinical translation of 3D printed modular scaffolds. *Annals of Biomedical Engineering*, *43*(3), 774–786. doi:10.100710439-015-1270-2 PMID:25666115

Huang, C. H. (2015). Continued evolution of automated manufacturing–cloud-enabled digital manufacturing. *International Journal of Automation and Smart Technology*, *5*(1), 2–5. doi:10.5875/ausmt.v5i1.861

Huang, Y., Leu, M. C., Mazumder, J., & Donmez, A. (2015). Additive manufacturing: Current state, future potential, gaps and needs, and recommendations. *Journal of Manufacturing Science and Engineering*, *137*(1), 014001. doi:10.1115/1.4028725

Krech, C., Rüther, F., & Gassmann, O. (2015). Profiting from invention: business models of patent aggregating companies. *International Journal of Innovation Management*, *19*(03), 1540005. doi:10.1142/S1363919615400058

Layne-Farrar, A., & Schmidt, K. M. (2010). Licensing complementary patents: Patent trolls, market structure, and excessive royalties. *Berkeley Technology Law Journal*, *25*, 1121–1143.

Leila, T., & Beaudry, C. (2015). Does government funding have the same impact on academic publications and patents? The case of nanotechnology in Canada. *International Journal of Innovation Management*, *19*(03), 1540001. doi:10.1142/S1363919615400010

Lu, J. (2012). The myths and facts of patent troll and excessive payment: have nonpracticing entities (NPEs) been overcompensated? *Business Economics (Cleveland, Ohio)*, *47*(4), 234–249. doi:10.1057/be.2012.26

McKinsey & Company. (2013). *Disruptive Technologies: Advances That Will Transform Life*. Business, and the Global Economy.

Olander, H., & And Hurmelinna-Laukkanen, P. (2015). Perceptions of Employee Knowledge Risks in Multinational, Multilevel Organisations: Managing Knowledge Leaking And Leaving. *International Journal of Innovation Management*, *19*(03), 1540006. doi:10.1142/S136391961540006X

Pearce, J. M. (2015). Applications of Open Source 3-D Printing on Small Farms. *Organic Farming*, *1*(1), 19–35. doi:10.12924/of2015.01010019

Piller, F. T., Weller, C., & Kleer, R. (2015). Business Models with Additive Manufacturing—Opportunities and Challenges from the Perspective of Economics and Management. In *Advances in Production Technology* (pp. 39–48). Springer International Publishing.

Pohlmann, T., & Opitz, M. (2013). Typology of the Patent Troll Business. *R & D Management*, *43*(2), 103–120. doi:10.1111/radm.12003

Royal Academy of Engineering. (2013). *Additive Manufacturing: Opportunities and Constraints*. London: Royal Academy of Engineering.

Seidle, R. (2015). Organizational Learning Sequences In Technological Innovation: Evidence From The Biopharmaceutical And Medical Device Sectors. *International Journal of Innovation Management*, *19*(03), 1540007. doi:10.1142/S1363919615400071

Shrestha, S. K. (2010). Trolls or Market-Makers? An Empirical Analysis of Nonpracticing Entities. *Columbia Law Review*, *110*, 114–160.

Smith, D. (2014). Finding the signal in the noise of patent citations: How to focus on relevance for strategic advantage. *Technology Innovation Management Review, 4*(9).

Smith, D. (2015). Disrupting the Disrupter: Strategic Countermeasures to Attack the Business Model of a Coercive Patent-Holding Firm. *Technology Innovation Management Review*, *5*(5), 5–16. Retrieved from http://timreview.ca/article/894

Stroh, P. J. (2015). Business Strategy—Creation, Execution and Monetization. *Journal of Corporate Accounting & Finance*, *26*(4), 101–105. doi:10.1002/jcaf.22055

Tekic, Z., & Kukolj, D. (2013). Threat of Litigation and Patent Value: What Technology Managers Should Know. *Research Technology Management*, *56*(2), 18–25. doi:10.5437/08956308X5602093

Turban, E., King, D., Lee, J. K., Liang, T. P., & Turban, D. C. (2015). Business-to-business E-commerce. In *Electronic Commerce* (pp. 161–207). Cham: Springer. doi:10.1007/978-3-319-10091-3_4

Van der Panne, G., Van Beers, C., & Kleinknecht, A. (2015). Success and Failure of Innovation: A Literature Review. *International Journal of Innovation Management*, *7*(3), 309–338. doi:10.1142/S1363919603000830

Vogel, C., Schindler, K., & Roth, S. (2015). 3D Scene Flow Estimation with a Piecewise rigid Scene Model. *International Journal of Computer Vision*, *115*(1), 1–28. doi:10.100711263-015-0806-0

Whalen, P. S., & Akaka, M. A. (2015). A dynamic market conceptualization for entrepreneurial marketing: The co-creation of opportunities. *Journal of Strategic Marketing*, *24*(1), 61–75.

Chapter 2
Addressing the Feasibility, Suitability, and Sustainability of the Blockchain

Renaud Redien-Collot
Institut Friedland, France

ABSTRACT

This chapter applies stakeholder theory in order to evaluate whether a blockchain community is demonstrating a communicational maturity in order to achieve its technical, social, and political agenda. The consistence of the mission of an organization or a community is clearly reflected in the eyes of its stakeholders. Therefore, the study adopts a qualitative lens in conducting 11 semi-structured interviews with experts that are prominent international stakeholders of the blockchain in order to gain a deeper understanding of their internalized perception about this technology and its social network. According to the results, the blockchain community members are ready to address the feasibility of their technology and its implications. They also address some aspects of the social suitability of their network. However, they do not fix clear conditions of communication and coordination to discuss the sustainability of the whole organization.

INTRODUCTION

Most of the work currently being done on the Blockchain is questioning its technological robustness and its optimization; other studies published by the media and social networks discuss the relevance of the business model promoted by Blockchain. However, the Blockchain also claims to have a social and political agenda. It is from its conflicting point of view that this agenda has emerged in the media. Following the exploitation of a loophole in the coding of the investment fund (The DAO), the crisis of the summer of 2016 showed that the only rule of the incorruptible code bent before that of its most influent contributors. The co-decisional process was distorted. In this context, the understanding of the Blockchain's social dimension cannot be reduced to the journalistic analysis of its conflicts of interest.

DOI: 10.4018/978-1-5225-7262-6.ch002

Blockchain is a complex and mutating entity that is, at the same time, a technology, a social network that exchanges crucial information and a challenging community. As blockchain establishes a control of users' behaviors in several contractual situations in order to guarantee and intensify economic exchanges, its members and stakeholders should open the proper space to discuss its social and political acceptability, that is, its feasibility, suitability and sustainability. Public media and public discourse randomly addresses these important issues. The contribution of this chapter is to determine whether the present community members of the blockchain are ready to discuss these issues, that is, to demonstrate stabilized social representations of their missions. Stakeholder theory emphasizes that an organization or an organized community may achieve their goals when they are ready to explore the different aspects of their future with the most appropriate addressees (Scherer & Palazzo, 2006; Scherer, Palazzo & Zeidl, 2012). In order to do so, they have to demonstrate stabilized believes in their ability to discuss and frame the future of their network and it has to be reflected in the eyes of their closer stakeholders (Voetglin & Scherer, 2017).

The existing literature that studies the blockchain is only examining the feasibility and the advantages and the pitfalls of this technology (Buterin, 2014; Rushkoff, 2016; Pilkington, 2016; Hileman and Rauchs, 2017). However, the blockchain is not only relying on a technology and a business model but this is also an organization that has to gain legitimacy and deploy a stakeholder approach in order to be sustainable. As the blockchain is not yet a common technology, its stakeholders who are well-informed are still very few. Therefore, we did not conduct a quantitative survey but, at the end of 2017, we used a qualitative approach with 11 semi-structured interviews with experts that are prominent international stakeholders of the blockchain in order to gain a deeper understanding of their internalized perception about this technology and its social network.

The results of this chapter show that, in their social representations, the members of the blockchain community attempt to stress their differences with the internet community. In this regard, we can observe two breaking points in their perception of the internet community and the formulation of a mission that may help to accomplish what present internet culture has failed to concretize. First, in their claim of technological disruption, they have constantly attempted to construct an asymmetrical model vis-à-vis internet. Secondly, blockchain people criticize the opportunistic mainstream posture of the geek generation and seek to re-articulate political and technological dimensions in their professional experience. More importantly, they promise to create a true society of the distributed information. However, we can observe clear discrepancies and even epistemological gaps in these shared and formulated social representations. This may endanger the ability of blockchain community members to address properly their stakeholders and assert its legitimacy both in society and cyber communities.

After a literature review that will examine the building of social legitimacy by novel organizations, the chapter will present its research design and methodology, then its results and their discussion.

LITERATURE REVIEW

Blockchain is both a technical and social innovation. However, there are no exhaustive studies concerning the social structure and interactions of the blockchain communities. It is true that blockchain is still an emerging social phenomenon that may be difficult to capture. In order to appreciate the robustness of its innovativeness, it is important to observe whether this social phenomenon is gathering the conditions of possibilities of its sustainability. According to the stakeholder theory, the sustainability of an organization or a community relies on its ability to formulate open scenario for the future with the appropriate stakeholders. We first synthetize the different academic descriptions of the blockchain. Then, we will see how the blockchain may perceive itself as an innovation community and what it implies. Finally, we will see how the stakeholder theory can help to evaluate the degree of social and organizational maturity of the blockchain.

The blockchain initiates a set of interactions that characterize an era of post-computer mediated communication. According to the specialized blog of the Wall Street Journal, blockchain corresponds to a chain of transactions collected in a registry shared and distributed on the internet. One speaks also of distributed ledger (DLT) technology (Hileman and Rauchs, 2017). Each transaction is a block of data. Each member of the network can create a new transaction with cryptography in a secure way and access the registry. The cryptocurrency is the asset of the blockchain.

Structurally, the blockchain constrains users' behaviors and stimulate members' motivation in processing record keeping and verification activities (Pilkington, 2016). However, beyond its technological and constraining aspects, the blockchain is frequently presented by the media[1], the business[2] and political[3] reports as a collaborative adventure that aims to simplify (including financial) transactions by removing any form of intermediary on the basis of a protocol developed by the blockchain community (Rushkoff, 2016). In solving algorithmic problems, blockchain members create a new space of financial and business transactions for people who may join their community and their values. Apparently, communication is not the core of this type of network. Direct value creation sustains its deployment (Pilkington, 2016).

Several papers stress that, in their efforts to concretize the blockchain, its founders are not only concerned by its structural dimension. They attempt to relate this technology with its various contexts in reflecting upon their practices and their choices (Buterin, 2014; Hileman and Rauchs, 2017). In doing so, though, their discourse is less reflective than self-celebratory. They accumulate examples of beneficial outcomes that can be withdrawn from the Blockchain. For example, they stress ad nauseam that the technology will create fully interactive and secure blockchains for people without banking infrastructure (Hileman and Rauchs, 2017), will facilitate the management of digital identity and will help to create networks of energy autonomy (photovoltaic, biomass or wind) that could sell their surplus and redistribute the benefits to the community concerned, or negotiate the purchase or borrowing of energy supplements for the same community, the years of low production (Leloup, 2017).

Moreover, this reflection takes frequently a messianic tone that describes the deployment of the blockchain as a categorical imperative and not as a social construct that is the result of several types of internal and external interactions. As it is a distributed system, the blockchain does not promote any form of central governance. (Leloup, 2017). Therefore, decision-making processes are collective. However, the different conflicts that have taken place in 2014 and during the summer 2016 have shown that the system of collegiate decision was very weak. If they refuse any type of governance and promote transparent debates in their networks, blockchain community members should clarify the dynamic that guides their innovation community. This can help them to maintain their cohesion in spite of their divergences. As Klerkx and Aarts (2013) point it out, oversimplified approaches of innovation communities as unified teams of champions are misleading. In their article, they contrast primary innovation communities that act as coordinated orchestrators of their own network and secondary innovation communities that have to struggle and negotiate over time in interaction in order to elaborate their own coordination. Blockchain communities seem to belong to the second category. However, in small group dynamic, coordination can be achieved if the community is able to construct collective social representations (Pinto et al., 2016). These representations nurture the formulation of specific norms and possibly the elaboration of a mission. At the same time, the formulation of norms will encourage a set of transgressions and will weaken the relevance of the mission. The members of the community have also to dedicate some time to the discussion and the reflection upon these transgressions.

Blockchain community members are deeply involved in socio-constructed interactions that have the power to produce social representations and reality (Berger and Luckman, 1971). However, in the context of intensive value production in an innovation community, social representations may proliferate and reveal themselves antagonist (Klerkx and Aarts, 2013). This proliferation can prevent its members to converge and to discuss with its stakeholders in order to find satisfying collective solutions and, more importantly, to establish legitimate decision-making processes that are supposed to replace central governance. Blockchain community is an organization that attempts to promote bottom-up approach. In this context, stakeholder theory stresses that this goal can be achieved if the participants have stabilized the social representation of the mission of their organization (Scherer, Palazzo & Zeidl, 2012). This stability is effective when it is reflected in the eyes of their closer stakeholders (Voetglin & Scherer, 2017).

RESEARCH DESIGN AND METHODOLOGY

The objective of this chapter is to determine whether the present community members of the blockchain are ready in 2017, after more than nine years of development, to discuss its feasibility, suitability and sustainability. As the blockchain is a distributed organization that does not have central governance, how do its members nurture a robust collective decision-making process? In other words, do they share common beliefs, that is, common and stabilized social representations in order to articulate efficiency and democracy, that is, to process their activity and welcome divergence and convergence in their debates?

As the blockchain is not yet a common technology, its stakeholders who are well-informed and well-engaged are still very few. Therefore, we did not conduct a quantitative survey but, at the end of 2017, we used a qualitative approach with 11 semi-structured interviews with experts that are prominent international stakeholders of the blockchain in order to gain a deeper understanding of their internalized perception about this technology and social network.

In the context of emerging innovations, Camagni (2017) suggests that the testimonials of the technology developers are either too technical or too evangelical to reflect on their effective social and organizational dynamics. Apparently, peripheral actors may elaborate more consistent and relevant social representations of a community that develops an emerging innovation. More importantly, this study wants to explore the blockchain stakeholders' perception. Therefore, our research design relies on a set of long interviews with two different categories of people who are facilitating the deployment of the blockchain technology but do not play a central role in its exploitation. We have extensively interviewed a panel of 5 investors and a group of 6 technological experts who are very familiar with the development of the blockchain. Investors were coded as follows: I1, 2, 3, 4 and 5. Technical experts were coded as TE 1, 2, 3, 4, 5, and 6. We have selected a sample of respondents that is demographically as diversified as possible in order to capture a comprehensive overview of the blockchain stakeholders' perceptions. The majority of the respondents show a clear expertise as they have been familiar with the blockchain since at least 6 years. As the interviewees come from very different backgrounds, the convergence of their points of view will reflect the present beliefs and social constructs at work in the blockchain community members' minds. In the following table, we can see demographic details concerning the sample of our respondents.

Table 1. Demographic characteristics of the sample of respondents

	Sex	Age	Nationality	Position	Exposition and Familiarity With the Blockchain
Investors					
I1	Female	42	UK	Independent	8 years
I2	Male	27	USA	Executive	7 years
I3	Female	33	Denmark	Senior Manager	6 years
I4	Male	54	France	Executive	8 years
I5	Female	25	Germany	Independent	7 years
Technical Experts					
TE1	Male	35	Spain	Engineer	6 years
TE2	Male	42	Greece	Senior Manager	7 years
TE3	Male	29	Australia	Independent	8 years
TE4	Female	63	USA	Executive	7 years
TE5	Female	38	Belgium	Engineer	7 years
TE6	Female	29	Canada	Independent	8 years

In our analysis of the results, we have adopted the point of view of the pragmatic sociology (Lemieux, 2012) that relies on an in-depth analysis of the social representations at work likely to determine the tendencies of its protagonists to act.

In order to identify the major beliefs of their stakeholders that guide the members of the blockchain community in their ability to construct their decision and their action, we have applied a discourse analysis on the discourse of their direct and even daily observers (Radu & Redien-Collot, 2013; Pinto, Marques, Levine & Abrams, 2016). The discourse analysis approach was used to understand how respondents attempted to objectify the blockchain members' behaviors and choices even though they were aware that the community was in permanent evolution. According to Alvesson and Karreman (2000), there are two principal techniques for analyzing discourse. This study adopted a mixed approach (Maingueneau, 1991; Redien-Collot and O'Shea, 2015). The first one takes into account the background context of the documents or descriptions to be analysed in order to focus on the meaning of the discourse in its situated dimension. Thus, we have selected two categories of observers in order to observe the convergence and divergence of their points of view. The second technique centres on the use of the language within a given discourse, without referring to the background context. The use of the mixed approach, whereby the research concentrated on the use of the verbatim language to be analysed and, at the same time, resituated each representation in the specificities of the observers' narrative.

RUPTURE BY INNOVATION

First, we will examine the social representations that may support the debates concerning the feasibility of the blockchain. According to their stakeholders' perceptions, blockchain members stress that their technology has already achieved the rupture by innovation. This core belief promotes the transparency of the blockchain that can become an organizational metaphor.

According to I2, I3, I4 and TE1, 5 and 6, blockchain technology demonstrates its robustness at least in four ways:

- It allows exchanges with its own cryptocurrency[4] whose flows are monitored by the community of participants to limit any form of inflation;
- It stores blocks of transaction validated and protected by a cryptographic solution (an algorithm); the clauses and documents associated with each transaction are collected in each block in a 'smart contract' [5]; This smart contract allows an inexorable application of the transaction on which the parties have agreed.
- It offers transparent information processing as a ledger stores all transaction blocks, once they are validated.
- It allows individual members to play several roles: user, developer, manager or data scientist, which, in the area of transactions, are often well compartmentalized.

The majority of respondents stress that the blockchain offers an innovative transaction modality. It is not guaranteed by a third party but by an algorithm. Seven respondents suggest that the blockchain value proposition is based on a conventional and social representation that aims to break from the 'restricted' model of innovation promoted by internet champions. As a convention that imposes a number of constraints on developers and speculators, the architecture of the Blockchain is asymmetrical to that of the internet. I2 notes: "For the internet, the protocol has no value; it belongs to the public domain (non-profit organizations, universities); on the other hand the value is in the application because it is there that are captured the data which are monetized - resold, treated, analyzed etc... For the Blockchain, the protocol holds the value: it is very efficient, integrates security features; it is a public good that is distributed to a small number of miners and founders of the Blockchain; the application and its data are automatically delivered via the ledger!"

As a common good, because it is not paying, the internet protocol is undervalued. The Blockchain tends to bring the research and development forces back to protocol, as a common good held by an enlightened oligarchy. For the first designers of the Blockchain, the protocol, if refined, must improve the analysis of this tool which is a place of registration of all transactions. According to Loubet and Epié (2015) who are prominent technical experts, we could better compare for a given activity the capture of value over a relatively short time (which remains the only indicator retained today) and the effective creation of value of the activity if it is subtracted from all the direct and indirect costs (which are hidden most often). Among the hidden costs of the flamboyant economic models of the net are, among other things, the rising energy costs.

To fully understand this break in the innovation model, TE 3 recalls that Blockchain communities frequently criticize the economic models of innovation giants whose certainly very efficient tools (accounting) promote scenarios of very patrimonial investments as well as temptations of frauds as old as time: "The economic models of the most innovative companies of our century are ultimately innovative but they are based on accounting rules which date from the 13th-14th century, the spirit of which was the most often forgotten... even to go back to very patrimonial strategies or even avoid paying taxes to preserve the interest, even though there is a social demand for more ethics from consumers and innovation leaders..." Transparency of the Blockchain would measure whether a model of economic celebrity is as innovative and virtuous as it claims. More rigorous than a balance sheet, the ledger could provide tangible evidence of a truly holistic approach to business innovation.

However, as three respondents point it out: "For many members, the frequent celebration of the transparency of the blockchain is a pretext. This technical transparency gives the illusion that any internal conflict can be easily solved!"

RUPTURE BY THE PROFESSIONAL AND SOCIAL MODEL

As it can be useful for private and public purposes (such as the management of digital identity), the economic social suitability of the blockchain seems to be obvious. However, blockchain community members question the suitability of their own social role. In order to justify their own social and political role, they contrast it with the professional choices of the mainstream cyber people. They want to stress a rupture with the existing models in their professions. This can give them legitimacy in addressing specific social and economic challenges such as the management of energy autonomy or the land registry.

In the eyes of their 11 stakeholders, the blockchain community is positioned in relation to a first social group - a professional group, that of the geeks of the internet galaxy who develop multiple projects, belonging to a community that claims to be universal but remains little or barely engaged (concerned) compared to what is developed in the long term on a site or a platform. The Blockchain actors also define their mission in contrast with that of the members of the social networks where it is a question of highlighting its point of view on the information, in general. According to I 3,4,5 and TE 1,4,6, for an actor of the Blockchain, there are two contributions which participate directly in the deployment of the network:

- A technical contribution that, through the validation of algorithms, validates transactions and makes them visible
- A political contribution which is supposed to determine the orientation of the Blockchain in its choices of (technological) developments and projects

Through this double technical and political contribution, respondents underline that Blockchain members seek to deepen their awareness and even their sense of responsibility for what is technologically deployed through them. By claiming an awareness at work in their professional purpose and standing apart from geeks and mere members of social networks, they seek to anchor legitimacy and authority to engage the whole of society to change with them. (Radu & Redien-Collot, 2008). This position of authority asserts a break away from the posture of the internet generation, which is more concerned with competitiveness than with the sustainability of technologies. At the same time, it marks an overcommitment of the individual by the technology (any relation is validated by an algorithm) and for the technology (any individual knows what he did in the Blockchain since this is recorded).

In three respects, this form of rupture proclaimed by the Blockchain community can be qualified in the social relation that binds the individual to technology. Stakeholders note that there are several pitfalls and discrepancies in the blockchain community members so-called social and political mission.

First, I4 notes that blockchain involvement is not a break from the modus operandi of the internet generation but an extension via tokenization: "The compensation of Blockchain members for their expertise, especially in mining and algorithmic resolution - assumes that the more you contribute to the

Blockchain, the more you are paid and the more the network and its tokens gain in value; it is an extension and a realization of the value of social networks. Hence the attractiveness of the Blockchain that brings together those who want to give more value and meaning to their interactions in the networks... "The militant posture of miners is therefore questionable in this field." (TE 5)

In addition, I3 recalls that mining is industrializing. It is deindividualized and allows the power of decision making to fall into the hands of a few: "The universe of mining which was historically that of individuals has been strongly professionalized! There is a race for the speed of resolution of the algorithms because there is a rise in the number of transactions and an ever-greater flow management. Miners are increasingly equipped to meet this demand and work is organized in the following way:

- Solitary miners equipped with cards / machines solving algorithms (sold by industrialists) and in search of cheap energy.
- Miners organised in pools
- A few major mining companies such as Bitfury in the West (B toB operator which holds 9% of the mining market and Bitmain (China) (B to B to C operator which covers 10% of the market)"

Finally, TE6 emphasizes the transitory nature of the current technological development of our societies where there is a striking gap between their technological potential and their coordination and decision-making practices: "The main principles on the autonomy of Blockchain organizations remain theoretical. And consensus-based decision-making experiments do not take place (for example) in Ethereum; we are in a technocracy; only a few very involved people can act on the protocol; this gives rise to confrontations that have spiced up the two years that have elapsed."

THE PROMISE OF AN INFORMATION SHARING SOCIETY

According to its stakeholders, the blockchain will prove itself sustainable if it opens an era of mutual trust in information exchange. In order to achieve this agenda, the majority of respondents stress that blockchain community members must develop a more comprehensive approach of the symbolic and political dimensions of monetary credibility. Otherwise, blockchain will be perceived as "a post-liberal toy that entertains the elites, may change the daily transactional behaviors but may also create more dissensions in society and ruin the business climate" (TE1).

As mentioned in a note from the French Ministry of Higher Education in 2011, the information sharing society should be characterized by a powerful reflexivity that causes the digital world to take into account the effects it necessarily produces in terms of cross-fertilization of information, production of

Table 2. A brief definition

MINING AND MINER Mining: Use of computing power to process transactions, secure the network and allow all users of the system to stay in sync. Miner: persons (individuals or companies) who connect to the network with one or more machines equipped to perform the mining. Each miner is remunerated in proportion to the computing power he brings to the network.

Source: https://blockchainfrance.net/le-lexique-de-la-blockchain/

knowledge and behavior to guide the development agenda of its processes and tools. Reflectivity on the effects of digital is certainly present everywhere but it seems difficult to set a governance. In this regard, TE5 notes: "The current property regime limits the total fulfillment of a shared information society! [...] There is not an information cadaster! If there is a massive loss of information, there is no truly legal recourse! [...] Google is, for example, sovereign over information. It can close an account at its discretion."

In the Blockchain, the ledger is an account-keeping cadaster of supposedly incorruptible, shared and archived information. This cadaster remains a tool; even if it can be combined with numerous tools for cross-referencing and data analysis, ideally, only the community of users, and well beyond that, the human community, remains sovereign to determine the interpretation of the distributed information and especially the modalities of dialogue to engage in any interpretation of what is recorded. The lack of debate on how to interpret data in the digital world leaves a hint of silence that many benefit from.

In order to progress towards the discovery of solutions acceptable to everyone via the Blockchain, TE1 underlines that the Blockchain community must question the gains and losses of individual and collective sovereignty "in a society where the system of calculation is incorruptible?" The advent of the information sharing society must be based on high-performance tools but also on coordination rules that the digital community may not yet be ready to formulate.

Coordination within the blockchain community is not only a managerial issue. TE 5, I2 and I3 stress that miners and founders have not yet clarified the precise nature of the bond of trust produced by the cryptocurrency. According to the majority of the respondents (especially I5), members of the blockchain community do not distinguish three forms of the notion of monetary credibility[6]: methodical trust, hierarchical trust and ethical trust (Gueydier, Pujos, Redien-Collot, 2018). Methodical confidence is the most obvious to admit, it is part of a mimetic behavior that makes each person routinely believe in the face value of money and its persistence over time: money is socially accepted because of the social consensus that everyone expects others to accept money as a means of stable exchange. Hierarchical trust (credibility) refers to a more political mechanism that establishes the credibility of the issuer of the currency by a legitimate representative of the account and payment community. Finally, ethical trust (trust) refers to the symbolic authority of the system of collective values that money embodies through its modes of emission, distribution and circulation that must appear capable of ensuring the reproduction of society with respect for its values and compliance with the principles of justice (unlike narco-dollars and "dirty money" for example). I6 stresses that blockchain community members do not open clear debates about the transformation of transactional behaviors: "The blockchain facilitates transactions but does not eliminate negotiations; however, as transaction will be accelerated, people may lose the sense of reality when they negotiate before finding an agreement. I am not sure whether people who promote the blockchain technology want to explore these types of issues and open debates about these behavioral mutations."

It is not only the human community and the notion of trust that must be taken into account in the development of the Blockchain. Nature and energy expenditure are directly impacted. TE4 remarks that a major issue must be the horizon of any form of technological research: "How much more energy must be allocated to progress technologically?" Of course, the advent of the information sharing society via a regulated and enriched ledger with a thousand and one applications and sensors has a disproportionate energy cost! The question is all the more relevant when the energy requirement for the operation of the Blockchain increases by 25% per month, for an annual consumption level today equivalent to that of Denmark. The projections are even more frightening when it appears that by 2020 the Blockchain will consume more energy annually than the whole world consumes today!

The current progress with the Blockchain should not lock us into the three inevitable debates of feasibility, suitability or sustainability that must, however, be courageously pursued. At the very heart of the notion of technological progress, the Blockchain debate reminds us that there will be no acceptable scientific progress if science does not make room for politics. TE1 notes: "[...] with the Blockchain, technology puts the political debate at the center of economic and financial concerns and indirectly demonstrates that political criticism is useful to progress technologically... Bitcoin is a composition with the economic / financial world more relevant than the posture to protest (Occupy Wall street); Bitcoin creates social situations that should not occur, encounters, conversations that could no longer happen in society, because of compartmentalization."

The information sharing society will therefore not be accomplished without excessive effort to develop methodologies and tools that encourage debates, integrate their conclusions both in the technology and in the management of the network and the people. TE3 emphasizes that the Blockchain is precious because it reintroduces the value of the debate within the world of technologies. Let us not forget, however, that the technical nature of regulatory concerns can also kill this debate. This is perhaps why I4 invites those interested in the Blockchain to adopt a very fluid perception of what it produces in terms of financial ramifications. In the case where segmentations occur in the Blockchain deployment chain (forks), one should not see it as a fragmentation of the cryptocurrency value. It is the Blockchain as a resource that presents a rupture that takes on an almost organic character: we speak of a mutant resource. I1 notes: "[...] owning a bitcoin and a cash bitcoin does not mean having two types of currency; it is to possess two seeds, one original and the other a seed in mutation, each opening on different resources and skills (mining and more). Through this invitation to not see in the Forks of the Blockchain the only mark of a human will, punished by the depreciation of a currency, but to pay attention to what mutates, on what unfolds, it is a question of engaging the human community to found a crucial political debate neither on a principle of precaution nor on a principle of speculation but on a sustained principle of observation.

CONCLUSION

According to the results of this study and to their stakeholders' point of view, the blockchain community members are ready to address the feasibility of their technology and its implications. They also address some aspects of the social suitability of their network. Proponents of the Blockchain claim a first rupture of a technological nature attributed not only to the Blockchain tool but also to the model of innovation that it proposes. They assume a second type of socio-professional breakdown that links work and political commitment. However, they do not fix clear conditions of communication and coordination to discuss the sustainability of the whole organization. The question of realizing a socio-political agenda - that of

Table 3. A brief definition

FORK
For a "cryptocurrency," a "fork" is a modification of the rules that govern it. Some transformations can be minor and retroactively compatible - we speak then of a "soft fork" - or more serious and without ascending compatibility, it is a "hard fork." A "soft fork" can be content with the support of the majority of nodes and miners to become functional. A "hard fork" on the other hand requires a very broad consensus, even a unanimity.

Source: https://bitcoin.fr/hard-fork-soft-fork/

a distributed information society – does not encompass the notion of monetary credibility. The majority of the stakeholders who were interviewed thought that, within the blockchain community, there was neither clear nor deep debate about the specific nature of the bond of trust produced by the cryptocurrency. This important dimension could help the members of the blockchain community to converge or, at least, to explore solutions that are less dramatic than forks.

Even though the media seem to echo the messages that some blockchain gurus wish to diffuse, we can see that this community has neither anchored an internal consistent dialog nor an external discussion with their stakeholders. In the interviews we can notice, for example, that the stakeholders of the blockchain identify gaps and discrepancies that are not shared with the community members. According to the stakeholder theory, this reflects the lack of communicational maturity in the blockchain community. In this context, during periods of crisis, this may encourage the triumph of technocracy and prevent the ultimate achievement of the distributed dimension of the blockchain.

At the same time, this study reveals how it is difficult for an innovation network to open a space of debates that articulate a concern for technical efficiency and another one for political relevance.

REFERENCES

Aglietta, M. & Orléan, A. (Eds.) (1998). *La monnaie souveraine*. Paris: Odile Jacob.

Alvesson, M., & Karreman, D. (2000). Varieties of discourse: On the study of organizations through discourse analysis. *Human Relations*, *53*(9), 1125–1149. doi:10.1177/0018726700539002

Berger, P., & Luckman, T. (1971). *The Social Construction of Reality*. Harmondsworth: Penguin.

Boyd, D., & Ellison, N. (2008). Social Network Sites: Definition, History, and Scholarship. *Journal of Computer-Mediated Communication*, *13*(1), 210–230. doi:10.1111/j.1083-6101.2007.00393.x

Buterin, V. (2014), A next generation smart contract and decentralized platform (White Paper). *Ethereum*. Retrieved from https://cryptorating.eu/whitepapers/Ethereum/Ethereum_white_paper.pdf

Camagni, R. (2017). Technological change, uncertainty and innovation systems: Toward a dynamic theory of economic space. In R. Capello (Ed.), *Seminal Studies in Regional and Urban Economics* (pp. 65–97). doi:10.1007/978-3-319-57807-1_4

CoinDesk. (2016). CoinDesk's State of Bitcoin.

Deloitte, L. L. P. (2016). Blockchain Enigma. Paradox. Opportunity. Retrieved from https://www2.deloitte.com/content/dam/Deloitte/uk/Documents/Innovation/deloitte-uk-blockchain-full-report.pdf

Gueydier, P., Pujos, A., & Redien-Collot, R. (2018), Blockchain, the Challenge of Trust, Optic Humana Technologia, http://optictechnology.org/index.php/fr/news-fr/144-revolution-technologique-white-papers-fr

Hileman, G., & Rauchs, M. (2017). *Global Cryptocurrency Benchmarking Study*. Cambridge Judge Business School.

Klerkx, L., & Aarts, N. (2013). The interaction of multiple champions in orchestrating innovation networks: Conflicts and complementarities. *Technovation*, *33*(6-7), 193–210. doi:10.1016/j.technovation.2013.03.002

Leloup, L. (2017), DLT: alternative aux blockchains Bitcoin et Ethereum. Retrieved from http://www.finyear.com/DLT-alternative-aux-blockchains-Bitcoin-et-Ethereum_a36182.html

Lemieux, C. (2012). Peut-on ne pas être constructiviste? *Politix*, *100*(4), 169–187. doi:10.3917/pox.100.0169

Loubet, N., & Epié, C. (2015). *Blockchain and Beyond*. Retrieved from .https://blockchainfrance.files.wordpress.com/2015/12/cellabz-blockchain-beyond.pdf

Maingueneau, D. (1991). *L'Analyse du Discours: Introduction aux Lectures de l'Archive*. Paris: Hachette.

Ministère de l'Education Nationale, de l'Enseignement Supérieur et de la Recherche (2011), *Stratégie de recherche pour le numérique*. Retrieved from https://cache.media.enseignementsup-recherche.gouv.fr/file/Strategie_Recherche/28/1/Rapport_atelier_7_314281.pdf

Pilkington, M. (2016). Blockchain technology principles and applications. In F. X. Olleros & M. Zhegu (Eds.), *Research Handbook on Digital Transformation*. Cheltenham, UK: Edward Elgar. doi:10.4337/9781784717766.00019

Pinto, I. R., Marques, J. M., Levine, J. M., & Abrams, D. (2016). Membership role and subjective group dynamics: Impact on evaluative intragroup differentiation and commitment to prescriptive norms. *Group Processes & Intergroup Relations*, *19*(5), 570–590.

Radu, M., & Redien-Collot, R. (2008). The Social Representation of Entrepreneurs in the French Press: Desirable and Feasible Models? *International Small Business Journal*, *26*(3), 259–298. doi:10.1177/0266242608088739

Redien-Collot, R., & O'Shea, N. (2015). Battling with institutions: How novice female entrepreneurs contribute to shaping public policy discourse. *Revue de l'Entrepreneuriat*, *15*(2-3), 57–80.

Rushkoff, D. (2016). *Reprogramming money – Bank vaults to blockchain. In Throwing Rocks at the Google Bus, How Growth Became the Enemy of Prosperity*. NY, NY: Penguin.

Scherer, A. G., & Palazzo, G. (2006). Corporate legitimacy as deliberation, A Communicative Framework. *Journal of Business Ethics*, *66*(1), 71–88. doi:10.100710551-006-9044-2

Scherer, A. G., Palazzo, G., & Seidl, D. (2012). Managing legitimacy in complex and heterogeneous environments: Sustainable development in a globalized world. *Journal of Management Studies*, *50*(2), 259–284. doi:10.1111/joms.12014

Voetglin, C., & Scherer, A. G. (2017). Responsible innovation and the innovation of responsibility: Governing sustainable development in a globalized world. *Journal of Business Ethics*, *143*(2), 227–243. doi:10.100710551-015-2769-z

ENDNOTES

1. CoinDesk, 2016
2. Deloitte, 2016
3. World Economic Forum Report, July 2016
4. The cryptocurrencies are a unit of digital account. The validity of each unit comes from recording the first transaction that generates this unit and all transactions that follow.
5. Smart contracts are computer programs that check if the terms of a transaction that have been defined beforehand are met in order to execute such transaction. They are infallible, and their application is inexorable. Those entering in transaction must agree on the conditions of its execution.
6. Aglietta and Orléan (1998) affirm it, saying that money is a bond of trust.

Chapter 3
Study of Intentions to Use Cloud CRM Within Saudi SMEs:
Integrating TAM and TPB Frameworks

Karim Mezghani
Al Imam Mohammad Ibn Saud Islamic University, Saudi Arabia & University of Sfax, Tunisia

Mohammed AbdulAziz Almansour
Al Imam Mohammad Ibn Saud Islamic University, Saudi Arabia

ABSTRACT

This chapter attempts to study the intentions to use cloud-based CRM applications through a combination between a Technology Acceptance Model (TAM) and a Theory of Planned Behavior (TPB). To test the different links identified in the research model, a research questionnaire was prepared and sent to marketing managers within Saudi SMEs in Saudi Arabia. A total of 41 useful questionnaires were collected. The authors opted to the structural equation modeling (SEM) using the Partial Least Squares (PLS) to analyze data. The tests are prepared with XLstat software since it integrates both factor analysis and PLS modules. Among the main statistical analyses, the authors conclude that the TPB-TAM is suitable to study cloud CRM. From a managerial perspective, the authors expect that cloud CRM is perceived with good impression and that this new technology should be implemented strongly and gradually in SMEs to improve the quality of services provided to customers and organizations.

INTRODUCTION

It is very important to notice that the current world is developing toward the use of nascent technologies in many fields. In addition to that, there are lots of changes in the perceived aspects of providing services to the customers. These changes refer to the customers themselves as they are moving toward using high technologies in their social interactions. As such, firms need to find additional tools to interact more closely with customers. The development of several disruptive technologies linked to cloud computing represents a unique opportunity to perform a digital transformation in the way of interactions with customers.

DOI: 10.4018/978-1-5225-7262-6.ch003

This study aims to clarify how managers in Saudi SMEs (Small and Medium-sized Enterprises) will perceive and deal with cloud CRM (Customer Relationship Management) technology by studying the intentions to use cloud CRM applications through SMEs and the main factors that could influence such intentions.

The main goal of CRM is to create a good and long run relationship with selected customers. Indeed, according to Baran and Galka (2017), "the purpose of CRM systems is not simply to retain customers" but to "identify, retain, and please the right kind of customer and to foster their repeat usage".

When presented as a service, Cloud CRM would generate the benefits of CRM combined with the multiple benefits of cloud computing (high scalability, accessibility, cost saving, etc.). However, even beneficial, every cloud technology is presented as a risky alternative. Moreover, previous studies showed that many managers were reluctant toward cloud adoption and use (Smaoui Hachicha and Mezghani, 2018).

Based on a literature review and a survey research, the current chapter attempts to propose a research model that could be used to study Saudi managers' intentions to use cloud CRM. In fact, despite the existence of several studies on cloud adoption and use, few of them focused on cloud CRM issues.

The proposed model is based on a combination between the Technology Acceptance Model (TAM) and the Theory of Planned Behavior (TPB). TAM "is an information systems theory developed in order to make predictions about technology acceptance. This model suggests that when users are presented with a new technology, a number of factors influence their decision about how and when they will use. TAM is based on the causal relationship belief - attitude - intention - behavior within the theory of reasoned action (TRA)" (Lala, 2014). TAM is "designed to provide an adequate explanation for and a prediction of a diverse user population's acceptance of a wide array of IT [Information Technologies] within various organizational contexts" (Hu et al., 1999).

As stated by Venkatesh and Davis (1996), TAM factors can be combined with other external variables. A literature review shows that the integration of TPB-linked factors with TAM provides a deep understanding of intentions to use IT (Lee, 2009; Safeena et al., 2013; Yu et al., 2018). Indeed, besides largely supported in several academic researches, TPB and TAM present common factors (intentions and attitudes) which facilitate their integration.

By following a quantitative approach, the current chapter aims to test the TAM-TPB integration in the case of cloud CRM. From a managerial perspective, this chapter attempts to formulate some recommendations to practitioners in the Saudi context regarding cloud CRM issues. Indeed, even this context is characterized by a rapid development in cloud computing adoption and use, very few experiences about cloud concerns are reported in academic research focusing on such context (Alharbi et al., 2016).

LITERATURE REVIEW

Customer Relationship Management and Digitalization Challenges

According to Baran and Galka (2017), CRM can be defined from different perspectives, but the holistic definition is proposed by Payne and Frow (2005). These authors consider CRM as "...a strategic approach that is concerned with creating improved shareholder value through the development of appropriate relationships with key customers and customer segments. CRM unites the potential of relationship marketing strategies and IT to create profitable, long-term relationships with customers and other key stakeholders."

From this definition, it is clear that CRM is deemed to be a strategy that focuses on customers. However, in general, CRM is a technologically driven concept (Baran and Galka, 2017). So, it is not strange to see the multiple developments of such concept in parallel with the digital transformations occurring in the business field due to the large use of disruptive technologies (cloud computing, Internet of Things, artificial intelligence, big data analytics, etc.).

It is very important to notice that the current world is developing toward the use of disruptive technologies in each aspect of life. In addition to that, we can see that there are lots of changes in the perceived aspects of providing services to the customers. These changes refer to the customers themselves as they are changing toward using a large set of technologies in their social interactions.

As an example, the massive use of social media by the firms to perform marketing efforts created a new term: the social CRM. This notion is related to the integration of the social media applications with the CRM capabilities in order "…to engage customers in collaborative conversations and enhance customer relationships…" (Trainor et al., 2014). These authors affirm that the technological and social shifts favored by the huge use of social media applications at the individual level (mainly via mobile devices), leading to the emergence of a "social customer" who look for more interaction with firms through social media. Baran and Galka (2017) report that "sixty-five percent of adults now use social networking sites." This can be considered as an additional challenge for managers pushing them to rethink their CRM systems and perform more digital transformations in the way of dealing with customers.

Such transformations would generate more challenges for marketing managers. The use of nascent technologies is accompanied by the explosion of structured and unstructured data volume, which will lead to Big Data concerns. Indeed, besides the challenge of storing data generated by interactions with customers, the firms need to find the suitable tools (data mining, text mining, etc.) to analyze costumers-linked data (emails, posts, tweets, etc.). Baran and Galka (2017) stipulate that one of the recent trends is the use of Real Time Analytics. Such analytics refer to the use of "technologies that deliver data between devices as the data is being generated", which allows firms "to implement customized and timely CRM tactics" (Baran and Galka, 2017). These authors affirm that several analytics and data storage alternatives are cloud-based.

Between social media applications and Big Data concerns, it is clear that the CRM-linked digitalization challenges are related to the use of cloud computing, as a new way to manage customers' interactions and data.

Customer Relationship Management in the Era of Cloud Computing

As a result of the emerging CRM trends, the increasing integration of cloud computing with CRM applications is taking a clear rule. Many companies are "running CRM on a mix of on-premise and SaaS[1] (cloud) systems" (Baran and Galka, 2017).

As such, we can notice that the combination of the CRM and the clouding technologies will generate the term which is called "cloud CRM". The most suitable definition of this new trend is presented by Sharma et al. (2014) as "Cloud CRM (Customer Relationship Model) is a dynamic application that is developed with latest technologies in order to mark a remarking presence of CRM process. This is built as a web application".

"Cloud Computing refers to a model of computing that provides access to a shared pool of computing resources (computers, storage, applications, and services), over a network, often the Internet" (Laudon and Laudon, 2012). When firms use computing as a service, they don't need to deploy additional IT infrastructure or to spend much money for servers and software (Handoko and Gaol, 2012; Mezghani and Ayadi, 2016). In fact, cloud computing is a cost saver alternative.

The use of the cloud computing has a very noticeable effect of increasing the performance of the services provided to customers. As many firms today are trying to provide good services to customers, cloud-based tools could play a crucial role in making these services available every time and everywhere.

With cloud computing, organizations "do not have to worry about storage capacity management when there are sudden surges in customer data capture activity, such as from a website or in response to an organization's CRM effort" (Baran and Galka, 2017). These authors add that "cloud computing also supports the relationship effort, as all relevant customer data can be easily captured and managed by the organization as well as the customer. Ease of access for both parties is a benefit to both". Indeed, the customer can easily access his data using many devices in a secured reliable mobile environment (Baran and Galka, 2017). This aspect is linked to "availability", considered as a one of the main features of the cloud CRM and one of the major characteristic that an organization focuses on (Sharma et al., 2014).

However, despites the multiple benefits of integrating CRM with cloud computing, many individuals and firms are still reluctant to use Cloud CRM applications because of the risks of cloud technologies (security, data confidentiality, network dependency, vendor lock-in, etc.). Baran and Galka (2017) stipulate that the mix of on-premise and SaaS will continue to be the dominant CRM model in the next five years.

Given such benefits and risks, SMEs are facing a true challenge of switching or not toward cloud CRM. On one hand, cloud CRM is a suitable alternative for such enterprises to perform digital transformations in their interactions with customers without adding substantial costs. On the other hand, using cloud-based CRM applications would generate the risks linked to cloud computing.

Hereafter, the authors propose a research model that could be beneficial to examine the main factors that lead to the intentions to use or not of cloud CRM.

INTEGRATING TPB AND TAM TO STUDY THE INTENTIONS TO USE CLOUD CRM

Why Integrating TPB and TAM?

The theory of planned behavior (TPB) is developed by Ajzen (1991) who considers the human behavior to be planned and determined by a set of factors linked to perceptions and motivations. A central factor that predicts directly the behavior is linked to intentions. "Behavioral intentions are indications of a person's readiness to perform a behavior" (Fishbein & Ajzen, 2010). Such intentions are influenced by three main factors:

- Attitudes "refers to the degree to which a person has a favorable or unfavorable evaluation or appraisal of the behavior in question" (Ajzen, 1991).
- Subjective norms which "express the perceived organizational or social pressures of a person who intends to perform the behavior in question" (Lee, 2009).
- Perceived behavioral control which "reflects a person's perception of the ease or difficulty of implementing the behavior in question" (Lee, 2009).

"When analyzing IT literature, it is possible to state that most previous studies and frameworks emphasize on intentions as the main antecedent of IT adoption and use" (Mezghani, 2014). However, limiting the study of intentions to the TPB-linked factors is not sufficient to get a deep understanding of IT use. That is why, several researchers attempted to integrate other factors with TPB as antecedents of intentions to adopt or use IT (Shih and Fang, 2004; Lee, 2009; Aboelmaged, 2010; Ekufu, 2012; Mezghani, 2014; Yu et al., 2018).

Another dominant framework largely applied to study the IT use is the Technology Acceptance Model (TAM). This model is developed by Davis (1989) and attempts to specify the IT related perceptions that most influence the attitudes and intentions toward IT adoption and use (Mezghani and Ayadi, 2016). In reference to this theory, two main factors influence, directly and indirectly, the attitudes and intentions toward IT use:

- **The Perceived Usefulness:** This factor refers to "the degree to which a person believes that using a particular system would enhance his or her job performance" (Davis, 1989).
- **The Perceived Ease of Use (PEOU):** This factor refers to "the degree to which a person believes that using a particular system would be free of effort" (Davis, 1989). This factor influences also the perceived usefulness "because technologies that are easy to use can be more useful" (Gangwar et al., 2015).

The use of TAM proved the robustness of this model in explaining the intentions to use different kinds of IT, even in case of disruptive technologies.

Rauniar et al. (2014) used the model of TAM as a main tool in explaining the effect of social media usage at the individual level. Through a survey study conducted with 398 users of Facebook, they confirmed the fact that PU and PEOU influence the intentions to use social media applications. However, Rauniar et al. (2014) revealed the importance to integrate other factors in order to get more explanation of intentions.

In order to study attitudes toward cloud computing, Mezghani and Ayadi (2016) applied TAM in a survey research within 74 Saudi firms operating in the financial sector. They found that the attitudes are closely linked to the PU and PEOU. As Rauniar et al. (2014), they affirm that the integration of other factors with TAM provides more explanation power to the research model.

When reviewing several studies applying TAM, it is clear to state that One of the most interesting integration is performed with TPB-linked factors. Indeed, besides derived from the same theory (theory of reasoned action), both TPB and TAM present "attitudes" and "intentions" as dominant factors to determine the personal behavior. Moreover, "The attitudinal beliefs of TAM can make up the precedent factors of attitude for TPB" (Lee, 2009).

In the study of Lee (2009), an interesting combination between TAM and TPB was performed. This combination has resulted in presenting the important factors of each model as enablers of the internet banking adoption. In their research, perceived benefits and perceived risks have been combined to show the importance of "positive" and "negative" factors in enhancing better implication for internet banking.

On the same path as Lee (2009) study, Safeena et al. (2013) attempted to study the internet banking use in India through a combination between TAM and TPB. Unlike Lee (2009) who kept "attitude" as a mediating variable, Safeena et al. (2013) used all factors of TAM and TPB as direct enablers of intentions to use internet banking. They found that all factors linked to perceptions have significant effects on intentions. Thus, they supported the usability of TAM-TPB integration.

Awa et al. (2015) has used the two models (TAM and TPB) to study e-commerce adoption within SMEs. In addition to those two models, this study has added the technology-organization-environment (TOE) to enhance the adoption of e-business technologies within SMEs. Awa et al. (2015) concluded that "the frameworks of extant TAM, TPB and TOE are insightful to the understanding of e-commerce adoption".

Yu et al. (2018) also combined TAM and TPB to study the intentions to use commercial bike-sharing systems. These authors noted that "TAM or TPB alone could not provide consistently superior behavior predictions… and there have been some empirical supports for the better exploratory power with the integration of TAM and TPB". Through a survey which yielded 286 persons and statistical analyses, Yu et al. (2018) supported the theoretical argument according to which the integration of TAM and TPB provides a greater explanatory power in predicting the use of commercial bike-sharing systems.

From the analysis of studies presented above, it is clear that the different integrations performed between TAM and TPB showed satisfactory results, regarding both classic and disruptive technologies. So, since cloud CRM is an emerging IT that is not yet largely used, it seems useful to adopt a TAM-TPB integration as a framework to study the cloud CRM use.

The Research Model and Hypotheses

As mentioned above, the research model developed in this chapter is based on the combination between TAM and TPB factors (Figure 1).

Regarding the TPB-linked factors, their effects on intentions to use cloud CRM can be justified as follow:

Figure 1. The research model

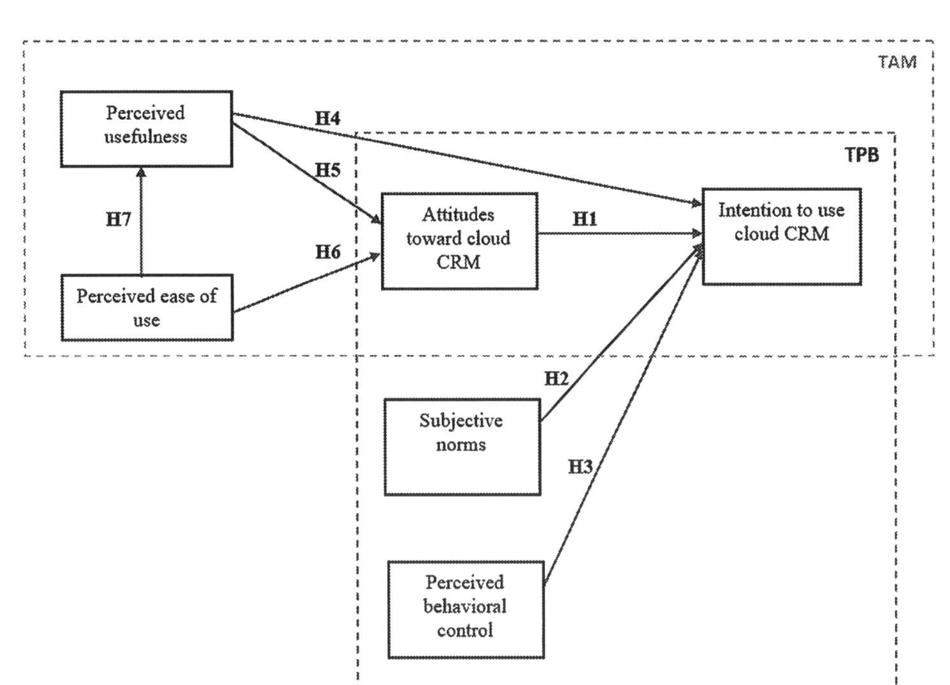

- **Attitudes:** In many studies deploying TPB to analyze IT use, it is argued that attitudes toward an IT are the main enablers of intentions. According to Yu et al. (2018), when people hold positive feeling toward the adoption of a specific IT, they tend to more frequently use it. In the case of a cloud-based system, Mezghani (2014) found that "when attitude refers to favorable or unfavorable evaluation, a manager would evaluate the opportunity of adopting cloud ERP to express his intention". Thus, it is possible to posit that attitudes toward cloud CRM positively influence the intentions to use a cloud-based CRM application.
- **Subjective Norms:** According to Payne and Frow (2005), CRM is considered as a strategic approach that aims to develop appropriate and long-term relationships with key customers. Regarding its strategic nature, the use of CRM applications could not be a usual decision, but rather a managerial choice imposed by the environment context and the strategic orientations of the firm. So, using CRM applications should be guided by organizational and environmental pressures. Such pressures would be greater when talking about cloud-based applications. From an organizational perspective, top managers are looking to use mobile technologies that are supported by cloud computing in order to get the benefits of mobile technologies (Fauscette, 2013; Mezghani, 2014). From an environmental perspective, Arnesen (2013) affirms that enterprise systems vendors "are in the process of developing hosted or cloud solutions as the market moves to a cloud environment". This can put more pressures on managers pushing them to use cloud-based solutions (Mezghani, 2014).
- **Perceived Behavioral Control:** Kidwell and Jewell (2003) state that the perceived behavioral control (PBC) can be externally and internally oriented. These authors affirm that a behavior is externally controllable "when it is perceived as easy to perform; that is, when it is relatively free of external or extrinsic influences that can act as a barrier toward behavioral performance" or when external helpful resources are available. A behavior is internally controllable "when an individual perceives that he or she possesses control over personal resources" (requisite skills, confidence,…) required to perform the behavior (Armitage et al. 1999; Kidwell and Jewell, 2003). Supporting this idea, Shih and Fang (2004) posit that "an individual with the self-assured skill to use a computer and the Internet is more inclined to adopt Internet banking". Based on these perspectives, as cloud CRM is an emerging technology based on Internet, a manager needs to assess his ability to use such technology before developing his intentions (internally oriented PBC). Besides, since cloud-based applications are deployed by a third party (the provider), a manager would intent to use cloud CRM if he feels supported by such party.

To sum up, the integration of the TPB factors allows to formulate the following hypotheses:

H1: The attitudes toward cloud CRM are positively related to the intentions to use a cloud-based CRM application.
H2: The subjective norms are positively related to the intentions to use a cloud-based CRM application.
H3: The perceived behavioral control is positively related to the intentions to use a cloud-based CRM application.

Regarding the TAM-linked factors, it is reported in Figure 1 that both PU and PEOU influence the intentions, mainly via the attitudes. More explanations are given below:

- **PU:** In line with TAM assumptions, PU is expected to favor the attitudes regarding a specific technology. It is important to note, however, that, when revising the original TAM, Davis (1989) removed the attitudes and found that PU could have a strong and direct link with intentions (Rauniar et al., 2014). Such perspective was adopted by many researchers in IT field (Thong et al., 2002; Venkatesh et al., 2002; Lee et al., 2007; Udoh, 2010; Rauniar et al., 2014; Gangwar et al., 2015). Although several studies limited their focus to only one effect (on attitudes or on intentions), the authors opted to keep both effects in the current chapter. Even the main objective is to study the enablers of intentions, considering the link between PU and attitudes is useful to get deeper explanations about cloud CRM use issues. According to Mezghani and Ayadi (2016), cloud based solutions are developed to facilitate doing business anytime and anywhere, meaning that managers would develop positive attitudes toward cloud solutions if they perceive that cloud computing is useful. Through a literature review on Internet-based technologies acceptance, Kalinic and Marinkovic (2016) stipulate that PU is considered as a direct predictor of intentions to use such technologies (Internet banking, m-commerce, m-health, etc.). Considering cloud CRM as an Internet-based application, it is possible to consider the direct link between PU and intentions to use cloud CRM.
- **PEOU:** In line with TAM, it is argued that when a person perceives that a specific IT is easy to use, he would consider it as useful and would develop positive attitudes toward it. On the subject of cloud-based solutions, Ekufu (2012) affirms that experiencing a friendly cloud application improves the positive attitude regarding using more cloud-based solutions. Through their survey research within Saudi managers, Mezghani and Ayadi (2016) supported empirically the idea that PEOU are linked positively to both PU and attitudes toward cloud computing. Indeed, these authors consider that, with the large use of Internet and mobile devices applications (email, social media, etc.), managers view any cloud-based application as easy to use, which helps them to develop positive attitudes toward it. Also, PEOU allows them to perceive more easily the benefits of cloud computing (Mezghani and Ayadi, 2016). In case of cloud CRM, Baran and Galka (2017) asserted that putting CRM on the cloud improves the relationships with the customers and allows to easily capture, analyze and access customers' data for both firms and customers. For such reasons, "Hi-tech organizations are rapidly adopting cloud computing services, and newly developed technology solutions rely heavily on this methodology" (Baran and Galka, 2017).

Based on these ideas, it is possible to formulate the following hypotheses:

H4: PU is positively related to the intentions to use a cloud-based CRM application.
H5: PU is positively related to the attitudes toward cloud CRM.
H6: PEOU is positively related to the attitudes toward cloud CRM.
H7: PEOU is positively related to the PU of cloud CRM.

RESEARCH METHODOLOGY

In order to test the research hypotheses, a survey research is conducted within a sample of Saudi SMEs selected randomly in the region of Riyadh (as the capital and the main attractive region in term of the economic weight with 25% of Saudi establishments[2]). Focusing on SMEs is justified by the fact that emerging technologies as cloud-based solutions "more effective for SMEs than large companies" (Gupta et al., 2018). Using cloud computing is more suitable for SMEs as a cost saver alternative.

Data collection was performed using a questionnaire built based on items (appendix) from previous researches (Venkatesh et al., 2003; Ekufu, 2012; Kalinic and Marinkovic, 2016; Mezghani and Ayadi, 2016) and addressed, after review and pretest, to the marketing managers of the selected SMEs. Before addressing the questionnaire, each manager is asked about his knowledge on Cloud CRM et if he already used such solution.

A total of 41 usable questionnaires were collected. The majority of respondents work for SMEs belonging to the services sector (43.9%). Regarding the small sample size, the authors opted for the XLSTAT software to analyze the collected data. In addition to its ability to perform factor analyses, this software allows to perform structural equation modeling (SEM) analyses with partial least square (PLS) technique. SEM analyses were preferred in this chapter since many dependent variables are reported in the research model. PLS is suitable when operating SEM on a small sample size bounded between 30 and 100 observations (Fernandes, 2012).

FINDINGS AND DISCUSSION

Findings of the Factor Analysis

To check the uniqueness of each variable and the robustness of the measurement model, a factor analysis with varimax rotation was performed.

From 29 items initially used, only 2 items were removed from the model (one item measuring subjective norms, another measuring perceived usefulness) due to their weak factors' loadings (less than 0.5).

The reliability of each factor was evaluated by assessing the internal consistency of items within each factor using Cronbach's α and Dillon-Goldstein (D-G) rho. The validity was verified by analyzing the Average Variance Extracted (AVE) value of each factor. The results are shown in Table 1.

Reliability estimates values higher than 0.7 suggest good reliability, "meaning that the measures all consistently represent the same latent construct" (Hair et al., 2010). Regarding convergent validity, these authors suggest using the AVE for the items loadings on each construct. An AVE of 0.5 or higher indicates a good convergence.

Table 1. Validity and reliability tests

Variables	Cronbach's α	D-G rho	AVE
Perceived ease of use	0.787	0.855	0.511
Perceived usefulness	0.807	0.863	0.513
Attitudes toward Cloud CRM	0.744	0.840	0.571
Subjective norms	0.802	0.886	0.709
Perceived behavioral control	0.770	0.857	0.587
Intention to use cloud CRM	0.890	0.925	0.758

When analyzing the values reported in the Table 1, it is clear that all obtained values from the factor analysis are in the acceptance range. Thus, it is possible to proceed to paths analyses within the structural model in order to test the research hypotheses. The structural equation modeling (SEM) technique using PLS with Xlstat software (suitable for small samples) permitted to test in the same time, the different links between variables (Path analysis).

Findings of the Paths Analysis

After verifying the factor structure, we built the structural model to test the research hypotheses. The paths analysis with PLS method, gave us the following fitted model:

"There is no overall fit index in PLS Path Modeling. Nevertheless, a global criterion of goodness of fit has been proposed by Tenenhaus, Amato and Esposito Vinzi (2004): the GoF index. Such an index has been developed in order to take into account the model performance in both the measurement and the structural model and thus provide a single measure for the overall prediction performance of the model" (Vinzi et al., 2010). The obtained GoF (Goodness of Fit) values after the model estimation are shown in Table 2.

According to Vinzi et al. (2010), "there is no inference-based threshold" to assess the statistical significance of GOF values. Nevertheless, Wetzels et al. (2009) propose a cut-off value of 0.5.

All obtained GoF values are high and indicate that the specified model (internal and external) represents well the data. So, it is possible to proceed to the hypotheses test. Such test is performed by analyzing the links' significance (critical ratios (CR)) and values (regressions).

The statistical analyses show that all links are significant at the level of 5% (CR > 1.96), meaning that all hypotheses are supported. These details are reported in the Table 3.

Figure 2. The fitted model (paths analysis)

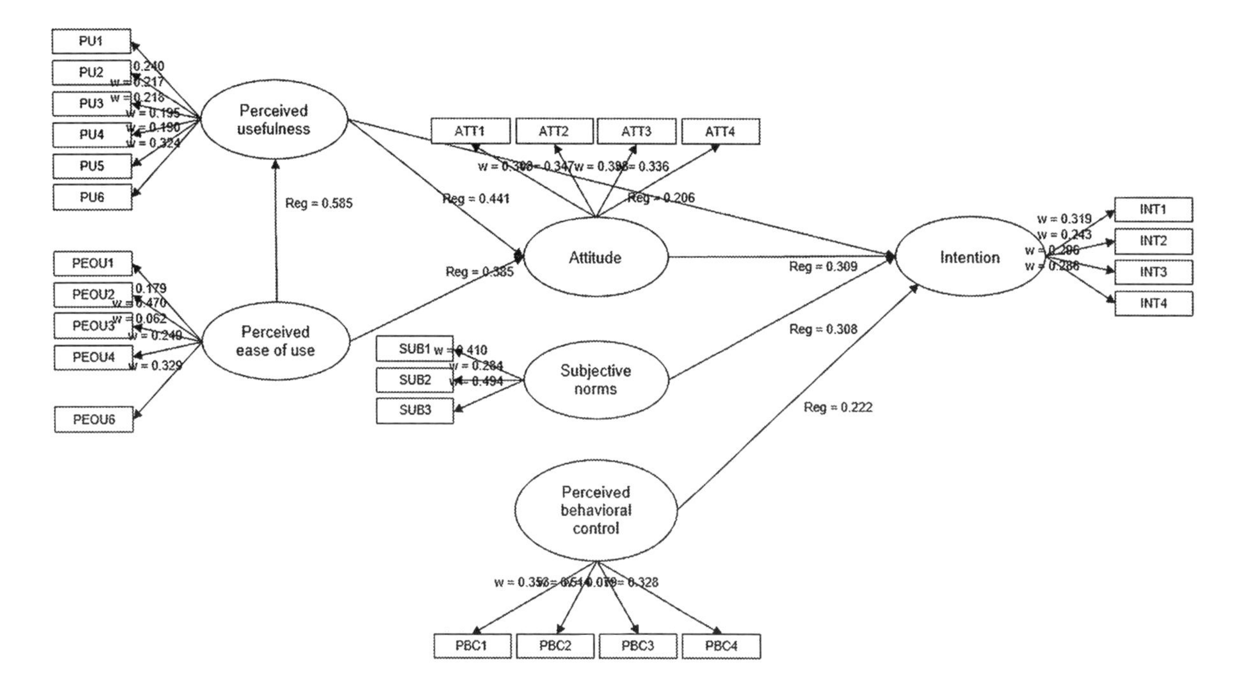

Table 2. Goodness of fit

	GoF
Absolute	0.550
Relative	0.809
External Model	0.978
Internal Model	0.827

Table 3. Significance of regression links

	Perceived Usefulness	Attitudes Toward Cloud CRM	Intentions to Use Cloud CRM
R^2 (CR)	0.342 (2.952)	0.541 (4.307)	0.647 (9.092)
Regression Links (CR)			
Perceived ease of use	0.585 (3.752)	0.385 (3.692)	-
Perceived usefulness	-	0.441 (8.297)	0.206 (5.159)
Attitudes toward Cloud CRM	-	-	0.309 (11.258)
Subjective norms	-	-	0.308 (11.419)
Perceived behavioral control	-	-	0.222 (2.912)

The results reported in the Table 3 show that the intentions to use Cloud CRM are largely influenced by the identified factors ($R^2 = 0.647$). This indicates that the TPB-TAM combination is a suitable alternative to study such intentions. Besides, it is clear that all hypotheses are supported in this research ($CR > 1.96$), which constitutes an additional argument to justify the suitability of the TPB-TAM combination.

DISCUSSION

The analysis of the means' values reveals that the surveyed managers strongly intent to use Cloud-based CRM solutions (means' values greater than 5 in a scale of 6 for each item). Their intentions are largely influenced by high levels of perceptions, norms and attitudes toward Cloud CRM.

Regarding perceptions, it is clear that the surveyed managers perceive many benefits from using Cloud-based CRM solutions. When comparing these results to previous studies conducted in the Saudi context, one can easily explain the positive effects found in the current research.

In a recent research aiming to study the attitudes toward cloud computing in Saudi firms, Mezghani and Ayadi (2016) found that the surveyed managers developed positive attitudes toward cloud-based solutions. They stated that these managers built their attitudes on "positive" perceptions (usefulness, ease of use and benefits) because they are well informed about the benefits of cloud computing.

Such results are also consistent with those of AlBar and Hoque (2017) who tried to identify the determinants of Cloud ERP adoption. Based on a survey study conducted within 136 Saudi firms, these authors found that the relative advantage is a key factor that favor the intentions toward Cloud-based ERP. This also supports the idea that many Saudi managers expect many benefits from cloud-based solutions, leading to develop more positive attitudes and intentions toward cloud computing in general.

Moreover, the surveyed managers believe that they have control over using cloud-based CRM, which encourage them to use such solution. This can be explained by the large use of cloud computing by individuals in Saudi Arabia leading to a certain familiarity with such computing. Indeed, many applications are used daily for personal and business issues thank to the development of smartphones. As an example, between 7 and 8.2 million uses of Snapchat are reported daily in Saudi Arabia during 2017 (Radcliffe and Lam, 2018). In their study, Mezghani and Ayadi (2016) affirmed also that many managers are largely engaged to cloud related information through seminars, literature and discussions (with peers, friends and family members). This helps to develop more knowledge and skills related to cloud solutions, leading to more perceptions about the ability to use similar solutions. Similarly, AlBar and Hoque (2017) affirm that IT skills are important predictors of intentions.

Regarding subjective norms, the statistical analyses show that the surveyed managers consider "influencers" opinions when to decide about using the cloud CRM. From previous researches focusing on Cloud ERP, senior managers are considered as key influencers regarding adoption and use (Mezghani, 2014; AlBar and Hoque, 2017). This can be explained by the fact that switching toward cloud-based solutions is a decisive change that requires particular support from senior management.

The external environment can also be a source of additional influences that favor the Cloud CRM. In fact, in line with the National Transformation Program and the Saudi Vision 2030, more and more firms are involved in digitalization efforts leading to more investments and use of cloud computing. According to Tahawultech (2018), "Saudi Arabia's IT spend in 2018 will reach the value of $40 billion" with a focus on cloud technologies and innovation.

Such results indicate a real aspiration to digital transformation in CRM within the surveyed Saudi SMEs.

RESEARCH IMPLICATIONS

In the light of the obtained results from this study, some research implications can be formulated.

From a theoretical perspective, the current chapter provides additional support to the robustness of the TAM-TPB integration. In fact, such alternative offers a deep explanation of the acceptance of IT innovations at the individual level, mainly the disruptive technologies as cloud CRM. Further research aiming to investigate more cloud CRM use issues can consider TAM-TPB integration as a basis to build suitable frameworks. Indeed, many authors integrate additional factors to explain the TAM or TPB factors in order to give more explanation to the attitudes and intentions toward IT use and adoption.

From a managerial perspective, as nascent technologies are more oriented toward remote access (cloud computing, social media, big data, Internet of things, etc.), cloud-based solutions are a suitable alternative to build new CRM strategies and technologies. Practitioners need appropriate tools to assess the readiness of managers for cloud CRM. Hence, the research model (Figure 1) developed in this research can be used by managers and consultants in Saudi context as a framework to examine practically issues linked to cloud CRM.

To our knowledge, this research is one of the first academic studies focusing on cloud CRM issues in Saudi Arabia. In fact, in line with Vision 2030, Saudi firms are expected to perform digital transformations in order to take part of the development of a knowledge-based economy. According to Nurunnabi (2017), information and communications technologies (ICT) are one of "key aspects which need to be considered in developing Saudi Arabia's knowledge economy". SMEs seem to be the main concerned as cloud solutions represent a suitable alternative regarding their benefits compared to their low costs.

CONCLUSION

With the large use of disruptive technologies in all fields, many firms are performing digital transformations in order to improve the way they do business. As the customers represent the core concern of every business, managers and consultants are continuously searching for innovative ways and technologies to build close relationships with their customers.

Quite a long time, CRM is considered as a suitable technologically oriented approach that helps to get a deep focus on customers. Recently, many alternatives are deployed to integrate CRM with new IT trends (social media, cloud computing, real time analytics, artificial intelligence, etc.) in order to get more knowledge about customers' needs and preferences. In fact, with the expansion of smartphones use at the individual level, a digital transformation in CRM practices represents a "must be" option to get more benefits from recent technologies.

From the new trends, the current chapter aimed to study the concept of cloud CRM and identify the main enablers of intentions to use cloud-based CRM applications. Through a literature review focusing on TAM and TPB (as the dominant theories in IT use literature), the authors attempted to propose a research model that could be useful to study the intentions to use cloud CRM. Such model is based on a TAM-TPB integration. This alternative was largely confirmed in previous studies focusing on IT innovations use.

A survey study was then conducted within 41 Saudi firms to test the research hypotheses. Considering the fact that cloud computing is, at the time, a suitable tool to perform digital transformations and a low-cost alternative, the authors limited their survey to the SMEs. Such category of firms is gaining more interest in the Saudi economy with an objective to rise the SMEs contribution to GDP from 20% to 35%[3].

The data, collected via a questionnaire addressed to the marketing managers, were analyzed using the SEM method with the PLS technique. The statistical analyses confirmed the factor structure of the research model and supported all formulated hypotheses indicating the suitability of the TAM-TPB integration. The means analyses indicated also that the surveyed managers show positive perceptions and intentions toward cloud CRM use. Such results are not surprising regarding the large use of cloud-based applications in Saudi context.

Hence, besides the theoretical contributions leading to confirm the TAM-TPB integration in the case of cloud CRM use, the results of this chapter could be helpful for practitioners to assess their readiness concerning cloud CRM. Cloud providers in Saudi context could also profit from these results in adjusting their solutions toward the integration of more CRM applications in the cloud.

Beyond contributions, two main limits need to be considered in this research. Firstly, no external variable was added to the TAM-TPB integration which limits the explanation of the obtained results. Adding more antecedents to intentions in future researches could provide deeper results. Thus, deploying additional frameworks as the unified theory of acceptance and use of technology (UTAUT) developed by Venkatesh et al. (2003) would be an interesting alternative. Secondly, the small sample size limits the possibility of extending the results and conclusions. It is important to keep in mind that PLS is considered as a predictive approach and that the linked statistical results need to be confirmed in a further research deploying larger sample size and a confirmatory approach as LISREL (Linear Structural Relations).

REFERENCES

Aboelmaged, M. G. (2010). Predicting e-procurement adoption in a developing country: An empirical integration of technology acceptance model and theory of planned behavior. *Industrial Management & Data Systems, 110*(3), 392–414. doi:10.1108/02635571011030042

Ajzen, I. (1991). The Theory of Planned Behavior. *Organizational behavior and human decision, 50*(2), 179-211.

AlBar, A. M., & Hoque, M. R. (2017). Factors affecting cloud ERP adoption in Saudi Arabia: An empirical study. *Information Development*. doi:10.1177/0266666917735677

Alharbi, F., Atkins, A., Stanier, C., & Al-Buti, H. A. (2016). Strategic Value of Cloud Computing in Healthcare organisations using the Balanced Scorecard Approach: A case study from A Saudi Hospital. *Procedia Computer Science, 98*, 332–339. doi:10.1016/j.procs.2016.09.050

Armitage, C. J., Armitage, C. J., Conner, M., Loach, J., & Willetts, D. (1999). Different perceptions of control: Applying an extended theory of planned behavior to legal and illegal drug use. *Basic and Applied Social Psychology, 21*(4), 301–316. doi:10.1207/S15324834BASP2104_4

Arnesen, S. (2013). Is a Cloud ERP Solution Right for You? *Strategic Finance*, (February), 45-50.

Awa, H. O., Ojiabo, O. U., & Emecheta, B. C. (2015). Integrating TAM, TPB and TOE frameworks and expanding their characteristic constructs for e-commerce adoption by SMEs. *Journal of Science & Technology Policy Management, 6*(1), 76–94. doi:10.1108/JSTPM-04-2014-0012

Baran, R. J., & Galka, R. J. (2017). *Customer Relationship Management: the foundation of contemporary marketing strategy* (2nd ed.). Routledge.

Davis, F. D. (1989). Perceived Usefulness, Perceived Ease of Use, and User Acceptance of Information Technology. *Management Information Systems Quarterly, 13*(3), 319–340. doi:10.2307/249008

Ekufu, T. K. (2012). *Predicting cloud computing technology adoption by organizations: an empirical integration of technology acceptance model and theory of planned behavior* [Doctoral Dissertation]. Capella University, MN.

Fauscette, M. (2013). ERP in the Cloud and the Modern Business (White paper). *International Data Corporation (IDC)*. Retrieved from http://resources.idgenterprise.com/original

Fernandes, V. (2012). (Re)discovering the PLS approach in management science. *M@n@gement, 15*(1), 101-123.

Fishbein, M., & Ajzen, I. (2010). Predicting and changing behavior: The reasoned action approach. New York: Psychology Press.

Gangwar, H., Date, H., & Ramaswamy, R. (2015). Understanding determinants of cloud computing adoption using an integrated TAM-TOE model. *Journal of Enterprise Information Management, 28*(1), 107–130. doi:10.1108/JEIM-08-2013-0065

Gupta, S., Misra, S. C., Kock, N., & Roubaud, D. (2018). Organizational, technological and extrinsic factors in the implementation of cloud ERP in SMEs. *Journal of Organizational Change Management, 31*(1), 83–102. doi:10.1108/JOCM-06-2017-0230

Hair, J. F., Black, W. C., Babin, B. J., & Anderson, R. E. (2010). *Multivariate data analysis* (7th ed.). Prentice Hall.

Handoko, I. P., & Gaol, F. L. (2012). Performance evaluation of CRM system based on cloud computing. *Applied Mechanics and Materials, 234*, 110–123. doi:10.4028/www.scientific.net/AMM.234.110

Hu, P. J., Chau, P. Y., Sheng, O. R. L., & Tam, K. Y. (1999). Examining the technology acceptance model using physician acceptance of telemedicine technology. *Journal of Management Information Systems, 16*(2), 91–112. doi:10.1080/07421222.1999.11518247

Kalinic, Z., & Marinković, V. (2016). Determinants of users' intention to adopt m-commerce: An empirical analysis. *Information Systems and e-Business Management, 14*(2), 367–387. doi:10.100710257-015-0287-2

Kidwell, B., & Jewell, R. D. (2003). An examination of perceived behavioral control: Internal and external influences on intention. *Psychology and Marketing, 20*(7), 625–642. doi:10.1002/mar.10089

Lala, G. (2014). The emergence and development of the technology acceptance model (TAM). In *The Proceedings of the International Conference" Marketing-from Information to Decision"* (p. 149). Babes Bolyai University.

Laudon, K.C. & Laudon, J.P. (2102). *Management Information Systems: Managing the digital firm* (12th ed.). New Jersey: Pearson Education.

Lee, M. C. (2009). Factors influencing the adoption of internet banking: An integration of TAM and TPB with perceived risk and perceived benefit. *Electronic Commerce Research and Applications, 8*(3), 130–141. doi:10.1016/j.elerap.2008.11.006

Lee, M. K., Cheung, C. M., & Chen, Z. (2007). Understanding user acceptance of multimedia messaging services: An empirical study. *Journal of the Association for Information Science and Technology, 58*(13), 2066–2077.

Mezghani, K. (2014). Switching toward Cloud ERP: A research model to explain intentions. *International Journal of Enterprise Information Systems, 10*(3), 48–64. doi:10.4018/ijeis.2014070104

Mezghani, K., & Ayadi, F. (2016). Factors explaining IS managers' attitudes toward Cloud Computing adoption. *International Journal of Technology and Human Interaction*, *12*(1), 1–20. doi:10.4018/IJTHI.2016010101

Nurunnabi, M. (2017). Transformation from an Oil-based Economy to a Knowledge-based Economy in Saudi Arabia: The Direction of Saudi Vision 2030. *Journal of the Knowledge Economy*, *8*(2), 536–564. doi:10.100713132-017-0479-8

Payne, A., & Frow, P. (2005). A strategic framework for customer relationship management. *Journal of Marketing*, *69*(4), 167–176. doi:10.1509/jmkg.2005.69.4.167

Radcliffe, D., & Lam, A. (2018). Social Media in the Middle East: The Story of 2017. Retrieved from https://papers.ssrn.com/sol3/papers.cfm?abstract_id=3124077

Rauniar, R., Rawski, G., Yang, J., & Johnson, B. (2014). Technology acceptance model (TAM) and social media usage: An empirical study on Facebook. *Journal of Enterprise Information Management*, *27*(1), 6–30. doi:10.1108/JEIM-04-2012-0011

Safeena, R., Date, H., Hundewale, N., & Kammani, A. (2013). Combination of TAM and TPB in internet banking adoption. *International Journal of Computer Theory and Engineering*, *5*(1), 146–150. doi:10.7763/IJCTE.2013.V5.665

Sharma, S., Padhy, S., & Verma, V. (2014). Multi-functional social CRM in cloud with cross-platform mobile application. *International Journal of Computers and Applications*, *93*(13), 9–15.

Shih, Y. Y., & Fang, K. (2004). The use of a decomposed theory of planned behavior to study Internet banking in Taiwan. *Internet Research*, *14*(3), 213–223. doi:10.1108/10662240410542643

Smaoui Hachicha, Z., & Mezghani, K. (2018). Understanding intentions to switch toward cloud computing at firms' level: A multiple case study in Tunisia. *Journal of Global Information Management*, *26*(1), 136–165. doi:10.4018/JGIM.2018010108

Tahawultech (2018). Saudi Arabia IT spend to reach $40 billion. https://www.tahawultech.com/news/saudi-arabia-spend-hit-40-billion/ (Retrieved: 14/05/2018).

Tenenhaus, M., Amato, S., & Esposito Vinzi, V. (2004). A global goodness-of-fit index for PLS structural equation modelling. In *Proceedings of the XLII SIS scientific meeting* (Vol. 1, pp. 739-742). CLEUP Padova.

Thong, J. Y., Hong, W., & Tam, K. Y. (2002). Understanding user acceptance of digital libraries: What are the roles of interface characteristics, organizational context, and individual differences? *International Journal of Human-Computer Studies*, *57*(3), 215–242. doi:10.1016/S1071-5819(02)91024-4

Trainor, K. J., Andzulis, J. M., Rapp, A., & Agnihotri, R. (2014). Social media technology usage and customer relationship performance: A capabilities-based examination of social CRM. *Journal of Business Research*, *67*(6), 1201–1208. doi:10.1016/j.jbusres.2013.05.002

Udoh, E. E. (2010). *The adoption of grid computing technology by organizations: A quantitative study using technology acceptance model* [Doctoral Dissertation]. Capella University.

Venkatesh, V., & Davis, F. D. (1996). A model of the antecedents of perceived ease of use: Development and test. *Decision Sciences*, *27*(3), 451–481. doi:10.1111/j.1540-5915.1996.tb01822.x

Venkatesh, V., Morris, M. G., Davis, G. B., & Davis, F. D. (2003). User acceptance of information technology: Toward a unified view. *Management Information Systems Quarterly*, *27*(3), 425–478. doi:10.2307/30036540

Venkatesh, V., Speier, C., & Morris, M. G. (2002). User acceptance enablers in individual decision making about technology: Toward an integrated model. *Decision Sciences*, *33*(2), 297–316. doi:10.1111/j.1540-5915.2002.tb01646.x

Vinzi, V. E., Trinchera, L., & Amato, S. (2010). PLS Path Modeling: From Foundations to Recent Developments and Open Issues for Model Assessment and Improvement. In *V.E. Vinzi, W.W. Chin, J. Henseler et al. (Eds.), Handbook of Partial Least Squares: Concepts, Methods and Applications*. Berlin: Springer-Verlag. doi:10.1007/978-3-540-32827-8_3

Wetzels, M., Odekerken-Schröder, G., & Van Oppen, C. (2009). Using PLS path modeling for assessing hierarchical construct models: Guidelines and empirical illustration. *Management Information Systems Quarterly*, *33*(1), 177–195. doi:10.2307/20650284

Yu, Y., Yi, W., Feng, Y., & Liu, J. (2018). Understanding the Intention to Use Commercial Bike Sharing Systems: An Integration of TAM and TPB. In *Proceedings of the 51st Hawaii International Conference on System Sciences*. 10.24251/HICSS.2018.082

ADDITIONAL READINGS

Ainin, S., Parveen, F., Moghavvemi, S., Jaafar, N. I., & Mohd Shuib, N. L. (2015). Factors influencing the use of social media by SMEs and its performance outcomes. *Industrial Management & Data Systems*, *115*(3), 570–588. doi:10.1108/IMDS-07-2014-0205

Makki, E., & Chang, L. C. (2015). Understanding the effects of social media and mobile usage on e-commerce: An exploratory study in Saudi Arabia. *International Management Review*, *11*(2), 98.

Mezghani, K., Ayadi, F., & Aloulou, W. (2014). Effects of Business Managers Skills: A Study of ERP Strategic Alignment in some Arab SME's. In K. Todorov & D. Smallbone (Eds.), *Handbook of Research on Strategic Management in Small and Medium Enterprises* (pp. 421–439). Hershey, PA: IGI Global. doi:10.4018/978-1-4666-5962-9.ch020

Park, S. C., & Ryoo, S. Y. (2013). An empirical investigation of end-users' switching toward cloud computing: A two factor theory perspective. *Computers in Human Behavior*, *29*(1), 160–170. doi:10.1016/j.chb.2012.07.032

Sharma, S. K., Al-Badi, A. H., Govindaluri, S. M., & Al-Kharusi, M. H. (2016). Predicting motivators of cloud computing adoption: A developing country perspective. *Computers in Human Behavior*, *62*, 61–69. doi:10.1016/j.chb.2016.03.073

KEY TERMS AND DEFINITIONS

Cloud CRM: Putting CRM applications on a cloud so CRM software is offered as a service. The cloud CRM is managed by the provider.

Theory of Planned Behavior (TPB): A theory developed by Ajzen (1991) who posits that the human behavior is planned and determined by a set of factors linked to perceptions and motivations.

Technology Acceptance Model (TAM): An information systems theory developed by Davis (1989) and attempts to study the IT acceptance and linked perceptions.

Partial Least Square (PLS): A variance-based structural equation modeling technique suitable for small sized samples.

ENDNOTES

1. Software as a service.
2. General Authority of Statistics: https://www.stats.gov.sa/en/16.
3. http://vision2030.gov.sa/en/goals.

Section 2

New Strategies Under Digital Transformation

Chapter 4
Enterprise Resource Planning (ERP) Systems and Multi-Organizational Enterprise (MOE) Strategy:

Ben Clegg
Aston University, UK

Yi Wan
Aston University, UK

ABSTRACT

This chapter critiques trends in enterprise resource planning (ERP) in respect to contemporary multi-organizational enterprise strategy in order to identify under-researched areas. It is based on the premise that multi-organization strategies and information systems span more than one legal company entity and are becoming increasingly important as digital Internet based systems become more prolific, and outsourcing and collaboration between companies becomes more widespread. This chapter presents a critique of literature covering theoretical, methodological and relational aspects of enterprise resource planning systems and multi-organizational enterprise strategy. The critique gives a unique perspective and highlights four major gaps in current research and points towards a trend which is referred to in this chapter as 'enterprization.' This research could help organizations make more effective use of their information and operations systems strategies when used across more than one company. It should interest researchers, teachers, IS developers and managers.

DOI: 10.4018/978-1-5225-7262-6.ch004

INTRODUCTION: WHY MULTI-ORGANIZATION ENTERPRISES AND MULTI-ORGANIZATION ERP SYSTEMS MATTER

This critique builds on two key definitions. First, Gable's (1998) comprehensive definition of "ERP" which is, "… a comprehensive package software solution [which] seeks to integrate the complete range of a businesses' processes and functions in order to present a holistic view of the business from a single information and IT architecture." Secondly and concomitantly this research builds on the European Commission's definition of an "enterprise"; where the term "enterprise" means, "… an entity including partnerships or associations that can be made up of parts of different companies" (European Commission, 2003). This research does not therefore consider manufacturing or service operations to be made up of a single legal company entity operating in isolation, but instead embodies multi-organizational enterprise management concepts, where parts of companies work with parts of other different companies to deliver complex product-service systems using a multi-organizational enterprise (MOE) (Binder and Clegg, 2007). An MOE will often result from a joint venture between companies as they focus on collaboratively delivering particular product-service systems (e.g. a family of cars, the construction of a building or bridge, the delivery of a complex integrated web-based shopping experience); critical interdependent and dynamic strategic relationships will develop between these company parts based on their relative core competencies.

Based on the above premises enterprise strategies and Enterprise Resource Planning (ERP) systems design and implementation, whether they are for single organizations or multi-organizational enterprises, need to go hand-in-hand. ERP systems per se have been extensively researched over recent decades and knowledge about how ERP systems work as single-company based systems are plentiful. For instance: single-company impact studies about manufacturing performance (Bose et al., 2008), single companies undertaking supply chain planning (Tarantilis et al., 2008), or single company implementation practice and business process re-engineering projects (Benlian and Hess, 2011) are easily found. These tend to be where ERP / ERPI systems are taken to be an integrated information management system supporting the operational transactions of a single company. In contrast there is a relative dearth of research into ERP systems development in a multi-organizational enterprises context, especially from dynamic and contingency perspectives, which leaves ERP systems and multi-organization enterprise strategy under-researched.

Firstly, in response to this dearth, this research critiques current literature and proposes that multi-organization ERP strategy should be better conceptualized and more clearly defined to help the evolution of ERP systems development and their deployment in multi-organization enterprises (MOEs). Secondly, this literature critique posits that multi-organization enterprise ERP systems are under-researched from a methodological perspective and as a result it is also unclear how some research methods could be effectively merged to investigate and shape ERP and multi-organizational enterprise management concepts. Thirdly, this critique posits that multi-organizational enterprise ERP systems are under-researched from a contingent perspective, as researchers need to better understand the synergies between ERP systems development and multi-organizational enterprise strategy over time (Clegg and Wan, 2013). For this purpose, this chapter determines the most frequently used keywords in relevant literature and classifies these publications according to their most commonly used (i) units of analysis (ii) theoretical perspectives, and (iii) research paradigms and techniques. The aim being to identify research gaps (Trauth et al., 1993) from literature pertaining to ERP and multi-organizational-enterprise strategy research to identify any prevailing trends.

The remainder of this chapter is structured as follows: first, the methodology for the literature critique is presented followed by sections which present identified gaps in the literature. Following these research gaps, the most relevant theoretical foundations for ERP systems development and multi-organizational enterprise management are discussed and the emerging concept of "enterprization" is proposed. Finally, the paper identifies some implications for future research based on prevailing trends.

APPROACH USED FOR THE CRITIQUE

The approach used for the review and interpretation of the literature is given in Figure 1.

A total of 53 highly rated peer reviewed scholarly journals were used, which were selected from five different discipline groups considered to be most relevant to this topic (as rated by the Chartered Association of Business Schools (C-ABS) Academic Journal Guide). 644 papers, published between 1987 and 2016 were found to cite either ERP systems or inter- and multi-organizational strategy for enterprises. These papers were selected by searching on the details given in papers' titles, abstracts and keywords. They were read in detail for further content analysis. The frequency and distribution of these 644 papers across these groups and journals are given in Table 1.

To gain a deeper understanding and potentially give a unique contemporary perspective, these 644 journal articles were read in detail to determine if they were directly relevant to both "ERP systems development" and "multi-organizational relationship and collaboration"; the outcome of this exercise reduced the number of papers to 255. The most relevant journals to this literature review were found to be: the International Journal of Operations & Production Management (IJOPM), the Journal of Enterprise Information Management (JEIM), the Information Systems Journal (ISJ), the Journal of Manufacturing Technology Management (JMTM) and the Journal of Operations Management (JOM). The remaining 60% of articles were spread over another 48 journals. This critique aimed to look at emerging trends through a non-traditional treatment of theory, that of a multi-organization enterprise, (Davison et al., 2012) in order to try and develop unique perspectives and new trends.

Critique of ERP and Multi-Organization Enterprise (MOEs) Literature

The content of the remaining most relevant 255 papers were reviewed by 'theoretical', 'methodological' and 'relational' themes, and then further by detailed elements of these themes, as summarized in Table 2.

Figure 1. Approach used for critique and interpretation of literature

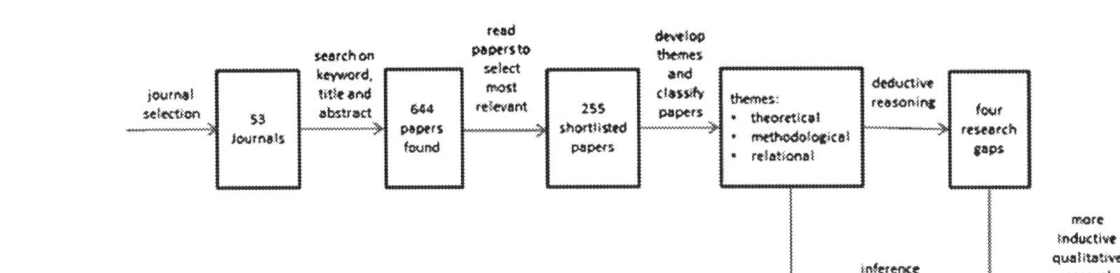

Table 1. Frequency distribution of papers per journal per category (note: only the highest frequency journals in each category are shown)

Discipline Groups	Selected Journals (Only Highly Cited Journals Within Each Group Are Named)	Number of Articles	
		Frequency	Percent
Operations, manufacturing & supply chain management	International Journal of Operations & Production Mgt.	60	9.32
	Journal of Supply Chain Management	40	6.21
	Journal of Operations Management	23	3.57
	International Journal of Production Economics	22	3.42
	Supply Chain Management: An International Journal	22	3.42
	Others	56	8.7
1st Group subtotal	12 journals	223	34.64
Information technology, engineering & innovation	Information Systems Journal	72	11.2
	Journal of Manufacturing Technology Management	48	7.45
	Journal of Enterprise Information Management	40	6.21
	European Journal of Information Systems	34	5.28
	International Journal of Information Management	22	3.42
	Others	109	16.95
2nd Group subtotal	26 journals	325	50.51
General strategic management & business practices	Business Process Management Journal	34	5.28
	Strategic Management Journal	6	0.93
	Academy of Management Review	4	0.62
	Harvard Business Review	4	0.62
	Others	9	1.41
3rd Group subtotal	9 journals	57	8.86
Organization & management science	Journal of Enterprise Transformation	8	1.24
	Organisation Science	4	0.62
	Management Decision	3	0.47
	Decision Sciences	2	0.31
4th Group subtotal	4 journals	17	2.64
Economic & marketing management	Journal of Business & Industrial Marketing	21	3.26
	Journal of Economics & Management Strategy	1	0.16
5th Group subtotal	2 journals	22	3.42
Total number	53 journals	644	100.00

THEME 1: THEORETICAL

Keywords

The most commonly found keywords, as taken from the 'keyword' sections in these papers were: "virtual enterprises" ($f = 85$), "extended enterprises" ($f = 70$), "interoperability" ($f = 66$), "information systems"

Table 2. Critique of ERP and MOE literature by theme

Themes	Themes' Elements	Rationale
1. Theoretical	Keyword	Determine the range of keywords, analytical units and theories used to explore ERP systems and multi-organizational enterprise management
	Unit of analyses	
	Theoretical perspective	
2. Methodological	Paradigmatic stance	Determine the methodological approaches used and the type of research techniques used to explore ERP systems and multi-organizational enterprise management
	Research technique	
3. Relational	ERP systems design and management	Critically review synergies in the extant literature between ERP and MOEs to direct future potential research
	ERP systems development	
	Multi-organizational enterprise design and management	
	Linkages between ERP systems and multi-organizational enterprise management	

($f = 60$), "supply chain integration" ($f = 55$), and "ERP" ($f = 54$) (see Table 3, left hand-side column). Keywords found between 1987 and 2016 suggest that research has been moving away from the traditional "ERP" system towards "ERPII" systems ($f = 47$) and on towards newer information systems terms relating to the next generation of ERP systems. ERPII systems are defined as web enabled technology information system with increasing transparency and potential for opening up new opportunities for (multi-organization) enterprises. ERPII systems are now increasingly being adapted towards collaborative e-business business models (Chen, 2001). By association a greater number of studies are therefore beginning to make the multi-organizational enterprise concept increasingly explicitly as strategic discussions shift from simplistic linear 'supply chain integration' and 'interoperability' towards 'enterprise management' (following the European Commission's definition); the latter work on 'enterprise management' includes an increasing amount of in-depth discussion on structural forms and strategic visions for different types of multi-organization enterprises - such as the 'vertically integrated enterprise (VIEs)' ($f = 48$), the 'extended enterprise (EE)' ($f = 70$) and the 'virtual enterprise' VEs ($f = 85$).

Units of Analyses

The units of analyses used in these 255 papers were determined by reading the methodological and analytical sections of each paper in detail. It was found that a single commonly used unit of analysis did not appear to exist in respect to how research was conducted into ERP systems, as different researchers used a myriad of different units of analysis for analyzing ERP related systems. However the most commonly used units of analyses are defined in Table 3 (middle column), namely: "ERP" itself ($f = 57$), "service-oriented architecture" (SOA) ($f = 13$), "ERPII" ($f = 10$), "cloud computing" ($f = 8$), "supply chain management systems" ($f = 6$), "inter-organizational information systems (IOIS)" ($f = 6$) and "open source" systems ($f = 5$). Currently there are many different units of analyses used for ERP systems analyses and a single unified term for the future generation of ERP systems (i.e. post ERPII) is still yet to be determined. If the future unit of analysis for investigating ERP related systems could be defined with clarity and uniformity it could help expedite future information systems research in multi-organization

Table 3. Classification of papers by theoretically themed elements

Keywords (*f*)		Units of Analyses (*f*)		Theoretical Perspectives (*f*)	
Virtual enterprises	5	ERP	7	Configuration theory	64
Extended enterprises	70	Supply chain	9	Relational view	51
Interoperability	66	Inter-firm network	5	Resource based view & resource dependency theory	45
Information systems	60	Virtual enterprises	2	Network theory	44
Supply chain integration	55	Dyadic relationships	1	IT and business alignment	36
ERP	54	Vertically integrated enterprises	7	IT and organizational capability	30
Vertically integrated enterprises	48	Extended enterprises	4	Competence theory	26
ERPII	47	Service-oriented architecture	3	Transaction cost economics	21
Dynamic capability	5	ERPII	0	Dynamic capabilities view	17
Enterprise transformation	24	Individual firms	9	Contingency theory	15
		Cloud computing	8	General strategic management	14
		Supply chain management systems	6	Critical successful factors-based view	10
		Inter-organizational Information systems	6	Value chain-based view	6
		Open source	5	Agent-based view	3
		Triadic relationships	3	Complex adaptive systems	3
Total	534		255		385

Note: 255 papers were reviewed. If more than one keyword or theory was addressed by an individual paper then the paper was placed in multiple categories

enterprises. This would require shifting the emphasis for the unit of analysis from intra-organization-wide ERP analyses towards inter-organization-wide ERP analyses. Thus, future ERP systems studies should use units of analyses spanning more than one legal organizational entity; and so would reflect contemporary ERP systems practice more closely.

Table 3 also shows the most commonly used suitable terms for inter-organizational units of analyses, which are probable candidates for future units of analysis for multi-organization information systems and multi-organizational strategy. For instance, terms such as "vertically integrated enterprises" (*f* =17), "extended enterprises" (*f* = 14) and "virtual enterprises" (*f* =22) were commonly used as suitable units of analyses where research into multi-organizational relationships had taken place. It seems that some researchers are realizing that "operations strategy has been developed from single firm level to dyadic and triadic relationships, and been upgraded into inter-organizational enterprise level [strategy]" (Eisenhardt and Martin, 2000). Even so, it should be noted that very few papers have pragmatically and rigorously applied the concepts of enterprise management (VIE, EE or VE) explicitly in the full sense of the European Commission's (2003) definition; notable exceptions include Binder and Clegg (2006) and Purchase et al., (2011). Most papers still prefer to use these newer terms in relatively general ways based on more traditional perspectives of 'supply chain' (*f* = 39), (e.g. Chen et al., 2008), 'inter-firm network' (f = 25) (e.g. Chi et al., 2010), 'dyadic and triadic relationships' (f = 21 and 3, respectively) (e.g. Choi and Wu, 2009) and simple contractual exchanges between 'individual firm(s)' (f = 9) (Mahoney, 1992). Perhaps it is time that researchers began to use multi-organizational enterprise concepts terms (e.g. VIE, EE, and VE) more specifically and explicitly - to more accurately reflect contemporary ERP (I, II and /or III use) and MOE practice.

Theoretical Perspective

Theoretical perspectives found in these 255 papers were determined by reading their theoretical sections. It was found that most of these papers built on a small number of closely related theoretical perspectives espoused from the following fields: economics (e.g. transaction cost economics, agent-based view), strategic management (e.g. configuration theory, resource-based view & resource dependency theory, competence theory, dynamic capabilities view, general strategic management, critical successful factors based view, value chain based view), sociology and organizational science (e.g. contingency theory, complex adaptive systems) and industrial marketing management (e.g. relational view, network theory) as shown in Table 3 (right-hand-side column). It is interesting to note that none of these articles claimed to use any totally new or original theories as they all used well known existing theories espoused from older business disciplines – and that were all popularized well before the advent of Internet based collaborative working. For instance, more detailed content analysis showed that configuration theory (f = 64), relational view (f = 51), resource-based view (RBV) / resource dependency theory (RDT) (f = 45), and network theory (f = 44) views were most popular. While articles grounded in dynamic capabilities view (DCV) (f = 17) and contingency theory (f = 15) were less well represented. Few publications attempted to use the value chain-based view (f = 6), the agent-based view (f = 3) or the complex adaptive systems (CAS) view (f = 3).

Perhaps new theoretical perspectives, developed since the popularization of the Internet, covering ERP and multi-organizational enterprises will emerge in the future as Internet based ERP solutions continue to pervade strategic and operational thinking. If so, these could encourage ERP developers and enterprise strategists to begin to place more emphasis on exogenous company factors and disruptive factors pertaining to external technical infrastructure developments; which in turn could better prepare industry in general for what is now commonly becoming increasingly referred to as the growth of 'Industry 4.0' (Brettel et al., 2014) or the growth of the 'Internet of Things' (IoT) (Cheng et al., 2016); which in this chapter's context, should perhaps be known as the 'Internet of Multi-Organizational Enterprise Operations'; as Industry 4.0 or the IoT is, in part, the coming together of companies, companies' IT systems and devices that link to these systems and processes together.

Further detailed content analysis on academic studies which explicitly acknowledged multi-organizational enterprise (MOE) management revealed that the same common theories, such as: transaction cost economics (TCE) (Cao and Zhang, 2011), RBV (Fawcett et al., 2012), the relational view (Cambra-Fierro et al., 2011) and the network theory view (Albani and Dietz, 2009) were used. Less commonly used theories were also revealed, such as: resource dependency theory (RDT), the extended resource-based view (ERBV), the CAS and the agent-based view. Further studies should perhaps be conducted to determine which theories have the greatest propensities for future innovative research, and which also take into account recent technological, societal and industrial changes.

Practitioner led ERP research papers were found to be light in theory but gave useful insights into ERP and MOE developments (e.g. Banker et al., 2010; Deep et al., 2008); they reported pragmatic approaches such as IT and business alignment, IT and organizational capability, configuration theory and critical successful factor (CSF) analyses, rather than explicit use of 'academic theories.'

In light of this critique about the use, or lack of use of, theoretical and practical perspectives' this study calls for researchers in these fields to be open-minded about developing new and/or adopting new theories for multi-organizational-enterprise ERP research (MOE-ERP) – especially those that aim to be more fitting for future ERP and MOE developments. As such researchers should perhaps loosen their pre-occupation with the few 'old favorite' theories which were popularized well before the advent of the Internet and become more cognizant of newly emerging business, social and technological requirements, to develop their theories accordingly.

THEME 2: METHODOLOGICAL

Paradigmatic Stance

Detailed content analysis classified these papers by paradigmatic stance into positivism, non-positivism, and multiple research methods as shown in Table 4 (see left-hand column). Table 4 shows that there were a moderate number of non-positivist research studies (*f* = 62) and very few multiple paradigm studies (*f* = 7) when compared to a large volume of positivistic studies (*f* = 186). The positivistic studies tended to test and retest known and accepted theories. The over-whelming propensity towards positivism in this field could possibly have prevented new and innovative theorizing about aspects of ERP systems development and multi-organizational enterprise conceptualization; which have omitted important aspects of multi-organization ERP and organization development as either unexplored, or unsupported, by rigorous research studies (Benlian and Hess, 2011). Therefore, this study suggest that there is potential for new research to develop clearer and more unified theories about multi-organization enterprise strategy and multi-organization ERP systems development.

Applied Research Techniques

Detailed content analysis also classified these 255 papers according to their prevailing research techniques (see Table 4, right-hand column). The classification distribution suggests that although a total of nine broadly different research techniques were recorded the majority of these studies used empirical case studies, analytical conceptual, empirical interviews or empirical questionnaire survey (*f* = 113, 81, 77 and 75, respectively). These techniques were followed in popularity by analytical mathematical methods (*f* = 41) and a handful of articles which used experimental design, analytical statistical, empirical sampling and Grounded Theory Method (GTM) (*f* = 18, 16, 14 and 10, respectively).

It can be seen that the distribution pattern of research techniques is skewed away from more qualitative and subjective approaches such as GTM. However, in information systems (IS) management and operations management (OM) fields GTM has proven itself to be extremely useful in developing context-based, process-oriented descriptions which can bridge information systems and operations management phenomena (Binder and Clegg, 2006; Nah et al., 2005; and Vannoy and Salam, 2010). GTM offers a well-signposted procedure for data analysis, and allows the emergence of original and rich findings tied

Table 4. Classification of methodological themed elements

Paradigmatic Stances		Applied Research Techniques	
Positivism (a.k.a. functionalist paradigm)	186	Case study	113
Non-positivism (e.g. interpretivism, phenomenalism, post-modernism)	62	Conceptual	81
Multiple research paradigmatic stances	7	Interviews	77
		Questionnaire survey	75
		Mathematical	41
		Experimental design	18
		Statistical	16
		Sampling	14
		Grounded theory method	10
Total	255		445

Note: While 255 paper were reviewed, some used more than one research method and so were placed in multiple categories

to data; despite this GTM is often omitted from information systems (IS) management and operations management (OM) studies. This is maybe due to some previous Grounded Theory based studies in IS management and OM being debunked for achieving relatively low levels of theory development. This study attributes this, in part, to GTM being used in limited coding-only ways for analysis (Fawcett et al., 2012) instead of being used as a purer, fully-fledged inductive theorizing methodology - as promoted by Glaser and Strauss (1967). Therefore, if GTM is used in a constricted form it will inevitably fall short of its full potential to theorize new combined theories about OM and IS management.

This study posits that today's complex and dynamic ERP and operations management research calls for more systematic observation to help managers deal with practical strategic and implementation OM and IS challenges. It seems likely that the use of mixed-methods research is the best way forward for information systems and operations management researchers, as the content analysis of these papers reveals that the majority have used more than one applied research technique. Thus, this study suggests that it would seem particularly beneficial for researchers in operations and information management to blend quantitative and qualitative methods.

THEME 3: RELATIONAL

This section explains the relational synergies between ERP systems and multi-organization enterprises (MOEs). It is the most significant section of this chapter as it identifies interrelated research gaps in the current literature on which useful future research could build.

ERP Systems Design and Management

ERP systems were originally designed and built to overcome problems associated with fragmented and incompatible information systems, and bring together inconsistent operating practices within organizations (Palaniswamy and Frank, 2000). By providing end-to-end connectivity ERP systems enabled companies to enhance the performance of their operations, manufacturing, and financial activities (Hendricks et al., 2007). For instance: order cycle times were reduced through cross-functional transaction automation giving improved throughput and customer response times (McAfee, 2002), also smoother and more streamlined data flows and real-time operational information was achievable. This enabled ERP vendors to claim dramatic gains in productivity for user organizations (Scott and Vessey, 2000).

Despite the widespread use of ERP systems, many companies have begun to realize that the real impact of ERP systems on management styles and practices often falls well below expectations. This might be explained by, "failed reconciliation between the technological imperatives of ERP and the business needs of the company itself" (Bingi et al., 1999). Therefore, some researchers have increasingly focused on the integration of applications (Samaranayake, 2009) to boost the impact of ERP systems, as integration is viewed as one of the most important issues in the successful design and deployment of business processes within ERP systems (Beretta, 2002). Other researchers have also pushed the notion of linking real-time information, provided by ERP systems, with tightly integrated business processes; but their arguments are still not radical enough. This may be because researchers are still overly concerned with single intra-organizational applications and individual business user's needs rather than being concerned with the dynamic reconfigurable inter-organizational integration of multi-organizational enterprises (Lin and Rohm, 2009). Static and intra-organizational views of ERP design and management will, by definition, entrench the status quo of development and limit the development and application of new information systems developments for production and operations management applications - and so limit the propensity for more radical advancements in associated IS strategy research.

To cope with these new challenges, arising from increasingly more Internet-based multi-organizational collaborations, Bose et al. (2008) and Lee et al. (2003) suggest that organizations need to improve the competitiveness of their systems by adding-on newly designed modules for processes such as supply-chain management (SCM), customer relationship management (CRM) and e-business. But once again, these studies focus too much on improving existing incumbent traditional core functionalities of ERP and operations, founded in supply chain management, rather than on more contemporary collaborative multi-organizational enterprises delivering complex product-service systems. These older more static intra-organizational views limit the investigatory, explanatory and developmental thinking of traditional ERP systems thinking, and are likely to inhibit the next generation of ERP systems development. Current theories, and those being developed, should therefore be increasingly focus on contingent, dynamic and multi-organizational enterprise management practices.

From a dynamic contingency perspective enterprise systems developments and implementations are often major change initiatives that cause high levels of disruption and potentially enable dramatic organizational transformation and process reengineering activities. Correspondingly, different multi-organizational enterprise structures and multi-organization strategies require different ERP solutions. However extant literature on ERP upgrade projects often fail to demark emerging and contemporary requirements from older and more traditional ERP critical success factors (Beatty and Williams, 2006). The reason for this could be that "ERP information systems have been developed for their own sake and not examined in the context of organization design as a whole" (Oura and Kijima, 2002). Therefore, there are distinct concerns about the alignment of ERP implementation with multi-organizational enterprise development activities. For instance, the scope of business processes reengineering during ERP adoption would need to be extended from intra-organizational processes to inter-organizational processes (Lu et al., 2006). One could then posit that ERP information systems are evolving towards a more inter-organizational enterprises perspective with increasing potential to adapt and reshape new business strategies in the Internet of Things (IoT) era; which is indicative of newer 'dynamic transformational views' of multi-organizational ERP systems management. Hence, in summary of this critique so far, Research Gap 1 is posited:

Research Gap 1: Research into designing and managing ERP systems should look beyond core ERP functionalities for single organizational level strategy and consider how ERP systems can become more supportive across multi-organization enterprises.

ERP Systems Development

It is commonly accepted that traditional ERP systems are internally integrated information systems used to gain operational competitive advantage (Blackstone and Cox, 2005, p. 38) by integrating core internal functions such as operations and production. ERP systems can then be further extended to other related support functions such as sales and distribution, accounting and finance (Al-Mudimigh et al., 2001), Product Data Management (PDM) and Decision Support Systems (DSSs) (Stevens, 2003; Themistocleous et al., 2001). Also, this more widespread adaption and connectivity is usually associated with higher degrees of proprietary in-house (re)development requiring, not only considerable financial commitment to implement, but also making further links to other organizations' ERP systems increasingly difficult - as more company-specific idioms become included.

Therefore, the most traditional forms of ERP systems (i.e. ERP I) do not necessarily support the increasing scope of future business requirements for Internet based commerce (Bond et al., 2000; Moller, 2005; Songini, 2002) and further functional modules are often developed as 'add-ons' to more closely approximate ERPII systems. Thus, it may be that traditional ERP (I) systems are slowly being usurped by ERPII systems as ERPII systems become more recognized as an integral part of business strategy; which will enable inter-organizational collaboration through the extension of operations to close and trusted partners (Bagchi et al., 2003). One might say that the first generation of ERP(I) systems was primarily concerned with supporting and coordinating activities across the inter-functional divisions and enhanced single organizational operations (Akkermans et al., 2003) whilst ERPII systems support, "… resource planning co-operations between different organizations at a meta-level" (Daniel and White, 2005). The mantra of "ERP is dead – long live ERPII" has been mooted by contemporary systems developers (Eckartz et al., 2009).

Currently ERPII systems are the dominant type of system to support modern enterprises. However, as competition increases, and markets become more turbulent, many manufacturers and service providers are attempting to re-design their operations and ERP systems to have even greater agility (Anussornnitisarn and Nof, 2003). As a result, information systems solutions increasingly encompass SOA (Service Orientated Architecture), Enterprise Application Integration (EAI), SaaS (Software as a Service) (Bass and Mabry, 2004; Sharif et al., 2005), cloud computing (Maurizio et al., 2007; Rappa, 2004; Sharif, 2010) and open-sources technologies (Benlian and Hess, 2011). These technologies have the potential to bring further flexibility, agility, efficiency, scalability and re-configurability to traditional ERP (I and II) systems and their associated operations; this is because they, amongst other features, can provide greater potential for multi-organizational connectivity (Torbacki, 2008; Wilkes and Veryard, 2004) within a multi-organizational enterprise context.

Therefore, the future of ERP systems is uncertain and unpredictable. New technologies and standards for SOA, SaaS, and openly-sourced enterprise applications are likely to bring about new challenges around features such as: the granularity of data-sharing, business privacy and de-centralization of strategic objectives (Candido et al., 2009; Xu et al., 2002). Despite this uncertainty, and the new challenges it brings, these emerging technologies are changing the way ERP systems are now being perceived, developed and deployed. New technical and conceptual IS developments may provide the catalyst for more sustainable competitive advantage and make the borderless multi-organization enterprise concept more common, more robust and more secure in the future.

In the absence of any unified or accepted definition of the next generation of ERP these authors refer to the next generation of ERP systems as 'ERPIII'. The authors define ERPIII as a flexible, powerful information system for operations management incorporating web-based technology which enables multi-organization enterprises to offer increasing degrees of connectivity, collaboration and dynamism through increased functional scope and scalability. Such future 'ERPIII' systems are of considerable current interest and are expected to play an increasingly significant role in new and different multi-organizational enterprise structures and strategies - as well as shaping the future for inter-organizational ERP systems. Therefore, in summary of this part of this critique, Research Gap 2 is posited:

Research Gap 2: Next generation ERP systems research (ERPIII) should emphasize how future ERP systems can react more dynamically and contingently as multi-organizational enterprise structures and strategies change.

Multi-Organizational Enterprise Design and Management

Enterprise management concepts may help innovate new multi-organizational thinking as they embrace new business partnerships and collaborative arrangements which can help contribute towards the commercial viability of a business. Parsimoniously this chapter focuses on three types of multi-organizational enterprise strategies which are: the vertically integrated enterprise (VIE), the extended enterprise (EE) and the virtual enterprise (VE).

Vertically integrated enterprises (VIE) operate as large single highly-integrated multi-functional firms as they strive for scales of economy in operations, VIEs typically have bureaucratic reporting hierarchies (Lynch, 2003) and evolve as, "a response to pre-existing market power problems or as a strategic move to create or enhance market power in upstream and downstream markets" (Joskow, 2003, p. 25). A VIE will typically process raw materials into end-consumer products and services and embed a company in an industry sector (Vallespir and Kleinhans, 2001). As a result, competitiveness may be gained through reduced transaction costs (Harrigan, 1985), stronger quality control, higher barriers to new entrants (Rothaermel et al., 2006) and rapid response to volume changes (Richardson, 1996). The downside to VIEs is that their structure and size can inhibit engagement with other organizations due to excessive bureaucracy.

The 'extended enterprise' (EE) concept, in contrast to the VIE, has been defined by Davis and Spekman (2004, p. 20) as "... the entire set of collaborating companies...which bring value to the marketplace..." and by Lyman et al. (2009) as "... a business value network where multiple firms own and manage parts of an integrated enterprise". This allows practices such as just-in-time (JIT) supply chain logistics (Sutton, 2006), collaborative innovation (Owen et al., 2008), and data warehouse interoperability (Triantafillakis et al., 2004) to be easily deployed across company boundaries; this is because an EE structure allows organizations to focus on their core business (usually the delivery of a complex product and or service) and technical activities whilst outsourcing non-core activities to other members in their extended enterprise (Thun, 2010). Thus, extended enterprises are conceived to be more agile than vertically integrated enterprises (VIEs) as they reduce cross-company barriers (Spekman and Davis, 2016). The key to effective EE management is to enlarge the scope of a focal firm's interactions with other businesses by partnering with other stakeholders, and nurturing the competences of EE neighbors, whilst simultaneously mitigating the risks inherited from having a myriad of multi-organizational relationships in operation. However even EEs cannot manage to follow very high economic turbulence and very high business unpredictability because they still operate in a partially restricted environment operating with only known and trusted members.

Highly turbulent and unpredictable market behavior is best coped with by the 'virtual enterprises' (VE) (Byrne and Brandt, 1993), rather than by EE or VIE, as VEs are the most agile type of enterprise. VEs are best thought of as a jigsaw of operations and information systems from more than one business entity that are very loosely governed by specific decentralized objectives (Martinez et al., 2001). Virtual multi-organizational relationships can facilitate innovative agile manufacturing and service delivery more easily (Sharp et al., 1999) and deal with dramatic dynamic market changes through Internet based information and communication technologies (ICTs), and by increasingly temporizing their structures and strategies (Madu and Kuei, 2004). This is because heterogeneity is a key characteristic of VE collaborations and operations, and VE infrastructure and coordination policies are by necessity flexible and reconfigurable; this in turn brings new challenges on how to best balance the 'autonomous activities' and 'coordinating activities' to every VE member. Therefore, in summary of this part of this critique, Research Gap 3 is posited:

Research Gap 3: Future research should develop, test and refine new theory for multi-organizational enterprise strategy and multi-organization information systems strategy.

Linkages Between ERP Systems and Multi-Organizational Enterprise Management

Although the importance of adopting ERP solutions to support multi-organizational relationships has begun to be addressed over the last decade or so, most studies still focus on improvements of ERP-enabled linear supply chain relationship management, which is essentially a mono-theoretic approach that is too un-innovative for the approach called for by the collaborative spirit of the European Commission's definition of an 'enterprise' (2003).

In addition most research dedicated to the ERP and multi-organizational collaboration still relies on traditional IT-business alignment theory (Vannoy and Salam, 2010), which is still far from satisfactory, as it has often been developed at either a single company level (Chengalur-Smith et al., 2012) using a traditional process-based inter-organizational IS (IOIS) approaches (Law and Ngai, 2007), or focused on information systems integration in mergers and acquisitions scenarios (Henningsson and Carlsson, 2011). Therefore, limited attention has been paid to investigate how ERP systems can best support multi-organizational enterprise strategy from a dynamic configuration and contingent perspective. This is a significant gap in understanding - as companies can experience different sets of endogenous and exogenous changes which require different degrees of integration of (multi-organization) ERP systems - as well as different multi-organizational enterprise practices to achieve desirable levels of competitiveness and agility. These pivotal issues should be addressed by newly formulated theoretical concepts, and their conceptual frameworks which are able to depict the alignment between ERP systems capabilities and multi-organizational enterprise management practice. Hence, in summary of this part of this critique, Research Gap 4 is posited:

Research Gap 4: Research should develop, test and refine conceptual contingency frameworks grounded in new theory and capable of providing useable decision-making capabilities for developers and managers of ERP systems in multi-organizational enterprises.

Towards Enterprization

These four gaps, identified through a critique of 255 papers, emphasizes that successful multi-organizational enterprise strategies rely upon appropriate implementation of ERP information systems in multi-organizational enterprises. Also, that ERP systems development and new multi-organizational enterprise paradigms and practice are inextricably linked. The authors refer to this trend as the trend of 'enterprization', which is summarized for the first time in Figure 2.

Figure 2 represents the overall findings from the literature and shows strong synergy between ERP(I) systems and VIEs, and between ERPII systems and EEs. The authors assert that VIE is the most appropriate multi-organizational enterprise structure for using a ERP(I) type systems. This is because ERP(I) type systems can adequately support all core process and some inter-departmental integration, give cycle time reduction, allow transaction costs to be minimized and maximize scales of economy; they can also facilitate speedy decision-making with real-time operating information as well as lowering conventional hierarchies within multi-functional units in highly integrated organizations. However, ERP(I) will tend to entrench current practice and can be relatively unreactive to strategic and environmental business changes by tending to over-emphasize internal transaction cost, strong product-service quality control, and market power.

Figure 2. The 'enterprization' trend of operations and ERP systems

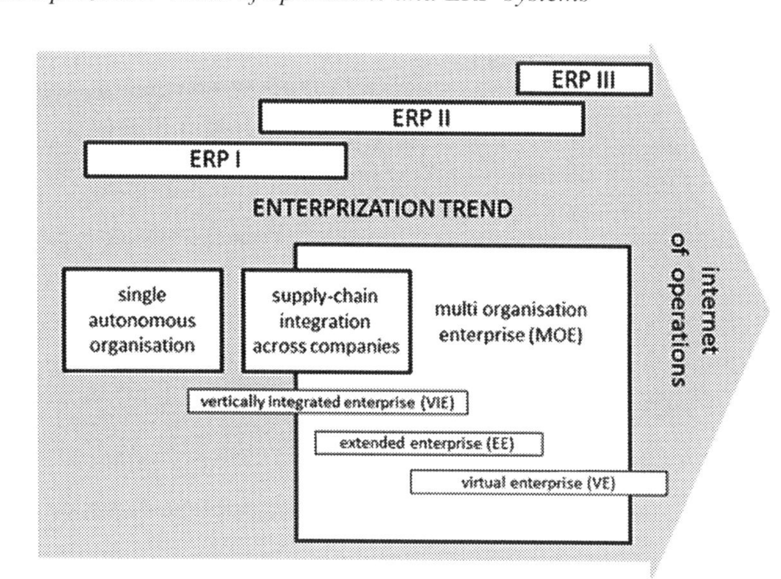

In contrast, ERPII systems should perform best in an EE serving medium-to-large sized operations aspiring to form closer partnerships within an extended value chain. Thus, ERPII systems extend ERP(I) capabilities to encourage more collaborative commerce potential (e.g. SCM, e-business) and encourage active participation from different entities. Specifically, ERPII can enable tighter integration between core supply chain components, optimize inter-firm operational processes, facilitate more accurate collaborative decision-making and exist in weaker bureaucratic authority environments. These features contribute directly towards a forward-thinking strategic vision for multi-organization enterprises which is both boundary-spanning and based increasingly upon distributed flatter structures for more adaptable and advanced information systems infrastructure.

It was also found by this study that fewer emerging publications on post-ERPII systems (a.k.a. ERPIII) were available; however, those found positioned ERPIII as most suited to VE implementations. Particularly interesting was the apparent transition towards VEs adoption via ERPIII use, which is maybe because the VE represents a more open, modular and reconfigurable structure combining multiple business entities which are quasi-interdependent and loosely coupled. Correspondingly, ERPIII solutions are expected to be cheaper, easier, quicker and more flexible to deploy for temporary, decentralized, and highly agile operations on a global scope and scale - whilst allowing the multi-organization enterprises to cope with these uncertainties with more informed and effective decision-making. This may be because technologies upon which they are based (e.g. SOA, SaaS, or PaaS) have become more mature in terms of security, robustness and usability (Rodon et al., 2011; Olsen and Sætre, 2007; Vathanophas, 2007). This is important because users of VEs and ERPIII systems are hoping for a quick-to-create and quick-to-dismantle multi-organizational enterprise whose operations can enable fast and accurate transactions in risky open environments.

Some research also made weaker synergies between traditional ERP and EEs (McAfee, 2002) and ERPII to VIEs (Weston, 2002; Eckartz et al., 2009), as well as between ERPII and VEs (Bond et al., 2000; Ericson, 2001); this reveals how a continuum of strategic operations, structural and ERP changes are observable in respond to changing factors in the digital business environment.

This critique suggest that different multi-organizational enterprise structures and supporting ERP information systems are not the result of different strategies, but are best considered as part of a strategic continuum with the same overall purpose of multi-organizational enterprise cooperation; thus the ERP-MOEs evolutional performance should be described as a path-dependent change towards what is now being labelled the 'Internet of Things' (or Industry 4.0) - which in this context means the 'Internet of multi-organizational enterprise operations'. This is because at different times and circumstances in an enterprise's lifecycle individual companies may prefer different multi-organizational enterprise structures and use different ERP system types (e.g. continuous upgrade or reconfiguration) to satisfy their exogenous and endogenous business requirements; which ultimately may grow and sustain their competitive advantage through the simultaneous strengthening of multi-organizational enterprise strategy, enterprise structures, and ERP systems interactivity; a pattern of behaviors which this study refer to here as the "enterprization of multi-organizational enterprise operations".

FUTURE RESEARCH DIRECTIONS

As noted above, there are growing trends to question why the synergies between ERP systems development and multi-organizational enterprise strategy are increasingly strengthening, and why they should be researched. Future research should examine the exogenous and endogenous factors driving ERP user companies to participate in ERP systems enabled multi-organizational relationship collaborations. Specific external challenges can be identified as global competition, increasing outsourcing tendencies, product-service proliferations with shorter life cycles, changing customer expectations, rapid technological development, and government regulations. Whilst internal key drivers involve expanding business process complexity, assets re-intermediation, and cost effectiveness. Literature indicates that companies can experience different sets of exogenous (i.e. industrial forces) and endogenous (i.e. intra-organizational issues) changes which critically affect the decision-making for selecting ERP systems types and multi-organization enterprise practices to achieve desirable levels of competitiveness. This future research direction should be followed to address Research Gaps 1 and 2.

Another important issue to be studied and developed is the role of ERP information systems and their providers as they shape multi-organizational structures. For instance, it will be important to understand how newly emerging digital technologies, such as cloud computing and service-oriented architecture impact ERP systems, enable future inter-company relationships. Researchers should also be aware of the trend for monolithic ERP solutions becoming a thing of the past, with many ERP vendors (e.g. SAP notably) beginning to address and achieve the aims of integrated, yet decoupled, non-monolithic enterprise systems. This provides potential research areas for future ERP systems design and deployment approaches, which are justified by the need for on-demand ERP solutions, which could benefit and enable organizations to access technologies without significant individual investment cost in Inter-organizational IS (IOIS) when collaborating virtually between their partners. This future research direction should also be followed to address Research Gaps 1 and 2.

It is also important to understand how to manage multi-organizational collaborations effectively, efficiently, and flexibly. This inevitably raises questions as to what extent organizational boundaries should be altered through ERP systems investment; and to what extent inter-firm integration could help sustain competitiveness (e.g. through high operational efficiency and quick responsiveness). This indicates that future organizational managers and ERP systems developers will both need multi-criteria decision-making, which will change over time, to cope with uncertainty and ambiguity in the collaborative business environment. These emerging issues need to be addressed in the future to extend knowledge on how to effectively manage these multi-organizational relationships and collaborations. This future research direction should be followed to address Research Gaps 2, 3, and 4.

ERP systems adoption and use within the context of multi-organizational collaboration is constantly posing new and challenging questions on whether, for instance, cloud-ERP, SaaS, and web-based system architectures are feasible and how these technical innovations should be managed in a fair and ethical manner as the IoT becomes increasingly omnipotent. Thus, there exists further interesting opportunities for researchers to provide insights about information reliability, security, manageability, and effectiveness - based upon flexible inter-organizational enterprise ERP infrastructures; as well as to re-evaluate strengths and weaknesses between 'best-of-breed' and 'one-size-fit-all' ERP solutions. Additionally, in the context of the cloud computing, inter-organizational-wide ERP systems (including business rules across heterogeneous software and middleware applications) are becoming increasingly managed by vendors (e.g. SAP, Oracle) or sophisticated third-party consulting company (e.g. IBM, WebMethods). Research effort is needed to investigate whether such trends will make traditional non-web-based ERP vendors lose their incumbent and influential positions over end-users, or whether they will integrate so effectively into the multi-organization enterprises of the future that they themselves become the most significant company in a multi-organizational enterprise – perhaps even greater than the IOIS user companies themselves. This future research direction should be followed to address all Research Gaps (1, 2, 3, and 4) identified by this critique.

CONCLUSION

In conclusion this chapter provides insights into the current state of research about future developments of ERP (I, II and III) systems for multi-organizational enterprise (MOE) strategies by presenting the results of a critique of 255 papers from 53 international peer-reviewed journals (as rated by the UK's Chartered Association of Business School's Journal list). This research has determining the frequency of keywords found and classified them according to their (i) units of analyses (ii) theoretical perspectives, and (iii) research paradigms and techniques applied. By doing this and through subsequent detailed content analyses four specific research gaps have been identified and discussed in depth. It is hoped that these research gaps give novel insights into current research in the operations and IS fields.

As a starting point the concept of 'enterprization of multi-organizational enterprises' has been proposed as an overarching term to encapsulate the emergent trending discussion relating to these gaps. Specifically, this concept tries to explain the fits between different types of ERP systems and different types of multi-organizational enterprises. It is hoped that the proposal of this concept will help catalyze inter-dependent operations and information systems research; which in turn should help key players in the 'Internet of Things' (or Industry 4.0) era to connect interdependent IS and operation management strategy paths.

Findings from this study are limited as only journals stated, within the timeframe defined, are covered in this critique. Reviews of other journals in other timeframes may have identified different research gaps which may have resulted in different trends and concepts being proposed.

REFERENCES

Akkermans, H., Bogerd, P., Yucesan, E., & Van Wassenhove, L. (2003). The impact of ERP on supply chain management: Exploratory findings from a European Delphi Study. *European Journal of Operational Research, 146*(2), 284–294. doi:10.1016/S0377-2217(02)00550-7

Al-Mudimigh, A., Zairi, M., & Al-Mashari, M. (2001). ERP software implementation: An integrative framework. *European Journal of Information Systems, 10*(4), 216–226. doi:10.1057/palgrave.ejis.3000406

Albani, A., & Dietz, J. L. G. (2009). Current trends in modeling inter-organisational cooperation. *Journal of Enterprise Information Management, 22*(3), 275–297. doi:10.1108/17410390910949724

Anussornnitisarn, P., & Nof, S. Y. (2003). E-work: The challenge of the next generation ERP systems. *Production Planning and Control, 14*(8), 753–765. doi:10.1080/09537280310001647931

Bagchi, S., Kanungo, S., & Dasgupta, S. (2003). Modeling use of enterprise resource planning systems: A path analytic study. *European Journal of Information Systems, 12*(2), 142–158. doi:10.1057/palgrave.ejis.3000453

Banker, R. D., Chang, H., & Kao, Y. (2010). Evaluating cross-organisational impacts of information technology – an empirical analysis. *European Journal of Information Systems*, *19*(2), 153–167. doi:10.1057/ejis.2010.9

Bass, T., & Mabry, R. (2004) Enterprise architecture reference models: a shared vision for Service-Oriented Architectures. In *Proceedings of the IEEE MILCOM* (pp. 1-8).

Beatty, R. C., & Williams, C. D. (2006). ERP II: Best practices for successfully implementing an ERP upgrade. *Communications of the ACM*, *49*(3), 105–109. doi:10.1145/1118178.1118184

Benlian, A., & Hess, T. (2011). Comparing the relative importance of evaluation criteria in proprietary and open-source enterprise application software selection – a conjoint study of ERP and Office systems. *Information Systems Journal*, *21*(6), 503–525. doi:10.1111/j.1365-2575.2010.00357.x

Beretta, S. (2002). Unleashing the integration potential of ERP systems. *Business Process Management Journal*, *8*(3), 254–277. doi:10.1108/14637150210428961

Binder, M., & Clegg, B. T. (2006). A conceptual framework for enterprise management. *International Journal of Production Research*, *44*(18/19), 3813–3829. doi:10.1080/00207540600786673

Binder, M., & Clegg, B. T. (2007). Enterprise management: A new frontier for organisations. *International Journal of Production Economics*, *106*(2), 406–430. doi:10.1016/j.ijpe.2006.07.006

Bingi, P., Sharma, M., & Godla, J. (1999). Critical issues affecting an ERP implementation. *Information Systems Management*, *16*(3), 7–14. doi:10.1201/1078/43197.16.3.19990601/31310.2

Blackstone, J. H. Jr. & Cox, J.F. (2005) APICS Dictionary (11th ed.). Chicago, IL: APICS: The association for Operations Management.

Bond, B., Genovese, Y., Miklovic, D., Wood, N., Zrimsek, B., & Rayner, N. (2000) ERP is dead - long live ERPII, Retrieved from www.pentaprise.de/cms_showpdf.php?pdfname=infoc_report

Bose, I., Pal, R., & Ye, A. (2008). ERP and SCM systems integration: The case of a valve manufacturer in China. *Information & Management*, *45*(4), 233–241. doi:10.1016/j.im.2008.02.006

Brettel, M., Friederichsen, N., Keller, M. and Rosenberg, M., How virtualisation, decentralisation and network building change the manufacturing landscape: an industry 4.0 perspective". *International Journal of Mechanical, Aerospace, Industrial, Mechatronic and Manufacturing Engineering*, *8*(1) 37-44.

Byrne, J. A., & Brandt, R. (1993, February 8). The virtual corporation. *Business Week*, 36-41.

Cambra-Fierro, J., Florin, J., Perez, L., & Whitelock, J. (2011). Inter-firm market orientation as antecedent of knowledge transfer, innovation and value creation in networks. *Management Decision*, *49*(3), 444–467. doi:10.1108/00251741111120798

Candido, G., Barata, J., Colombo, A. W., & Jammes, F. (2009). SOA in reconfigurable supply chain: A research roadmap. *Engineering Applications of Artificial Intelligence*, *22*(6), 939–949. doi:10.1016/j.engappai.2008.10.020

Cao, M., & Zhang, Q. (2011). Supply chain collaboration: Impact on collaborative advantage and firm performance. *Journal of Operations Management*, *29*(3), 163–180. doi:10.1016/j.jom.2010.12.008

Chartered Association of Business Schools (C-ABS). (n.d.). Academic Journal Guide, 2014 and 2018. Retrieved from https://charteredabs.org/academic-journal-guide-2018/

Chen, D., Doumeingts, G., & Vernadat, F. (2008). Architectures for enterprise integration and interoperability: Past, present and future. *Computers in Industry*, *59*(7), 647–659. doi:10.1016/j.compind.2007.12.016

Chen, I. J. (2001). Planning for ERP systems: Analysis and future trend. *Business Process Management Journal*, *7*(5), 374–386. doi:10.1108/14637150110406768

Cheng, Y., Tao, F., Xu, L., & Zhao, D. (2016). Advanced manufacturing systems: Supply-demand matching of manufacturing resource based on complex networks and Internet of Things. *Enterprise Information Systems*, 1751–7575.

Chengalur-Smith, I., Duchessi, P., & Gil-Garcia, J. R. (2012). Information sharing and business systems leveraging in supply chain: An empirical investigation of one web-based application. *Information & Management*, *49*(1), 58–67. doi:10.1016/j.im.2011.12.001

Chi, L., Ravichandran, T., & Andrevski, G. (2010). Information technology, network structure, and competitive action. *Information Systems Research*, *21*(3), 543–570. doi:10.1287/isre.1100.0296

Choi, T. Y., & Wu, Z. (2009). Triads in supply networks: Theorizing buyer-supplier-supplier relationships. *The Journal of Supply Chain Management*, *45*(1), 8–25. doi:10.1111/j.1745-493X.2009.03151.x

Clegg, B., & Wan, Y. (2013). ERP systems and enterprise management trends: A contingency model for the enterprization of operations. *International Journal of Operations & Production Management*, *33*(11/12), 1458–1489. doi:10.1108/IJOPM-07-2010-0201

Daniel, E. M., & White, A. (2005). The future of inter-organisational system linkages: Findings of an international delphi study. *European Journal of Information Systems*, *14*(2), 188–203. doi:10.1057/palgrave.ejis.3000529

Davis, E. W., & Spekman, R. E. (2004) *Extended Enterprise: Gaining Competitive Advantage through Collaborative Supply Chains*. New York, NY: Financial Times Prentice-Hall.

Davison, R. M., Powell, P., & Trauth, E. M. (2012). ISJ inaugural edition. *Information Systems Journal*, *22*(4), 257–260. doi:10.1111/j.1365-2575.2012.00417.x

Deep, A., Guttridge, P., Dani, S., & Burns, N. (2008). Investigating factors affecting ERP selection in made-to-order SME sector. *Journal of Manufacturing Technology Management*, *19*(4), 430–446. doi:10.1108/17410380810869905

Eckartz, S., Daneva, M., Wieringa, R., & Hillegersberg, J. V. (2009) Cross-organisational ERP management: How to create a successful business case? In *SAC'09 Proceedings of the 2009 ACM Symposium on Applied Computing*. Honolulu, HI.

Eisenhardt, K. M., & Martin, J. A. (2000). Dynamic capabilities: What are they? *Strategic Management Journal*, *21*(10/11), 1105–1121. doi:10.1002/1097-0266(200010/11)21:10/11<1105::AID-SMJ133>3.0.CO;2-E

Ericson, J. (2001). What the heck is ERPII? Retrieved from http://www.line56.com/articles/default.asp?ArticleID=2851

European Commission. (2003). Commission recommendation of 6 May 2003 concerning the definition of micro, small and medium sized enterprises. *Official Journal of the European Union, L, 124*(1422), 36–41.

Fawcett, S. E., Fawcett, A. M., Watson, B. J., & Magnan, G. M. (2012). Peeking inside the black box: Toward an understanding of supply chain collaboration dynamics. *The Journal of Supply Chain Management, 48*(1), 44–72. doi:10.1111/j.1745-493X.2011.03241.x

Gable, G. (1998). Large package software: A neglected technology? *Journal of Global Information Management, 6*(3), 3–4.

Glaser, B. G., & Strauss, A. L. (1967). *The Discovery of Grounded Theory: Strategies for Qualitative Research*. New York, NY: Aldine.

Harrigan, K. R. (1985). Vertical integration and corporate strategy. *Academy of Management Journal, 28*(2), 397–425.

Hendricks, K. B., Singhal, V. R., & Stratman, J. K. (2007). The impact of enterprise systems on corporate performance: A study of ERP, SCM, and CRM system implementations. *Journal of Operations Management, 25*(1), 65–82. doi:10.1016/j.jom.2006.02.002

Henningsson, S., & Carlsson, S. (2011). The DySIIM model for managing IS integration in mergers and acquisitions. *Information Systems Journal, 21*(5), 441–476.

Joskow, P. L. (2003). Vertical integration. In *Handbook of New Institutional Economics*. Boston, MA: Kluwer.

Law, C. C. H., & Ngai, E. W. T. (2007). An investigation of the relationships between organisational factors, business process improvement, and ERP success. *Benchmarking: An International Journal, 14*(3), 387–406. doi:10.1108/14635770710753158

Lee, J., Siau, K., & Hong, S. (2003). Enterprise integration with ERP and EAI. *Communications of the ACM, 46*(2), 54–60. doi:10.1145/606272.606273

Lin, F., & Rohm, C. E. T. (2009). Managers' and end-users' concerns on innovation implementation: A case of an ERP implementation in China. *Business Process Management Journal, 15*(4), 527–547. doi:10.1108/14637150910975525

Lu, X. J., Huang, L. H., & Heng, M. S. H. (2006). Critical success factors of inter-organisational information systems: A case study of Cisco and Xiao Tong in China. *Information & Management, 43*(3), 395–408. doi:10.1016/j.im.2005.06.007

Lyman, K. B., Caswell, N., & Biem, A. (2009). Business value network concepts for the extended enterprise. In P. H. M. Vervest, D. W. Liere, & L. Zheng (Eds.), *Proc. of the Network Experience*. Berlin: Springer. doi:10.1007/978-3-540-85582-8_9

Lynch, R. (2003). *Corporate strategy (3rd ed.)*. Harlow: Prentice-Hall Financial Times.

Madu, C. N., & Kuei, C. (2004). *ERP and Supply Chain Management*. Fairfield, CT: Chi Publishers.

Mahoney, J. T. (1992). The choice of organisational form: Vertical financial ownership versus other methods of vertical integration. *Strategic Management Journal*, *13*(8), 559–584. doi:10.1002mj.4250130802

Martinez, M. T., Fouletier, P., Park, K. H., & Faurel, J. (2001). Virtual enterprise: Organisation, evolution and control. *International Journal of Production Economics*, *74*(1-3), 225–238. doi:10.1016/S0925-5273(01)00129-3

Maurizio, A., Girolami, L., & Jones, P. (2007). EAI and SOA: Factors and methods influencing the integration of multiple ERP systems (in an SAP environment) to comply with the Sarbanes-Oxley Act. *Journal of Enterprise Information Management*, *20*(1), 14–31. doi:10.1108/17410390710717110

McAfee, A. (2002). The impact of enterprise information technology adoption on operational performance: An empirical investigation. *Production and Operations Management*, *11*(1), 33–53. doi:10.1111/j.1937-5956.2002.tb00183.x

Moller, C. (2005). ERPII: A conceptual framework for next-generation enterprise systems? *Journal of Enterprise Information Management*, *18*(4), 483–497. doi:10.1108/17410390510609626

Nah, F. F., Tan, X., & Beethe, M. (2005) End-users' acceptance of Enterprise Resource Planning (ERP) Systems: an investigation using grounded theory approach. *Paper presented at the Eleventh Americas Conference on Information Systems*, Omaha, NB.

Olsen, K. A., & Sætre, P. (2007). IT for niche companies: Is an ERP system the solution? *Information Systems Journal*, *17*(1), 37–58. doi:10.1111/j.1365-2575.2006.00229.x

Oura, J., & Kijima, K. (2002). Organisation design initiated by information system development: A methodology and its practice in Japan. *System Research and Business Science*, *19*(1), 77–86. doi:10.1002res.415

Owen, L., Goldwasser, C., Choate, K., & Blitz, A. (2008). Collaborative innovation throughout the extended enterprise. *Strategy and Leadership*, *36*(1), 39–45. doi:10.1108/10878570810840689

Palaniswamy, R., & Frank, T. (2000). Enhancing manufacturing performance with ERP systems. *Information Systems Management*, *17*(3), 1–13. doi:10.1201/1078/43192.17.3.20000601/31240.7

Purchase, V., Parry, G., Valerdi, R., Nightingale, D., & Mills, J. (2011). Enterprise transformation: Why are we interested, what is it, and what are the challenges? *Journal of Enterprise Transformation*, *1*(1), 14–33. doi:10.1080/19488289.2010.549289

Rappa, M. A. (2004). The utility business model and the future of computing services. *IBM Systems Journal*, *43*(1), 32–42. doi:10.1147j.431.0032

Richardson, J. (1996). Vertical integration and rapid response in fashion apparel. *Organization Science*, *7*(4), 400–412. doi:10.1287/orsc.7.4.400

Rodon, J., Sese, F., & Christiaanse, E. (2011). Exploring users' appropriation and post-implementation managerial intervention in the context of industry IOIS. *Information Systems Journal*, *21*(3), 223–248. doi:10.1111/j.1365-2575.2009.00339.x

Rothaermel, F. T., Hitt, M. A., & Jobe, L. A. (2006). Balancing vertical integration and strategic outsourcing: Effects on product portfolio, product success, and firm performance. *Strategic Management Journal*, *27*(11), 1033–1056. doi:10.1002mj.559

Samaranayake, P. (2009). Business process integration, automation, and optimization in ERP: Integrated approach using enhanced process models. *Business Process Management Journal*, *15*(4), 504–526. doi:10.1108/14637150910975516

Scott, J. E., & Vessey, I. (2000). Implementing enterprise resource planning systems: The role of learning from failure. *Information Systems Frontiers*, *2*(2), 213–232. doi:10.1023/A:1026504325010

Sharif, A. M. (2010). It's written in the cloud: The hype and promise of cloud computing. *Journal of Enterprise Information Management*, *23*(2), 131–134. doi:10.1108/17410391011019732

Sharif, A. M., Irani, Z., & Love, P. E. D. (2005). Integrating ERP with EAI: A model for post-hoc evaluation. *European Journal of Information Systems*, *14*(2), 162–174. doi:10.1057/palgrave.ejis.3000533

Sharp, J. M., Irani, Z., & Desai, S. (1999). Working towards agile manufacturing in the UK industry. *International Journal of Production Economics*, *62*(1-2), 155–169. doi:10.1016/S0925-5273(98)00228-X

Songini, M. L. (2002). J.D. Edwards pushes CRM, ERP integration. *Computerworld*, *36*(25), 4.

Spekman, R., & Davis, E. W. (2016). The extended enterprise: A decade later. *International Journal of Physical Distribution & Logistics Management*, *46*(1), 43–61. doi:10.1108/IJPDLM-07-2015-0164

Stevens, C. P. (2003). Enterprise resource planning: A trio of resources. *Information Systems Management*, *20*(3), 61–71. doi:10.1201/1078/43205.20.3.20030601/43074.7

Sutton, S. G. (2006). Extended-enterprise systems' impact on enterprise risk management. *Journal of Enterprise Information Management*, *19*(1), 97–114. doi:10.1108/17410390610636904

Tarantilis, C. D., Kiranoudis, C. T., & Theodorakopoulos, N. D. (2008). A web-based ERP system for business service and supply chain management: Application to real-world process scheduling. *European Journal of Operational Research*, *187*(3), 1310–1326. doi:10.1016/j.ejor.2006.09.015

Themistocleous, M., Irani, Z., & O'Keefe, R. (2001). ERP and application integration: Exploratory survey. *Business Process Management Journal*, *7*(3), 195–204. doi:10.1108/14637150110392656

Thun, J. H. (2010). Angles of integration: An empirical analysis of the alignment of internet-based information technology and global supply chain integration. *The Journal of Supply Chain Management*, *46*(2), 30–44. doi:10.1111/j.1745-493X.2010.03188.x

Torbacki, W. (2008). SaaS – direction of technology development in ERP/MRP systems. *Archives of Materials Science and Engineering*, *31*(1), 57–60.

Trauth, E. M., Farwell, D. W., & Lee, D. (1993). The IS expectation Gap: Industry Expectations versus Academic Preparation. *Management Information Systems Quarterly*, *17*(September), 293–307. doi:10.2307/249773

Triantafillakis, A., Kanellis, P., & Martakos, D. (2004). Data warehousing interoperability for the extended enterprise. *Journal of Database Management*, *15*(3), 73–82. doi:10.4018/jdm.2004070105

Vallespir, B., & Kleinhans, S. (2001). Positioning a company in enterprise collaborations: Vertical integration and make-or-buy decisions. *Production Planning and Control*, *12*(5), 478–487. doi:10.1080/09537280110042701

Vannoy, S. A., & Salam, A. F. (2010). Managerial interpretations of the role of information systems in competitive actions and firm performance: A grounded theory investigation. *Information Systems Research, 21*(3), 496–515. doi:10.1287/isre.1100.0301

Vathanophas, V. (2007). Business process approach towards an inter-organisational enterprise system. *Business Process Management Journal, 13*(3), 433–450. doi:10.1108/14637150710752335

Weston, F.C., Jr. (2002). *A vision for the future of extended enterprise systems* [Presentation]. In *J.D. Edwards FOCUS Users Conference*, Denver, CO, June 12.

Wilkes, L., & Veryard, R. (2004) Service-oriented architecture: considerations for agile systems. *Microsoft Architect Journal,* (April). Retrieved from www.msdn2.microsoft.com

Xu, W., Wei, Y., & Fan, Y. (2002). Virtual enterprise and its intelligence management. *Computers & Industrial Engineering, 42*(2-4), 199–205. doi:10.1016/S0360-8352(02)00053-0

ADDITIONAL READING

Bickof, A. (2018). Netsuite ERP for Administrators: Summarising NetSuite ERP in simple terms with practical tips to make you an effective administrator. Birmingham, UK: Pakt Publishing.

Binder, M., & Clegg, B. (2010). *Sustainable supplier management in the automotive industry: leading the 3rd revolution through collaboration*. New York: Nova Science Publishers Inc.

Callaway, E. (2000). ERP – the next generation: ERP is web enabled for e-business. Computer Technology Research Corporation.

Davis, E. W. (2003). *The extended enterprise: gaining competitive advantage thorough collaborative supply chains*. London, UK: Financial Times Prentice Hall.

Gross, P. (2018). *Milestone deliverables: ERP project management methodology*. Toronto, Canada: Pemeco Financial Holdings Inc.

Gutteridge, L. (2018). *Avoiding IT disasters: fallacies about enterprise systems and how you can rise above them*. Vancouver, Canada: Thinking Works Inc.

Johnson, B., & de Rouw, L.-P. (2017). *Collaborative business design: improving and innovating the design of IT-driven business services*. UK: IT Governance Publishing. Ely.

Paton, S., Clegg, B., Hsuan, J., & Pilkington, A. (2011). *Operations Management*. Maidenhead, UK: McGraw-Hill.

Pelphrey, M. W. (2015). *Directing the ERP implementation: a best practice guide to avoiding program failure traps while tuning system performance*. Boca Raton, LA: CRC Press. doi:10.1201/b18278

Skinner, P. (2018). *Collaborative Advantage: How collaboration beats competition as a strategy for success*. London, UK: Robinson. Little Brown Book Group.

KEY TERMS AND DEFINITIONS

Enterprization: This refers to the trend where future ERP systems development / implementation and new multi-organizational enterprise paradigms and practice (e.g. strategy and operations) are inextricably linked.

Enterprization of Multi-Organizational Enterprises: This is a pattern of behavior that explains how, at different times and circumstances in an enterprise's lifecycle, individual companies may prefer different multi-organizational enterprise structures and use different ERP system types (e.g. continuous upgrade or reconfiguration) to satisfy their exogenous and endogenous business requirements. This may lead to the 'internet of multi-organizational enterprise operations.

ERP Systems: A comprehensive package software solution that integrates the complete range of a businesses' processes and functions in order to present a holistic view of the business from a single information and IT architecture, these can vary in maturity (e.g. from ERPI, to ERPII, to the highest maturity and technologically advanced ERPIII). ERP systems were originally designed and built to overcome problems associated with fragmented and incompatible information systems, and bring together inconsistent operating practices within organizations.

ERP I: An Integrated information management system supporting the operational transactions of a single company.

ERP II: Web enabled information system for managing operations in organizations, these can increase potential transparency of data and operations between organizations and open up new opportunities for (multi-organization) enterprises. ERP II systems can facilitate resource planning and co-operations between different organizations at a meta-level.

ERP III: A flexible, powerful information system for operations management incorporating web-based technology which enables multi-organization enterprises to offer increasing degrees of connectivity, collaboration and dynamism through increased functional scope and scalability across parts of different organizations.

Extended Enterprise: Semi-permanent group of parts of different organizations working towards joint strategic objectives. Extended enterprises (EEs) tend to be lean and agile based on technical and social competence.

Multi-Organization Enterprise: An organization in which parts of companies work with parts of other different companies to collaboratively deliver complex product-service systems.

Vertically Integrated Enterprise: Almost permanent and extremely well-integrated group of parts of organizations; very similar to a single legal entity. Vertically integrated enterprises tend to be lean.

Virtual Enterprise: Temporary group of parts of different organizations exploiting a short-term high-risk opportunity. Virtual enterprises (VEs) tend to be very agile and based on new innovative technology.

Chapter 5
Toward a Conceptualization of Big Data Value Chain:
From Business Problems to Value Creation

Rim Louati
University of Sfax Higher Institute of Business Administration, Tunisia

Sonia Mekadmi
University of Carthage Institute of Higher Commercial Studies, Tunisia

ABSTRACT

The generation of digital devices such as web 2.0, smartphones, social media and sensors has led to a growing rate of data creation. The volume of data available today for organizations is big. Data are produced extensively every day in many forms and from many different sources. Accordingly, firms in several industries are increasingly interested in how to leverage on these "big data" to draw valuable insights from the various kinds of data and to create business value. The aim of this chapter is to provide an integrated view of big data management. A conceptualization of big data value chain is proposed as a research model to help firms understand how to cope with challenges, risks and benefits of big data. The suggested big data value chain recognizes the interdependence between processes, from business problem identification and data capture to generation of valuable insights and decision making. This framework could provide some guidance to business executives and IT practitioners who are going to conduct big data projects in the near future.

INTRODUCTION

The increasing digitalization of organizations, coupled with the advances in the capabilities of technology, has led to the explosion of data in different formats from various digital sources. The volumes of data amassed by organizations are "big" (McDonald and Lévéillé, 2014). According to estimates, the volume of business data, across almost all companies worldwide, doubles every 1.2 years (McKinsey Global Institute, 2012; Chen and Zang, 2014). The volume of data available today is measured in zettabytes (ZB) – a measure equal to one trillion gigabytes (GB) and equivalent to the data storage capacity

DOI: 10.4018/978-1-5225-7262-6.ch005

of about 250 billion DVDs (Alharthi *et al.*, 2017). As a consequence, companies in many industries are increasingly interested in how to leverage on these "big data" to draw insights from the various kinds of data available to them and gain an in-depth understanding of the hidden values in order to exploit new opportunities (Tan et al., 2015; Raguseo, 2018). The International Data Corporation (IDC, 2017) forecasted that "Big data and business analytics worldwide revenues, which reached $49.1 billion worldwide in 2016, are expected to maintain a compound annual growth rate (CAGR) of 11.9% through 2020 when revenues will be more than $210 billion".

Big data is a nascent concept introduced to describe the tremendous quantity of data that requires to be managed in organizations. The proliferation of digital devices such as web 2.0, smartphones, social media and sensors has led to a growing rate of data creation and is driving an increasing need for real-time analytics and evidence-based decisions. Notwithstanding the emerging nature of big Data, the origin of the concept is not new. It was introduced by the Gartner analyst Doug Laney in a research note from 2001 in which he noted: "*While enterprises struggle to consolidate systems and collapse redundant databases to enable greater operational, analytical, and collaborative consistencies, changing economic conditions have made this job more difficult. E-commerce, in particular, has exploded data management challenges along three dimensions: volume, velocity and variety*" (p. 1).

Big data is a new concern for organizations. The objectives of big data initiatives are generally focused on data value (McDonald and Lévéillé, 2014). With big data, firms can extract new insights about their markets, customers and products which are important to innovation. However, as noted by McAfee et al. (2012), businesses are collecting more data than they know what to do with. In this way, managers should define a clear strategy for how to use big data to respond to business problems and support firms' innovation capabilities. The main challenges for ensuring this strategy are both related to the development of skills and a new decision-making culture to turn all this data into a competitive advantage (Raguseo, 2018), and to the establishment of a robust IT architecture that enables acquisition, storage and analysis of very huge data sets (Barton and Court, 2012; Kumar et al., 2013; Wamba et al., 2015).

The purpose of this chapter is to provide an integrated view of big data management. A conceptualization of big data value chain is proposed as a research model to help firms understand how to cope with challenges, risks and benefits of big data. Value chain, a concept introduced by Porter (1985), sees an organization as a series of processes and each process adds value to the product or service for their customers (Chou, 2014). Value delivered by actions is therefore a foundation of decision-making. Value chain architecture helps to analyze the sources of value creation by identifying the main activities and processes a firm should perform and how they interact. Miller and Mork (2013) applied Porter's value chain theory into a reference model for big data value creation. However, the model is limited to the big data management process and neglected other processes like big data architecture and business process reengineering.

The purpose of defining an integrated big data value chain is to provide a holistic approach to big data management that begins with the identification of business problems explaining the need to carry out big data projects, and ends with the value creation by generating and leveraging deep customer insights. From the beginning to the end of the value chain, three interrelated processes are positioned which are related to big data management process, big data architecture and business process reengineering. The suggested value chain focuses on the interrelations between strategic, organizational and technological elements in order to extract value from big data. Such a definition is needed for strategic planning and alignment of the elements, and thus should not only be of academic interest, but also have important managerial implications.

Based on a literature review, this paper first describes the big data concept and its dimensions and characteristics, and then proposes a research model conceptualizing the main issues that need to be considered when managing big data. After this discussion, the different components of the big data value chain are described in detail. Finally, the conclusion section and directions for future research are given.

BIG DATA CONCEPT AND DIMENSIONS

The term big data is used to characterize the massive volume of data generated by digital machines and computing devices, cloud-based solutions, business management processes, etc. This volume of data is very difficult to manage using traditional data analysis tools. The importance of big data can be explained by the fact that data are produced extensively every day in many forms and from many different sources. Organizations have now more data to use than ever before. This data may be internal or external, structured or unstructured (Raguseo, 2018). Companies are searching for new technological solutions in order to harness this 'big' quantity of data and therefore to understand patterns of consumer activity and to deliver new business insights.

Laney (2001) suggested that Volume, Variety, and Velocity (or the three V's) are the three dimensions of challenges in data management. The Three V's have emerged as the main characteristics to describe big data (Gandomi and Haider, 2015; Gartner, 2017). *Volume* refers to the size of data. The huge quantity of data currently gathered by organizations is reported in multiple terabytes and petabytes. But the concept of big data goes beyond volume. *Variety* which is the multidimensionality and heterogeneity of data formats and sources, including various types of structured, semi-structured, and unstructured data. *Velocity* is related to the frequency or the speed of data generation and/or analysis and delivery. Drawing on the ubiquity of the Vs definition, others include two additional Vs (5 Vs) namely value and veracity. *Value* highlights the need of creating economic benefits or business value from the available big data. *Veracity* emphasizes the importance of data quality and the level of trust in various data sources (White, 2012; Lee, 2017).

There are also other definitions that stress the role of advanced technologies to enable the capture, storage, management and analysis of big data (Wamba et al., 2015). The term big data designates not only data, but also the tools and techniques used for capturing, analyzing, processing, and managing these massive, complex, and rapidly evolving data sets (Alharthi *et al.*, 2017).

In this way, big data can be defined as a holistic strategy and a set of cost-effective and innovative techniques and technologies to manage the volume, variety, velocity, value and veracity of data in order to get enhanced insights for decision-making and business value creation. Big data requires taking into account the main characteristics of data (the 5 Vs), the desired outcomes from its analysis and the means to achieve those benefits. Big data must bring a chain value for an organization.

CONCEPTUALIZING BIG DATA VALUE CHAIN

Value creation is a challenge for every firm seeking a sustainable competitive advantage. The value added to their product or service plays an important role while business executives are deciding to invest in a vital project. It is therefore necessary to target high value-adding processes within an organization for enhancing the final value delivered to its customers. In that sense, Porter (1985) proposed the value-chain concept, describing an organization as a series of activities and processes that culminate in value creation. Whereas applying value analysis into IT context, Smith and McKeen (2003) identified three stages of generating an IT value in business processes, including identification of potential opportunities for adding value, conversion of those opportunities into effective applications of technology for the process, and realization of that applications for developing IT value (Chou, 2014).

Big data value creation may also be examined through a value chain model. Miller and Mork (2013) adapted Porter's value chain theory for big data management. The aim of this framework was to examine how to bring disparate data together and manage them in an organized process from data capture to valuable information creation that can inform decision making at the enterprise level. Although interesting, the suggested value chain is restricted to the big data management process without taking into account other processes like big data architecture and business process reengineering. Then, we propose an integrated big data value chain, which aims to manage big data across a continuum from business problem identification to value creation. As illustrated in Figure 1, big data value creation model consists of five components: business problem identification, big data management process, big data architecture, business process reengineering, and big data value creation. The suggested big data value chain gets closer to the Smith and McKeen (2003) logic's. The only difference is that the conversion phase has been divided into three processes linked to the management and application of big data technologies. The framework targets simultaneously strategic, organizational and technological aspects related to effective management of big data projects. The individual components of the research model are discussed below.

Big Data and Business Problems

Big data is characterizing firms whose resource requirements for data management exceed their capabilities of traditional computing environments, and who are searching for innovative and cost-effective techniques to solve existing and future business problems (Loshin, 2013). Therefore, a deep understanding of the business process or problem behind the implementation of big data analytics projects is the first critical step. Senior management and business stakeholders of the project must set the right expectations about what the management of big data can achieve for their organization, and the possible insights that can be generated from it (Dutta and Bose, 2015).

Figure 1. Big data value chain

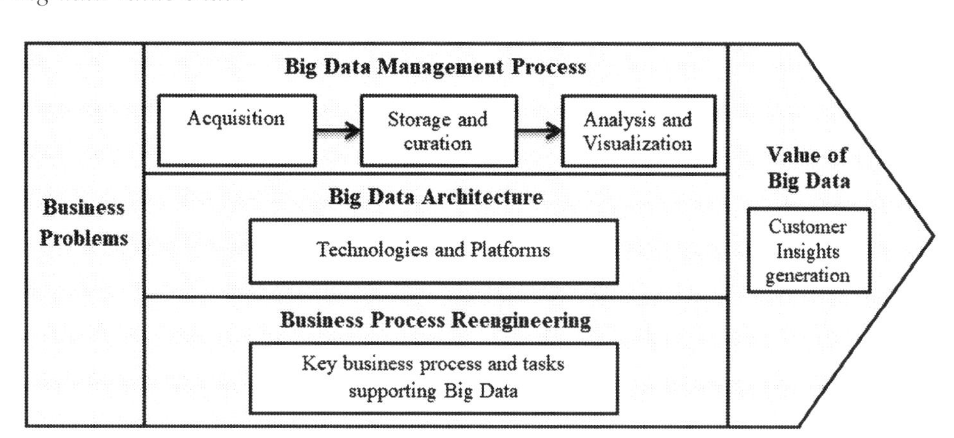

There are many examples of business problems explaining the need to carry out big data projects in multiple companies belonging to different industries. Besides the giants of the web, like Amazon, Google and Facebook, who are multiplying initiatives to exploit their big data, a significant number of firms, both in traditional sectors and e-commerce, are trying to understand their business needs in order to set targets for achievement in a big data project. The most noteworthy case is the retail sector which is searching to build a culture of customer data-driven decision making. Confronted with the competition of e-commerce, retailers must innovate in stores and rely on big data to cope with online business competitors. All the signs are now leading a strategy of digitalization of the retail outlet and try to mix the digital, the mobile and the store to collect as much information as possible. For a big retailer, the objective is naturally to convert the occasional visitors into customers. Analysis of the behavior is thus crucial to react at best to the abandonments of shopping cart, to bring back the visitor on the site and urge him to convert its visit in act of purchase with a multi-channel approach, whether it is on mobile, in shop, on the Web or via the call center. Hence, big data offer possibilities to track customers during their customer journey, and optimize advertising campaigns and budgets. Technical analysis of customer journeys has become an important feature for retailers, who follow the customer when he or she seeks information, compares products, and ultimately takes the decision to purchase a product and buy it. For example, the number one Retailer in UK "Tesco" has created a powerful data collection engine through the combination of data obtained from loyalty cards, scanners, Web sites, and market research, in order to systematically turn big data into customer insights and insights into business decisions (Leeflang et al., 2014).

Other business problems can be encountered in the financial sector, telecommunications, transport, etc. For instance, credit card companies need to analyze fraudulent behavior with stolen credit cards. Based on this analysis, they can automatically block credit cards that show payment patterns frequently displayed with stolen credit cards, in order to reduce the financial risk for both the customer and the credit card company. Yet another example is Telecommunications companies who are submerged by big data from their call center records. The results of data analysis can help them determine what segments of their market are experiencing particular technical issues, and develop enhanced technical support services targeted to specific demographic segments (McDonald and Lévéillé, 2014). Railway companies can use their own travel data and data provided by navigation systems to provide information to customers on their expected travel time by train and by car. This helps customers make more grounded decisions on their choice of travel mode.

In another instance, the tremendous growth of social media and consumer-generated content on the Internet had impacted hospitality industry. Websites including TripAdvisor.com, and online travel agencies (OTAs) such as Expedia and Travelocity, had allowed consumers to post their ratings and reviews regarding their experiences with hotel properties they have stayed at in the past. Customer reviews reflect the way consumers describe, relive, reconstruct, and share their experiences. Because other consumers are tapping into this information for travel planning purposes, customer reviews can generate a huge impact on travel planning and subsequently attitudes and behavioral intentions. Hotels can indeed apply big data analytic techniques, like text mining approach, to a large quantity of consumer reviews extracted from social websites to deconstruct hotel guest experience and examine its association with satisfaction ratings (Xiang et al., 2015).

One last example is related to airline companies. In airports, every minute counts, including schedules of arrival of the flights: if a plane lands before the ground staff is ready to accommodate it, the passengers and the crew find themselves blocked. If the plane is presented later that envisaged, staff remains seated doing nothing and the costs rise. Also, airline companies have commonly the issue of waiting times for landing with a variation between the estimated hour and the effective hour of landing. In search of a better customer service, those companies must offer the right estimated hour of landing (ETA). They need to collect instantaneously a large set of information by combining the weather conditions data, the schedules of flights, and diverse other factors with data coming from the airplane pilot and radar tracking stations. Big data analytics can allow an airline company to know when its planes are going to land and organizing itself accordingly, saving as a result several costs in every airport (McAfee and Brynjolfsson, 2013).

Note that all examples assume that data-driven customer insights are gaining the most challenge in different businesses such as retail, financial, high-tech, and telecom companies. The same challenge is ubiquitous in every Internet company. Big data allows better planning, and thus to better decision making.

Big Data Management Process

Considering the complexity of big data management, due to its high-volume, high velocity and high variety, initiatives taken by organizations must involve the development of processes and systems designed to capture, examine, store, analyze, visualize and otherwise exploit data from existing systems and databases.

Typically, as shown in Fig. 1, valuable insights and knowledge extracted from big data require the management of a series of stages or sub-processes, beginning with the acquisition of data from one or several sources, and proceeding through steps that include the examination and storage of data, data analysis and modeling, data visualization and the eventual production of statistics, reports and other forms of information (McDonald and Lévéillé, 2014; Chen and Zang, 2014). Different challenges are facing firms in each sub-process:

- **Data Acquisition:** Data requirements are different due to different organizations' needs and problems. First, it is essential for a company to understand what information it needs in order to create as much value as possible (Tan et al., 2015). This helps identify data sources and capture the right data. A better management of big data depends on the integration of multiple sources of data, including structured data such as historical business transactions, Internet traffic (e.g., clickstreams), mobile transactions, customer surveys, etc. and unstructured data obtained from social media sources, sensor networks, customer interaction emails, user-generated content, etc. Often, these unstructured data give additional insights that can augment insights generated from structured data. Thus, the accessibility of Big Data sources is on the top priority of the knowledge creation process. Big Data should be accessed easily and rapidly for further analysis (Chen and Zang, 2014).
- **Data Storage and Curation:** It is important for organizations to meet their bulk storage requirements in big data management process for experimental databases, array storage for statistical computations, and large output files (Tan et al., 2015). However, challenges in Big Data storage include data inconsistence and incompleteness, scalability, timeliness, redundancy and data security. The pre-processing of the data is necessary before it is stored to improve its quality and

simplify its authentication, retrieval, reuse and preservation over time. Thus, a number of data pre-processing techniques, including data cleaning, data integration, data transformation and data reduction, can be used to remove noise and correct inconsistencies from datasets (Chen and Zang, 2014). The unstructured data may require going through text tagging and annotation for creation of metadata, using new variables like product, customer, location, etc. This metadata in turn could serve as a dimension of analysis for the structured data (Dutta and Bose, 2015). Hence, a well-constructed data is the prior step to data analysis.

- **Data Analysis and Visualization:** The main characteristic of Big Data is its volume. So, the biggest and most important challenge is analysis of the large amount of data. Data analytics can be used to help managers generate lots of useful information. Analytics is the practice of using and interpreting data to extract insights that can help firms make better fact-based decisions with the ultimate aim of driving strategy and improving performance (Tan et al., 2015). Various data mining techniques can be used on the collected data in order to generate lots of useful information, identify trends in the data and discover underlying reasons explaining the trend. The choice of quantitative modeling technique depends on the type of business problem which is solved as well as the nature of data. For structured data analysis, statistical approaches such as factor analysis, regression, structural equation modeling, etc. as well as machine learning approaches such as decision trees, neural networks, clustering, among others, can be used. For unstructured data analysis, text mining can be used for identification of important concepts and their interrelationships as well as extraction of important reviews and terms related to concepts and analysis of sentiments expressed about these concepts (Dutta and Bose, 2015). Furthermore, data visualization techniques can be used to represent knowledge more intuitively and effectively. Different graphs with both aesthetic form and functionality are necessary to transmit information easily by providing knowledge hidden in the complex and big data sets (Chen and Zang, 2014).

Big Data Architecture

With big data, firms can extract new insights about their markets, customers and products which are important to innovation. However, managers must define a clear strategy for how to use big data to support firms' innovation capabilities. The main challenge for ensuring this strategy is the establishment of a robust IT architecture that enables acquisition, storage and analysis of very huge data sets (Barton and Court, 2012; Kumar et al., 2013; Wamba et al., 2015). Organizations need to use and develop new techniques and technologies to manage volume, variety and velocity of big data that often remain beyond traditional IT capabilities. IT infrastructure is especially required to support the two main sub-processes of managing big data: data storage and data analytics. As shown in figure 2, data storage involves technologies to capture and store data and to prepare and retrieve it for analysis. Data analytics, on the other hand, refer to techniques and technologies used to analyze and make sense from big data (Gandomi and Haider, 2015).

Architecture for Big Data Storage

An efficient accessible architecture is required to the storage and management of large data sets and for achieving availability and reliability of big data. However, the rapid growth of data has restricted the capability of existing storage technologies to manage big data. Even though traditional database systems

Figure 2. Big data architecture

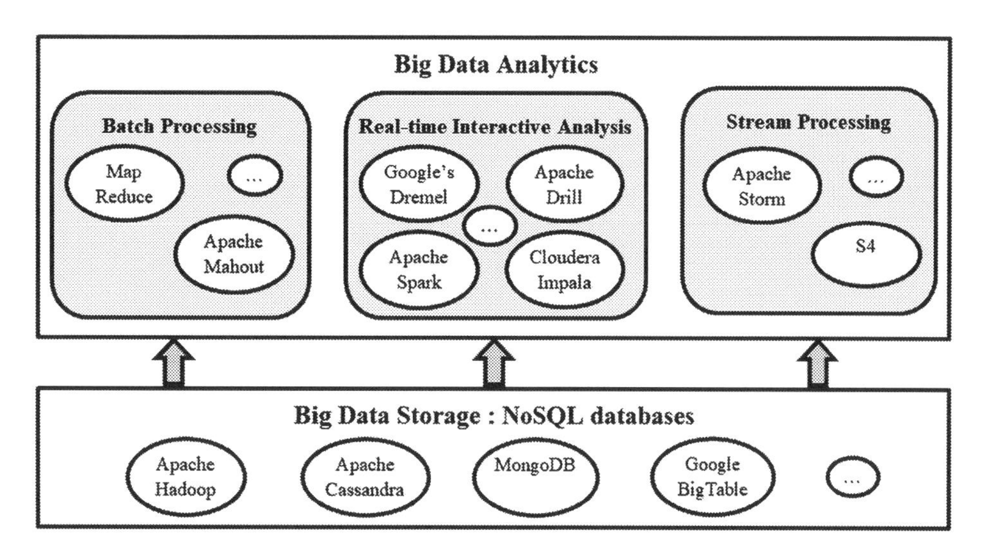

are able to handle the Volume associated with Big Data, they are inadequate when dealing with the Variety and Velocity of Big Data (Madden 2012, Dutta and Bose, 2015).

Generally, traditional storage systems are utilized to store data through structured Relational Database Management Systems (RDBMS) (Hashem et al., 2015). For example, a data warehouse is a popular relational database system that is used to store and manage large-scale datasets. The data warehouse is Standard Query Language (SQL) based database system which is mainly responsible to store structured data that is sourced from the operational systems. Although some Big Data analytic platforms, like SQL stream and Cloudera Impala, still use SQL in their database systems, because SQL is more reliable and simpler query language with high performance in stream Big Data real-time analytics, the most popular big data architectures use NoSQL database, also called "Not Only SQL", to store and manage unstructured data or non-relational data for large and distributed data management and database design (Chen and Zang, 2014; Self and Voorhis, 2015).

The most famous used NoSQL database is Apache Hadoop which is an open-source Software framework written in Java that allows the distributed processing of large datasets across clusters of commodity. Hadoop provides high reliability and a high fault tolerance to applications by maintaining multiple working copies of data and redistributing the failed node (Tian and Zhao, 2015). Its scalability is enhanced by the parallel and fast processing of petabytes of data (Lam, 2010; Tan et al., 2015). The most significant feature of Hadoop is that the storage system is not physically separated from the processing system. Other NoSQL implementations include Apache Cassandra, MongoDB, Google BigTable, etc.

It should be noted that big data utilizes distributed storage technology based on cloud computing rather than local storage attached to a computer or electronic device. Big data analysis is driven by using virtualized technologies like Hadoop clusters. Virtualization is a process of resource sharing and isolation of underlying hardware to increase computer resource utilization, efficiency, security and scalability. Therefore, cloud computing not only provides facilities for the computation and processing of big data but also serves as a service model for minimizing infrastructure maintenance cost (Hashem et al., 2015).

Architecture for Big Data Analytics

Big data analytics is a new research paradigm that can help organizations to analyze a mix of structured, semi-structured and unstructured data in search of new business insights. A diverse set of analytical tools can be used to transform large data sets containing a variety of data types into valuable information and then action. The analytical findings enable companies to understand business environment, customer needs, competitors, strategic stakeholders, market characteristics, products and more generally to predict future trends from the data and make more informed business decisions (Aiden and Michel, 2014; Xiang et al., 2015).

There are an increasing number of tools and platforms developed to optimize the time of processing on distributed databases and make sense of Big Data. Current technologies for big data computing can be divided into three categories, namely, batch processing technologies, real-time interactive analysis technologies and stream processing technologies (Chen and Zang, 2014; Tian and Zhao, 2015):

- **Batch Processing Technologies:** Batch processing is the execution of frequently used programs ("jobs") with minimum human interaction. Jobs can run to completion without any manual intervention. Batch process jobs are designed so that all input data are pre-defined through scripts or command-line parameters. A job takes a set of data files as input, processes the data, and produces a set of output data files. The input data are collected into *batches* or sets of records and each batch is processed as a unit. The output is another *batch* that can be restated for computation. Batch processes are used when a real-time response is not critical in data transmission. The files to be transmitted are gathered over a period and then sent together as a batch. The major benefit of batch processing is the ability to share computer resources between users and programs and shift job processing time to other resources that are less busy. Most batch processing technologies are based on the Apache Hadoop framework which provides infrastructures and platforms for other analytical Big Data applications, such as MapReduce and Apache Mahout. These systems are built on Hadoop, and have specific usages in different domains like data mining and machine learning;
- **Real-Time Interactive Analysis Technologies:** Nowadays, many organizations need the continuous access and processing of events and data in real time to gain constant awareness and take immediate action. The most important requirement of real-time analysis is the response to user needs in real time. The interactive analysis processes the data in an interactive environment, allowing users to undertake their own analysis of information. The user is directly connected to the computer and so can interact with it in real time. The data can be reviewed, compared and analyzed in tabular or graphic format or both at the same time (Chen and Zang, 2014; Tian and Zhao, 2015). Hadoop does well in processing large amount of data in parallel. However, Hadoop is designed for batch processing, but is not a real-time and high performance tool. In recent years, Open source Big Data systems have emerged to address the need not only for scalable batch processing, but also real-time interactive processing. Google's Dremel, Apache Drill, Apache Spark and Cloudera Impala are the most prominent technologies based on real-time and interactive analysis;
- **Stream Processing Technologies:** When the data volume is large, the requirement for real-time response is high, and the data sources are continuous, the data is considered as a data stream. A data stream is an accumulation of data records that are unbounded in number and time distribution. Data streams are now very common. Log streams, click streams, message streams, and event streams are some examples. Stream processing technologies are designed to process large

real-time streams of data. They enable applications such as real-time trading in financial services, fraud detection, process monitoring, location-based services in telecommunications, ad delivery, real-time searches and analysis of social networks. Actually, there are a few stream processing frameworks. Storm from Twitter and S4 (Simple Scalable Streaming System) from Yahoo! Inc. are two notable ones that are designed for big data streams (Tian and Zhao, 2015). Both frameworks are written in Java but their programming model is different.

To capture Big Data value and benefit their specified purposes, firms need to develop high-level computer architecture for data storage management and data analysis. Big data techniques and technologies are still emerging and will certainly grow much more in the future so that businesses need to wade in or risk being left behind.

Business Process Reengineering

In the last few years, a lot has been written about Big Data either in academic literature or in professional magazines (Gandomi and Haider, 2015; Wamba et al., 2015; Wamba and Mishra, 2017) and also about BPR (Business Process Reengineering), BPM (Business Process Management), and BPI (Business process Improvement) (Anand et al., 2013). However, few research efforts have been done to review and explain the relationship between both concepts. How can they co-exist and how can they be mutually supportive? How could an organization combine them to deliver real business benefits? Big data is expected to help organizations improve substantially their business processes and redefine their business model (Mishra et al., 2016) through BPR (Business Process Reengineering) and BPM (Business Process Management).

Originally pioneered in the early 1990's by Hammer and Champy, BPR is an approach to redesign management by radically transforming core processes, often using Information technology, to enable sustainable improvements and thus successfully create long lasting costumer value (Jha et al., 2016). BPR has been criticized as too risky and too radical. More flexible and less risky than BPR, BPM (Business Process Management) is a business practice that covers techniques and methods aiming at studying, identifying, changing and monitoring business processes to ensure they run smoothly and can be improved over time. It is about to manage performance of existing processes in a smooth and continuous manner rather than removing and replacing them by "perfect processes" at one time.

Companies willing to successfully integrate a Big Data technology will face issues related not only to their information and data architecture but also to how to make their business processes more intelligent. First of all, companies should identify business processes which are using unstructured and semi structured data. They should then determine the critical processes that need to be radically changed in order to release quick wins in the area of business performance improvement and cost reduction undertakings. The way decisions are made has known a fundamental change, by relying on multi-source and semantically enriched data. Traditionally, business processes generate structured data. This kind of data is indeed helpful in solving numerous business issues but when isolated from other data sources, it may improve existing processes that are on the way to become obsolete and unprofitable. Besides, decisions based on these processes overlook unstructured and often external data that, undoubtedly, contain valuable business information which can be analyzed to improve overall performance. The real concern is not to rely on a sole data source overlooking the other ones, but to understand what variety of data is required for analysis and what additional variables will enhance or refine the analytical result. Only in this way, can the organization reach BPR outcomes.

Big data databases are supported by systems which are themselves supported by business processes. Consequently, "Any effort to understand the organizational context from the perspective of big data on the one hand and records retention and disposition on the other should be based on the business process that generated the data in the first place" (McDonald and Lévéillé, 2014, p. 107). There is an obvious relation between big data and business processes. Big data analytics need to be combined with business processes to improve operations and offer innovative services to customers. Business processes need to be reengineered for big data analytics.

Big Data strategy could be utilized as a mean to support reengineering and thus to transform business processes into more intelligent. Based on the Six Sigma tool (Tennant, 2001) for process improvement, Jha et al. (2016) proposed a methodology to develop a Big Data strategy for reengineering. They identified six steps for applying this strategy: data from different sources (internal, external, social media…); key business process supporting the key tasks; key tasks that need to be accomplished to be successful; desired outcome and critical success factors; business initiatives that support the business strategy; and finally, big data strategy.

Reengineering is in turn an imperative step to make Big Data strategy feasible (figure 3). Park et al. (2017) proposed a modeling framework for BPR using big data analytics and goal orientation, called IRIS. This model consists of two parts, business modeling part and big data part. The authors defined a modeling language which helps diagnose as-is processes according to business goals and transform them into to-be processes with insights supported by big data. The model shows why, what and how concretely big data analytics may serve BPR and BPM initiatives.

Big Data Value

"*You can't manage what you can't measure!*" and "*If you can't measure it you can't improve it!*" These famous and very old management philosophies frequently attributed to both Perter Drucker and Edwards Deming (McAfee and Brynjolfsson, 2013), are more accurate and recent than ever. They represent the key driving force behind the emergence of Big Data solutions.

Not very long ago, companies have been making business decisions based on transactional data stored on relational databases. With the advent of big data, they are using less structured data (weblogs, social media, email, sensors, and photographs) which have more volume, velocity and variety than any data used so far. Even though very useful, they are still characterized by relatively " low value density " especially due to their volume (Gandomi and Haider, 2015). Volume is not Value. Having more data does not mean delivering more customer value. Customer value results from new insights into customers and from building deeper relationships with them. Also, velocity is not value because companies must take action on these insights in order to identify new opportunities which might increase customer engagement and then value creation. The velocity of big data is pushing companies to make real-time decisions. The real value of big data depends heavily on the speed of the company's execution ability.

Figure 3. Relationship between big data and business process reengineering

One of the key factors to survive in today's competitive business context is the ability of firms to extract value from big data so that they can understand their customer behavior trends, customize more their products and accordingly gain meaningful customer insights (Wong, 2012; Tan et al., 2015). If well filtered, deeply mined, analyzed and operationalized on time, big data may reveal a real treasure trove. The main challenge for companies is to have the maturity and the readiness to turn big data into insights, into smart decisions and finally into business value.

To extract value from Big Data, one fundamental step consists of taking a consumer-centric focus. Sophisticated and powerful quantitative tools may be used to switch from big data to customer value. To build long lasting advantage, companies must use big data to answer questions like "what will be the customer's lifetime value?" or "what will prevent a customer from switching out when a competitor offers a better price?" rather than " what price is he willing to pay?" or "what will trigger his next purchase?" Put simply, firms must shift from asking what big data can do for them, to what it can do for customers. Advanced analytic techniques such as Business Intelligence tools, data mining and predictive analytics can help answering these new questions. These tools can turn very large volumes of data derived from many different types and formats, into predictive insights for optimal marketing decision making.

Data is only as useful as it can inform metrics which might be later combined to provide the "ardently desired" insights to managers seeking to understand their customer behavior. For example, Kumar and Reinartz (2006) list marketing metrics in three categories: (1) traditional marketing metrics; (2) primary customer based metrics; and (3) strategic customer-based value metrics. Notwithstanding their aid in determining the total value that a single buyer could provide a firm, the above-mentioned metrics do not provide much deeper insights about future customer purchasing behavior because they assume that a customer's past buying behavior and future buying behavior will be similar. A complementary metric more accurate to predict the future profitability of a customer is the CLV "customer Lifetime Value" metric (Kumar and Reinartz, 2006; Kumar and Rajan, 2012; Kumar et al., 2013). It is probably the most powerful metric for measuring the true value of a company's customer and for giving weight to customer relationship. This forward-looking metric does not prioritize loyalty over profitability, it rather makes certain whether a consumer is going to be profitable and for how long. It allows firms to learn about the consumer's personal behavior and then to develop personalized initiatives that improve profit and retention. Calculating customer's lifetime value is not a trivial task. It involves an incredible amount of data which need to be thoroughly tracked and analyzed. Big data sets can help calculate CLV and other significant metrics by creating extremely precise pictures that take into account a variety of environmental variables such as demographics.

All key metrics are then brought into a single display called "dashboard", thereby avoiding potential problems such as data overload, disseminated data, managerial biases and lack of transparency and accountability (Pauwels et al., 2009). In big data analytics, reporting the insights gathered from the analysis of large and disparate data sources is the ultimate step of the process. Before big data, dashboard and data visualization offered limited benefits because organizations were unable to access large data sets without applying batch processing which may take sometimes several days. Big data applications make data more accessible for all company's levels and actors including consumers.

CONCLUSION

Today, big data has been recognized by the IT industry and business executives as an opportunity for any organization that knows how to capitalize on it. More and more managers view data as a significant driver of innovation and an important source of value creation and competitive advantage. Firms can use likes, tweets, click streams, videos and other unstructured sources to extract new insights about their products, customers, and markets. The challenge here is to develop new ideas and products or improve customer service processes using several data sources.

Various types of benefits can be gained from the use of big data. Its practice can bring substantial value in such areas as product and market development, market demand predictions, customer experience and loyalty, and operational efficiency (Yin and Kaynak, 2015; Raguseo, 2018). In other words, this new concept adds new value chain to companies by offering them the possibility to get strategic data which are relevant to decision making and action. However, this value may not be easy to be discovered. To get the most out of the big data, firms must identify its value realization processes to manage data holistically from business problem identification and data capture to decision making.

Data volume, variety, velocity and veracity are significant impediments to value creation. So, the paper's primary focus has been on the conceptualization of a big data value chain for helping managers in their quest to take advantage and gain valuable insights from huge sets of data. The paper first defines what is meant by big data by describing its various dimensions. Then, a big data value chain model is proposed by identifying five components: (1) business problem, (2) big data management process, (3) big data architecture, (4) business process reengineering, and (5) big data value creation.

After giving some examples of business problems underlying the need to implement big data projects, this paper lays down a number of challenges that big data management comes across, including the choice of support technologies for big data acquisition, storage, analysis and visualization; the management and improvement of business processes that generate this data; and the setting of appropriate metrics and KPIs to control value creation. The suggested big data value chain recognizes the interdependence between processes, from business problem identification and data capture to generation of valuable insights and decision making. This framework could provide some guidance to business executives and IT practitioners who are going to conduct big data projects in the near future.

To illustrate the usefulness of our research model for business and to better explain the relationships between its components, we can take fund distribution industry as a practical example. Financial Companies operating in fund management industry are facing a fragmented and complex market burdened by the weight of many intermediaries. This causes slow transactions and too many relationships. Moreover, clients or fund buyers, in order to select the most appropriate investment product, they require reliable financial information about products offered by fund promoters. Therefore, there is a need for a global solution to digitalize and enhance the efficiency of the entire fund distribution chain. The main driver here is clients' needs for relevant, accurate and up-to-date information about products. In order to achieve this goal, financial institutions need to ensure the management of a series of big data levels, starting with capture and storage of a wide range of data (static, transactional & historical fund data, fund marketing documentation, etc.) from internal and external data sources like market data vendors and web scraping, and ending with processing and analysis of large amount of data to perform custom analytics from reporting to predictive. For architecture, fund managers can leverage Big Data technologies to make the fund distribution process easier and more efficient. They can first create a central distributed platform for big data storage and dissemination enabling an easy access to fund data for both distributors and investors. They can then use big data analytics tools to help investors, distributors, and fund manufacturers explore fund data to compare and select funds according to multiple criteria. The digital platform could transform fund liability management and support functions like commercial, marketing, legal and risk functions. It could also simplify processes for order execution and support new and optimized straight through processing (STP) transaction placement and management functions in a controlled and streamlined process across the fund distribution chain, addressing thus the transparency challenge raised by new regulations. This initiative will significantly improve customer knowledge and the new investor on-boarding experience, for both the investor and the promoter, in terms of speed, simplicity, and transparency. It will also enable to transform the revenue model by optimizing costs structure and providing added-value distribution services through online monitoring with dashboard and big data analytics technologies.

Looking beyond this research, there is a need for more practical research analyzing challenges to big data within business, as well as a need for organizational culture changes to encourage data-based decision making. Companies need to cultivate a data-driven culture. A culture in which the business is empowered to make better use of data, in which business teams are more autonomous in the ways they use data, and people are aware of how data can be beneficial and how it can lead to new opportunities.

Future studies should be carried out to validate the structure of this conceptual model by testing it among companies belonging to different industries such as financial institutions, retailers, hotels, etc. Conducting case studies in companies who are adopting or willing to adopt big data technologies is necessary in order to understand the dynamics of their adoption and confirm or not the different components of the proposed big data value chain and their eventual interrelationships.

REFERENCES

Aiden, E., & Michel, J. B. (2014). The Predictive Power of Big Data. *Newsweek*. Retrieved from http://www.newsweek.com/predictive-power-big-data-225125

Alharthi, A., Krotov, V., & Bowman, M. (2017). *Addressing barriers to big data*. Business Horizons.

Anand, A., Wamba, S. F., & Gnanzou, D. (2013). A Literature Review on Business Process Management, Business Process Reengineering, and Business Process Innovation. In *Workshop on Enterprise and Organizational Modeling and Simulation* (pp. 1-23). Springer.

Barton, D., & Court, D. (2012). Making advanced analytics work for you. *Harvard Business Review*, *90*(10), 78. PMID:23074867

Chen, Ph., & Zhang, Ch.-Y. (2014). Data-intensive applications, challenges, techniques and technologies: A survey on Big Data. Information Sciences, 275(10), 314-347.

Chou, D. C. (2014). Cloud Computing: A Value Creation Model. *Computer Standards & Interfaces*. doi:10.1016/j.csi.2014.10.001

Dutta, D. & Bose, I. (2015). Managing a Big Data Project: The Case of Ramco Cements Limited. *International Journal of Production Economics: Manufacturing Systems, Strategy & Design*, 1-51.

Gandomi, A., & Haider, M. (2015). Beyond the hype: Big data concepts, methods, and analytics. *International Journal of Information Management*, *35*(2), 137–144. doi:10.1016/j.ijinfomgt.2014.10.007

Gartner. (2017). Big data. Retrieved from http://www.gartner.com/it-glossary/big-data

Hashem, I. A. T., Yaqoob, I., Anuar, N. B., Mokhtar, S., Gani, A., & Khan, S. U. (2015). The rise of "big data" on cloud computing: Review and open research issues. *Information Systems*, *47*, 98–115. doi:10.1016/j.is.2014.07.006

IDC. (2017). Worldwide Semiannual Big Data and Analytics Spending Guide. Retrieved from https://www.idc.com/getdoc.jsp?containerId=prUS42371417

Jha, M., Jha, S., & O'Brien, L. (2016). Combining Big Data Analytics with Business Process using Reengineering.

Kumar, V., Chattaraman, V., Neghina, C., Skiera, B., Aksoy, L., Buoye, A., & Henseler, J. (2013). Data-driven services marketing in a connected world. *Journal of Service Management*, *24*(3), 330–352. doi:10.1108/09564231311327021

Kumar, V., & Rajan, B. (2012). Customer lifetime value management: strategies to measure and maximize customer profitability. In V. Shankar & G.S. Carpenter (Eds.), Handbook of Marketing Strategy (pp. 107-134). Edward Elgar Publishing.

Kumar, V., & Reinartz, W. J. (2006). *Customer Relationship Management: A Databased Approach*. Hoboken, NJ: Wiley.

Lam, C. (2010). *Hadoop in Action*. Greenwich, CT, USA: Manning Publications Co.

Laney, D. (2001). *3D Data Management: Controlling Data Volume, Velocity and Variety, Application Delivery Strategy. Gartner*. Retrieved from http://blogs.gartner.com/doug-laney/files/2012/01/ad949-3D-Data-Management-Controlling-Data-Volume-Velocity-and-Variety.pdf

Lee, I. (2017). Big data: Dimensions, evolution, impacts, and challenges. *Business Horizons*. doi:10.1016/j.bushor.2017.01.004

Leeflang, P. S. H., Verhoef, P. C., Dahlström, P., & Freundt, T. (2014). Challenges and solutions for marketing in a digital era. *European Management Journal*, *32*(1), 1–12. doi:10.1016/j.emj.2013.12.001

Loshin, D. (2013). *Market and Business Drivers for Big Data Analytics*.

Madden, S. (2012). From databases to Big Data. *IEEE Internet Computing*, *16*(3), 4–6. doi:10.1109/MIC.2012.50

McAfee, A., & Brynjolfsson, E. (2013). Le Big Data, une revolution du management. *Harvard Business Review*, (Avril-Mai), 1–9.

McAfee, A., Brynjolfsson, E., Davenport, T. H., Patil, D. J., & Barton, D. (2012). Big data: The management revolution. *Harvard Business Review*, *90*(10), 61–67. PMID:23074865

McDonald, J., & Léveillé, V. (2014). Whither the retention schedule in the era of big data and open data? *Records Management Journal*, *24*(2), 99–121. doi:10.1108/RMJ-01-2014-0010

McKinsey Global Institute. (2012). *Big data: The Next Frontier for Innovation*, Competition, and Productivity.

Miller, H. G., & Mork, P. (2013). *From data to decisions: a value chain for big data. In IT Pro* (pp. 57–59). IEEE.

Mishra, D., Gunasekaran, A., Papadopoulos, T., & Childe, S. (2016). Big Data and Supply Chain Management: A Review and Bibliometric Analysis. *Annals of Operations Research*. doi:10.100710479-016-2236-y

Park, G., Chung, L., & Khan, L. (2017). A Modeling Framework for Business Process Reengineering Using Big Data Analytics and A Goal-Orientation.

Pauwels, K., Ambler, T., Clark, B. H., LaPointe, P., Reibstein, D., Skiera, B., ... Wiesel, T. (2009). Dashboards as a service: Why, how, and what research is needed? *Journal of Service Research*, *12*(2), 175–189. doi:10.1177/1094670509344213

Porter, M. E. (1985). *Competitive Advantage: Creating and Sustaining Superior Performance*. New York: The Free Press.

Raguseo, E. (2018). Big data technologies: An empirical investigation on their adoption, benefits and risks for companies. *International Journal of Information Management*, *38*(1), 187–195. doi:10.1016/j.ijinfomgt.2017.07.008

Self, R. J., & Voorhis, D. (2015). Tools and technologies for the implementation of big data, In Application of Big Data for National Security (pp. 140-154).

Smith, H. A., & McKeen, J. D. (2003). Developments in practice VII: Developing and delivering the IT value proposition. *Communications of the Association for Information Systems*, *11*, 25.

Tan, K.H., Zhan, Y.Z., & Ji, G. Ye F. and Chang, Ch. (2015). Harvesting big data to enhance supply chain innovation capabilities: An analytic infrastructure based on deduction graph. *International Journal of Production Economics*, 1–11.

Tennant, G. (2001). *Six Sigma: SPC and TQM in Manufacturing and Services*. Gower Publishing, Ltd.

Tian, W., & Zhao, Y. (2015). *Big Data Technologies and Cloud Computing*. In *Optimized Cloud Resource Management and Scheduling* (pp. 17–49). doi:10.1016/B978-0-12-801476-9.00002-1

Wamba, F. S., Akter, S., Edwards, A., Chopin, G., & Gnanzou, D. (2015). How 'big data' can make big impact: Findings from a systematic review and a longitudinal case study. *International Journal of Production Economics*, 1–33. doi:10.1016/j.ijpe.2014.12.031

Wamba, F. S., & Mishra, D. (2017). Big data integration with business processes: A literature review. *Business Process Management Journal*, *23*(3). doi:10.1108/BPMJ-02-2017-0047

White, M. (2012). Digital workplaces: Vision and reality. *Business Information Review*, *29*(4), 205–214. doi:10.1177/0266382112470412

Wong, D. (2012). *Data is the Next Frontier, Analytics the New Tool: Five Trends in Big Data and Analytics, and Their Implications for Innovation and Organisations*. London: Big Innovation Centre.

Xiang, Z., Schwartz, Z., Gerdes, J. H. Jr, & Uysal, M. (2015). What can big data and text analytics tell us about hotel guest experience and satisfaction? *International Journal of Hospitality Management*, *44*(January), 120–130. doi:10.1016/j.ijhm.2014.10.013

Yin, S., & Kaynak, O. (2015). Big Data for Modern Industry: Challenges and Trends. *Proceedings of the IEEE*, *103*(2), 143–146. doi:10.1109/JPROC.2015.2388958

Chapter 6
Strategic and Business-IT Alignment Under Digital Transformation:
Towards New Insights?

Nabyla Daidj
Institut Mines-Télécom Business School, France

ABSTRACT

Digital transformation is at a very early stage. Digital transformation has several impacts on business, on organization and process and raises several questions. Over the years, the aims of strategic fit and IT-business alignment have remained constant but the environment in which companies operate has changed significantly becoming more dynamic, very competitive and global. This chapter attempts to analyse how the digital transformation could affect more specifically strategic and IT-business alignment.

INTRODUCTION

The digital economy marks a historical shift in the approach to businesses and potential commercial activities. In the digital environment, a competitive firm should constantly be able to reposition itself in terms of its value proposition, resources and competences, financial capacity in light of the changing entry (and exit) barriers, the disruptive technologies and business models and periods of evolving competitive dynamics. Hypercompetitivity and digitalization presuppose permanent transformation of competitive advantages and critical success factors (CSFs). D'Aveni et al. (2010) thus have proposed "the age of temporary advantage" as an alternative concept to sustainability. This idea is also shared by McGrath (2013) who has called into question the relevance of sustainability in today's fast moving and hypercompetitive marketplaces.

DOI: 10.4018/978-1-5225-7262-6.ch006

The digital transformation has several impacts on business, on organization and process and raises several questions. "A digital transformation strategy impacts a company more comprehensively than an IT strategy and addresses potential effects on interactions across company borders with clients, competitors and suppliers" (Hess et al., 2016, p. 1). This chapter attempts to analyse how the digital transformation could affect more specifically strategic and IT-business alignment.

LITERATURE REVIEW: DIGITIZATION, DIGITALIZATION AND DIGITAL TRANSFORMATION

Digital transformation is a polysemous, buzz word (Sugahara et al., 2017, p.71). There are many dimensions of digital transformation sometimes confused with other terms such as digitization or digitalization (Table 1).

Digital transformation is often considered as the next step of digitization. Digitization is the process of converting information from an analog to digital format. For example, in the entertainment industry, the dematerialization of content has had repercussions not only on production, but also importantly on exploitation, broadcasting and distribution. In the context of convergence, these technological changes are also accompanied by changes in common practices adopted by consumers and strategies of the main firms involved (Daidj, 2015; Danowski & Choi, 1998).

The term "digitalization' closely related with computerization was used first by Robert Wachal in 1971 who discussed the social implications of the "digitalization of society". In the Oxford English Dictionary (OED), digitalization refers to "the adoption or increase in use of digital or computer technology by an organization, industry, country, etc." Gartner's IT glossary defines digitalization as "the use of digital technologies to change a business model and provide new revenue and value-producing opportunities; it is the process of moving to a digital business."

As regards digital transformation, definitions are numerous elaborated by both practitioners and academic scholars. Keyur Patel and Mary Pat McCarthy (2000) have been among the first authors to address the digital transformation issue by highlighting three questions as follows: What is digital transformation? How does it change business? What are the challenges for B2B companies and B2C companies in making a digital transformation?

Table 1 presents an overview of definitions of digital transformation summarizing key focus words contained in the definitions.

The digital transformation was first associated with technologies. "Within an enterprise, digital transformation is defined as an organizational shift to big data, analytics, cloud, mobile and social media platforms. The current business environment is witnessing a radical altering of the business landscape fueled by the emergence of digital innovations and opportunities. Firms are increasingly adopting various opportunities such as analytics, big data, cloud, social media and mobile platforms in a bid to build competitive digital business strategies" (Nwankpa & Roumani, 2016, p. 2). Then, according to most definitions, emphasis has been placed on one or two main concepts developed both in the strategic area and the IS/IT literature including (IT) capabilities, process, value (creation and capture), business model and organization. Several scholars have underlined the impact of the digital transformation on performances (Westerman et al., 2011; Peppard, 2016). Very little information and analysis can be found in academic papers on the relevance of new KPIs. Kotarba (2017) presented an analysis of metrics used to measure digitalization activities. Digital KPIs are numerous and there are already too many metrics and

Table 1. Overview of definitions of digital transformation

Authors	Definition	Key Features/Words
Dörner and Edelman (2015)	Digital transformation is "creating value at the new frontiers of the business world, create value in the processes that execute a vision of customer experiences and building foundational capabilities that support the entire structure". (p. 1)	Marketplace Capabilities Agility
Hess et al. (2016)	"Digital transformation is concerned with the changes digital technologies can bring about in a company's business model, which result in changed products or organizational structures or in the automation of processes. [...] (The) conceptual framework for formulating a digital transformation strategy identifies the four key dimensions of every digital transformation endeavor: *The use of technologies* reflects a firm's approach and capability to explore and exploit new digital technologies. *Changes in value creation* reflect the influence of digital transformation on a firm's value creation. *Structural changes* refer to the modifications in organizational structures, processes and skill sets that are necessary to cope with and exploit new technologies. The *financial aspects* dimension relates to both a firm's need for action in response to a struggling core business as well as its ability to finance a digital transformation endeavor." (p. 124)	Strategy Changes in business models organization, process, skills Value creation
Schuchmann, and Seufert (2015)	"Digital transformation is one of the major challenges in all industries. It embraces the realignment of technology and new business models to more effectively engage digital customers at every touchpoint in the costumer experience lifecycle. Therefore, successful digital transformation begins with an understanding of digital consumer behavior, preferences and choices. It then leads to major consumer-centric changes within the organization that address these needs". (p. 31)	Realignment New business models Consumer experience

Source: elaborated by the author, based on analysis of the articles cited.

targets. As Kotarba (2017, p. 136) mentioned it, "the overall number of digital KPIs already exceeds 100 items raising a problem of selecting the best metrics to monitor with limited control budgets (cost of data acquisition and processing). Further work should be conducted to select the metrics with the highest descriptive/statistical potential to pronounce the development of digital performance". A number of key metrics should be identified according to industries and should be followed by companies accordingly.

The digital transformation is a multidimensional phenomenon that can be studied from many different perspectives and thus requires the decompartmentalization of disciplines, particularly of economics, strategic management, digital marketing and information systems. As the digital transformation may take a variety of forms for company and have diverse (positive or adverse) consequences, several levels of analysis are required as shown in Table 2.

The digital transformation has a direct impact on the enterprise both internally (Abrell et al., 2016; Libert et al., 2016; Porter & Heppelmann, 2015; Yoo et al., 2010) and externally (Markus & Loebbecke, 2013; Porter & Heppelmann, 2014; Von Leipzig et al., 2017). In addition, its business relationships (with its main suppliers, competitors, and customers), partnerships and practices (for example within business ecosystem and/or networks a company belongs to) will change because of the transformation. Value creation mechanisms, value chains and business models (Bock et al., 2017) are expected to be renewed (even disrupted) enabled by the increasing development of cloud computing, data analytics, machine learning, artificial intelligence (AI) and IOT (Internet of Things).

What will drive competitive advantage in the future? The 'traditional' sustainable competitive advantage (based on cost reduction and/or differentiation) is already questioned. The notion of "transient advantage" is more and more used (McGrath, 2013). 'Classical' CSFs such as financial resources, in-

Table 2. Digital transformation levels: what major issues for companies?

<table>
<tr><td rowspan="4">Strategic & IT business-IT Alignment</td><td colspan="2">Internal Level</td><td rowspan="2">Business Ecosystems / Networks Technological Platforms</td><td colspan="2">External Level</td></tr>
<tr><td>Organization* Information Systems (IS) Process</td><td>Company** (Corporate Level)</td><td>Market</td><td>BtoB / BtoC Suppliers Competitors</td></tr>
<tr><td colspan="5">Reconfiguration of value chains
Value creation (new mechanisms)
Redefinition of business models and revenue models</td></tr>
<tr><td colspan="5">Towards disruption ?</td></tr>
</table>

Note:
* Organization includes SBUs (strategic business units) and/or main functions (Research and Development; Production; Purchasing; Logistics; Marketing; Sales; Human Resources (HR); DevOps etc.).
The black arrow corresponds to the concept of business-IT alignment analyzed in the following section.
** The corporate level refers to the overall scope of an organization, its portfolio of businesses.
Source: elaborated by the author.

novation, marketing and promotion should evolve towards an increasingly customer satisfaction and user experience thanks in particular to a more efficient strategy and IT alignment.

As the digital transformation is a very broad notion, in this chapter,the authors have therefore decided to analyze and to focus on its impact on strategic and business-IT alignment. This issue has not yet been fully explored by scholars. On the business side, companies are aware of the fact that strategic and business-IT alignment could support successful digital transformation. This view has been expressed in several reports and studies. In 2016, MIT Sloan Management Review and Deloitte Digital published their Digital Business Global Executive Study and Research Report (July 2016). Based on a survey of 3,700 executives across 131 countries and 27 industries, the report entitled 'Aligning the Organisation for Its Digital Future', highlighted the key concern of IT alignment called also congruence. "Digital congruence is the crux: To navigate the complexity of digital business, companies should consider embracing what we call *digital congruence* — culture, people, structure, and tasks aligned with each other, company strategy, and the challenges of a constantly changing digital landscape". This is the focus of the next sections.

FROM STRATEGIC FIT AND IT-ALIGNMENT TO DIGITAL BUSINESS STRATEGY

As the notion of alignment has several meanings according to the approach adopted (strategic focus or IT oriented), it is helpful to review how this concept was created and then adopted by scholars from various disciplines.

Strategic alignment dates back to the 1980s and 1990s with the work of several researchers in strategic management (Hamel & Prahalad, 1989; Porter, 1996). Hamel and Prahalad (1989) began to address the issue of strategic intent question then leading them to the analysis of the concept of strategic fit. Creating competitive advantages and (re)gaining competitiveness requires a 'strategic intent' that drives the entire corporation. They made a distinction between 'strategic intent' and 'strategic fit' in their model of strategic leadership and strategic vision. Traditional strategic management involves a search for the strategic fit between business portfolios, market niches and products, customers and distribution channels. Strategic fit is related to the degree to which a company is matching its resources and competencies with the opportunities in the external environment. Strategic fit is based on the resource-based view (RBV) of the firm which explains that the key to profitability is rather through unique characteristics of the company's resources and competencies in order to develop a sustainable competitive advantage. The idea of considering firms as a large set of resources goes back to the seminal work of Penrose (1959). This concept received renewed attention in the 1980s in particular by Wernerfelt (1984, 1989) and has become an influential framework for analysing corporate strategy (Barney, 1991; Hoopes et al., 2003; Peteraf, 1993). In 1996, Porter has pursued Hamel and Prahalad' work (1989) and has explained that strategic fit drives both competitive advantage and sustainability. According to him, the success of a business is explained often by the implementation of strategic fit. "Fit locks out imitators by creating a chain that is as strong as its *strongest* link (...). One activity's cost, for example, is lowered because of the way other activities are performed. Similarly, one activity's value to customers can be enhanced by a company's other activities. That is the way strategic fit creates competitive advantage and superior profitability." (p. 70).

The well-known strategic alignment model (SAM) proposed by Henderson and Venkatraman (1993) was the first to clearly describe the relationship between business strategies and information technology strategies. IT has evolved from its traditional role as a support function to a strategic role. The SAM model relies on four key domains of strategic choice: business strategy, information technology strategy, organizational infrastructure and processes, and information technology infrastructure and processes. They illustrated this model in terms of two fundamental characteristics for strategic management: strategic fit (the interrelationships between external and internal components) and functional integration (integration between business and functional domains).

IT-business alignment has been expressed using various words such as 'harmony' (Luftman et al., 1993), 'fusion' (Smaczny, 2001), 'integration' (Weill & Broadbent, 1988), and 'linkage' (Henderson & Venkatraman, 1993). Several IT alignment models have been elaborated and different methodologies have been adopted (Avison et al., 2004; Coltman et al. 2015; Gerow et al., 2014; Gerow et al., 2015; Henderson & Venkatraman, 1993; Luftman et al., 1993; Maes et al, 2000; Marchand et al., 2001; Reich & Benbasat, 1996, 2000) to identify key elements that need to be aligned for business IT alignment. Luftman (2000) developed a framework called the Strategic Alignment Maturity Model (SAMM) based on the twelve components of Business/IT-Alignment, which can be recognized in the model of Henderson and Venkatraman (1993). "Alignment addresses both how IT is in harmony with the business, and how the business should, or could be in harmony with IT. Alignment maturity evolves into a relationship where the function of IT and other business functions adapt their strategies together. Achieving alignment is evolutionary and dynamic. IT requires strong support from senior management, good working relationships, strong leadership, appropriate prioritization, trust, and effective communication, as well as a thorough understanding of the business and technical environments". (Luftman, 2000, p. 6-7).

IT has played a fundamental and powerful role in facilitating business activities and has become a catalyst of fundamental changes in the structure, operations, and management of organizations (Brown & Magill, 1994; Kearns & Sabherwal, 2006; Luftman *et al.*, 2006). This place is being strengthened today in the context of digital transformation. Strategy and business-IT issues are closely related and should be more combined in order to achieve a competitive advantage that might be called a 'digital competitive advantage'. Several scholars have stressed the need for companies to take into account all these dimensions in a digital strategy (Bharadwaj et al., 2013; Drnevich & Croson, 2013; Mithas et al., 2013; Pagani, 2013). As explained by Bharadwaj et al. (2013, p. 471), [they] "argue that the time is right to rethink the role of IT strategy, from that of a functional-level strategy—aligned but essentially always subordinate to business strategy—to one that reflects a fusion between IT strategy and business strategy. This fusion is herein termed digital business strategy". This is the purpose of the next section.

Managerial Contributions and Recommendations

The importance of strategic alignment (business-IT) has been recognized for a long time. "The best performing companies are often the best aligned" (Trevor, 2018, p. 2). Digital transformation does not imply the end of alignment but requires addressing new challenges in relation with digital business strategy. As mentioned previously, digital transformation is not only digital upgrades (Libert et al., 2016). A real digital business strategy aligned requires redesigning the "big picture" (Mintzberg, 1994, p. 111) and rethinking the strategic vision in relation with several key pillars as follows:

- **Value Creation & Capture:** "Firms investing in digital transformation are able to align digital insights about customers with innovative processes and investments leading to improved customer experience and performance" (Nwankpa & Roumani, 2016, p. 10). Approaches like co-conception, co-design, co-creation, co-production all emerged in the early 2000s and have grown without interruption ever since. In varying degrees, they all imply a collaborative, collective and interdisciplinary dimension. Instead of relying upon a "technology push" philosophy, which prevailed for a long time, they confide in a "market pull" outlook, which seeks to integrate market needs with the expectations of final users. Co-conception and co-design specifically acknowledge the context in which the goods are consumed to collect information and thus be able to improve and enhance customers´ experience. These collaborative practices transform the relations between the company and its customers, because the latter are now involved in the design of new products, services, applications, information systems before or during the development stage. These different players become "co-producers" of value (Prahalad & Ramaswamy, 2004; Vargo & Lusch, 2004).
- **Organization/Culture:** "Digital business strategy is different from traditional IT strategy in the sense that it is much more than a cross-functional strategy, and it transcends traditional functional areas (such as marketing, procurement, logistics, operations, or others) and various IT-enabled business processes (such as order management, customer service, and others). Therefore, digital business strategy can be viewed as being inherently transfunctional. All of the functional and process strategies are encompassed under the umbrella of digital business strategy with digital resources serving as the connective tissue." (Bharadwaj et al. (2013, p. 473). This new organization requires to hire "digital" workers and managers. It means also the development of an agile culture. The "agility" concept was created in 1991 by a group of researchers at the Iacocca Institute (Lehigh University, USA). The agility concept is defined as "a manufacturing system with extraordinary capabilities (Internal capabilities: hard and soft technologies, human resources, educated management, information) to meet the rapidly changing needs of the marketplace (speed, flexibility, customers, competitors, suppliers, infrastructure, responsiveness). A system that shifts quickly (speed and responsiveness) among product models or between product lines (flexibility), ideally in real-time response to customer demand (customer needs and wants)." Principles of agility have been then applied to other functions of enterprise, and the "agile enterprise" concept has been launched (Goldman et *al*., 1995; Nagel, 1992).

The "traditional business-IT alignment" framework should progressively evolve (Table 3).

FUTURE RESEARCH DIRECTIONS

Digital transformation is at a very early stage. Multi-disciplinary approaches will be required to address this issue in a comprehensive manner. Further work is required to understand the impact of digital transformation on IT-business alignment and related performances. It includes also case studies with different types of impact assessments (value chains, dynamic capabilities, organizational processes, business processes and managerial models). Among other consequences, digital transformation reshapes also business models. Advanced analytics tools ("business analytics" or "big data analytics") enable an easier access to infinite data and could lead to the adoption of business intelligence (BI) strategies for many kinds of public and private organizations. Advanced business intelligence tools thus contribute to the value creation for the organization and, in addition, an incremental transformation (Loebbecke and Picot, 2015). Business models which are increasingly digitalized and 'BI-Based' (Davenport, 2006) should replace less efficient business models.

CONCLUSION

The 'old' debate on strategic fit is not closed. Since the beginning of the 1980s, the alignment has been more and more linked with IT/IS facilitating the business operations and ensuring that the strategy was implementable. Most of scholars presented previously have stressed the role of IT/IS in creating

Table 3. Towards a renewed alignment framework in the context of digital transformation

	Previous Model **Strategy Drives Technology Towards Alignment**	**Renewed Model** **Technology Drives Strategy Towards the Implementation of a Digital Business Strategy Leading to Alignment**
	Internal Level (Organization)	
Organization (structure, management, resources and competencies)	Silos / Functional structure	"Transfunctional structure" Emergence of new functions (e.g. DEV/OPS) Agility Digital skills (chief information officers - CIO)
IT	Information system / Process	Development of dynamic IT capabilities (architecture, data management services, application platforms)
	External Level	
Market	Mainly competition	Co-existence of rivalry and cooperation
Relationships with other external stakeholders	Customers Suppliers (subcontracting relationships)	Customers (co-conception, co-creation and co-design practices) Suppliers (real partnerships)
	Business Models	
		Data-driven business models

Source: elaborated by the author.

value. Over the years, the aims of strategic fit and IT-business alignment have remained constant but the environment in which companies operate has changed significantly becoming more dynamic, very competitive and global. Competitors are more innovative and new entrants are more digitally focused (Von Leipzig et al., 2017). As mentioned by Smaczny (2001, p. 797), "…the alignment concept is about a sequential development of strategies and a sequential operationalisation. Unfortunately, now that reliance on IT has increased to the level where a lack of IT agility and responsiveness can create a competitive disadvantage, the sequential model is too slow. [We] promote a notion of fusion. Today's chief information officers are increasingly becoming company strategists who help their chief executive officer colleagues to steer companies. Business and IT strategies have to be developed simultaneously and implemented simultaneously." The digital transformation requires higher value at each chain of the internal and external value chain for the organizations.

REFERENCES

Abrell, T., Pihlajamaa, M., Kanto, L., Brocke, J., & Uebernickel, F. (2016). The Role of Users and Customers in Digital Innovation: Insights from B2B Manufacturing Firms. *Information & Management*, *53*(3), 324–335. doi:10.1016/j.im.2015.12.005

Avison, D., Jones, J., Powell, P., & Wilson, D. (2004). Using and validating the strategic alignment model. *The Journal of Strategic Information Systems*, *13*(3), 223–246. doi:10.1016/j.jsis.2004.08.002

Barney, J. B. (1991). Firm resources and sustained competitive advantage. *Journal of Management*, *17*(1), 99–120. doi:10.1177/014920639101700108

Bharadwaj, A., El Sawy, O. A., Pavlou, P. A., & Venkatraman, N. (2013). Digital Business Strategy: Toward a Next Generation of Insights. *Management Information Systems Quarterly*, *37*(2), 471–482. doi:10.25300/MISQ/2013/37:2.3

Bock, R., Iansiti M., & Lakhani, K. (2017, January 31). What the companies on the right side of the digital business divide have in common. *Harvard Business Review Digital Articles*.

Brown, C. V., & Magill, S. L. (1994). *Alignment* of the IS functions with the enterprise: Toward a model of antecedents. *Management Information Systems Quarterly*, *18*(4), 371–403. doi:10.2307/249521

Coltman, T., Tallon, P., Sharma, R., & Queiroz, M. (2015). Strategic IT alignment: Twentyfive years on. *Journal of Information Technology*, *30*(2), 91–100. doi:10.1057/jit.2014.35

D'Aveni, R. A., Dagnino, G. B., & Smith, K. G. (2010). The age of temporary advantage. *Strategic Management Journal*, *31*(13), 1371–1385. doi:10.1002mj.897

Daidj, N. (2015). *Developing Strategic Business Models and Competitive Advantage in the Digital Sector*. Hershey, PA: IGI Global. doi:10.4018/978-1-4666-6513-2

Danowski, J. A., & Choi, J. H. (1998). Convergence in the information industries. Telecommunications, broadcasting and data processing 1981-1996. In H. Sawhney & G.A. Barnett (Eds), Progress in Communication Sciences (pp. 125-150). Stamford, CT: Ablex Publishing

Davenport, T. H. (2006). Competing on Analytics. *Harvard Business Review*, *84*(1), 98–107. PMID:16447373

Dörner, K., & Edelman, D. (2015). *What 'digital' really means. McKinsey & Company: Insights & Publications*. Retrieved from http://www.mckinsey.com/insights/high_tech_telecoms_internet/what_digital_really_means?cid=digital-eml-alt-mip-mck-oth-1507

Drnevich, P. L., & Croson, D. C. (2013). Information Technology and Business-Level Strategy: Toward an Integrated Theoretical Perspective. *Management Information Systems Quarterly*, *37*(2), 483–509. doi:10.25300/MISQ/2013/37.2.08

Gerow, J. E., Grover, V., Thatcher, J. B., & Roth, P. L. (2014). Looking toward the future of IT-business strategic alignment through the past: A meta-analysis. *Management Information Systems Quarterly*, *38*(4), 1059–1085. doi:10.25300/MISQ/2014/38.4.10

Gerow, J. E., Thatcher, J. B., & Grover, V. (2015). Six Types of IT-Business Strategic Alignment: An investigation of the constructs and their measurement. *European Journal of Information Systems*, *24*(5), 465–491. doi:10.1057/ejis.2014.6

Hamel, G., & Prahalad, C. K. (1989). Strategic intent. *Harvard Business Review*, *67*(3), 63–78. PMID:10303477

Henderson, J., & Venkatraman, N. (1993). Strategic alignment: Leveraging information technology for transforming organizations. *IBM Systems Journal*, *32*(1), 4–16. doi:10.1147j.382.0472

Hess, T., Matt, C., Wiesböck, F., & Benlian, A. (2016). Options for Formulating a Digital Transformation Strategy. *MIS Quarterly Executive*, *15*(2), 103–119.

Hoopes, D. G., Madsen, T. L., & Walker, G. (2003). Guest Editors' Introduction to the Special Issue: Why is there a Resource-Based View? Toward a Theory of Competitive Heterogeneity. *Strategic Management Journal*, *24*(10), 889–902. doi:10.1002mj.356

Iacocca Institute. (1991). *21st century manufacturing enterprise strategy*. Bethlehem, PA: Lehigh University.

Kearns, G. S., & Sabherwal, R. (2006). Strategic alignment between business and information technology: A knowledge-based view of behaviors, outcome, and consequences. *Journal of Management Information Systems*, *23*(3), 129–162. doi:10.2753/MIS0742-1222230306

Kotarba, M. (2017). Measuring Digitalization - Key Metrics. *Foundations of Management*, *9*(1), 123–138. doi:10.1515/fman-2017-0010

Libert, B., Beck, M., & Wind, Y. (2016). 7 Questions to ask before your next digital transformation. *Harvard Business Review*. Retrieved from https://hbr.org/2016/07/7-questions-to-ask-before-your-next-digital-transformation

Loebbecke, C., & Picot, A. (2015). Reflections on Societal and Business Model Transformation arising from Digitization and Big Data Analytics: A Research Agenda. *The Journal of Strategic Information Systems*, *24*(3), 149–157. doi:10.1016/j.jsis.2015.08.002

Luftman, J. (2000). Assessing business-information technology alignment maturity. *Communications of the Association for Information Systems*, *4*(1), 1–49.

Luftman, J., Kempaiah, K., & Nash, E. (2006). Key issues for information technology executives 2005. *MIS Quarterly Executive*, *5*(2), 81–99.

Luftman, J., Lewis, P., & Oldach, S. (1993). Transforming the Enterprise: The Alignment of Business and Information Technology Strategies. *IBM Systems Journal*, *32*(1), 198–222. doi:10.1147j.321.0198

Maes, R., Rijsenbrij, D., Truijens, O., & Goedvolk, H. (2000). Redefining business: IT alignment through a unified framework. (PrimaVera working paper; No. 2000-19). Amsterdam: Universiteit van Amsterdam, Department of Information Management.

Marchand, D., Kettinger, W., & Rollins, J. (2001). *Information orientation: The link to business performance*. Oxford: Oxford University Press.

Markus, M. L., & Loebbecke, C. (2013). Commoditized digital processes and business community platforms: New opportunities and challenges for digital business strategies. *Management Information Systems Quarterly*, *37*(2), 649–654.

McGrath, R. G. (2013). *The End of Competitive Advantage: How to Keep Your Strategy Moving as Fast as Your Business*. Boston, Mass: Harvard Business Review Press.

Mintzberg, H. (1994). The fall and rise of strategic planning. *Harvard Business Review*, *72*(1), 107–114.

Mithas, S., Tafti, A., & Mitchell, W. (2013). How a firm's competitive environment and digital strategic posture influence digital business strategy. *Management Information Systems Quarterly*, *37*(2), 511–536. doi:10.25300/MISQ/2013/37.2.09

Nagel, R. N. (1992). *21st Century Manufacturing Enterprise Strategy. An Industry-Led View. Prepared for the Office of Naval Research Arlington, VA*. Bethlehem, PA: Iacocca Institute. Lehigh University. doi:10.21236/ADA257032

Nwankpa, J. K., & Roumani, Y. (2016). IT Capability and Digital Transformation: A Firm Performance Perspective. In Proceedings of the *Thirty Seventh International Conference on Information Systems, Dublin*. Retrieved from https://pdfs.semanticscholar.org/e8c4/16395a5d6690550b4aa74d81950eaa28bd84.pdf

Pagani, M. (2013). Digital Business Strategy and Value Creation: Framing the Dynamic Cycle of Control Points. *Management Information Systems Quarterly*, *37*(2), 617–632. doi:10.25300/MISQ/2013/37.2.13

Patel, K., & McCarthy, M. P. (2000). *Digital Transformation: The Essentials of E-Business Leadership*. New York: McGraw-Hill Professional.

Penrose, E. G. (1959). *The Theory of the Growth of the Firm*. New York: Wiley.

Peppard, J. (2016). A Tool for Balancing Your Company's Digital Investments, *Harvard Business Review Digital Articles*, October 18, 2016, 2-5.

Peteraf, M. A. (1993). The cornerstone of competitive advantage: A resource-based view. *Strategic Management Journal*, *14*(3), 179–191. doi:10.1002mj.4250140303

Porter, M. E. (1996). What is Strategy? *Harvard Business Review, 74*(6), 61–78. PMID:10158474

Porter, M. E., & Heppelmann, J. E. (2014). How Smart, Connected Products Are Transforming Competition. *Harvard Business Review, 92*(11), 64–88.

Porter, M. E., & Heppelmann, J. E. (2015). How Smart, Connected Products Are Transforming Companies. *Harvard Business Review, 93*(10), 98–114.

Prahalad, C. K., & Ramaswamy, V. (2004). Cocreation experience: The next practice in value creation. *Journal of Interactive Marketing, 18*(3), 5–14. doi:10.1002/dir.20015

Reich, B., & Benbasat, I. (1996). Measuring the linkage between business and information technology objectives. *Management Information Systems Quarterly, 20*(1), 55–81. doi:10.2307/249542

Reich, B., & Benbasat, I. (2000). Factors that influence the social dimension of alignment between business and information technology objectives. *Management Information Systems Quarterly, 24*(1), 81–113. doi:10.2307/3250980

Schuchmann, D., & Seufert, S. (2015). Corporate Learning in Times of Digital Transformation: A Conceptual Framework and Service Portfolio for the Learning Function in Banking Organisations. *International Journal of Corporate Learning, 8*(1), 31–39. doi:10.3991/ijac.v8i1.4440

Smaczny, T. (2001). Is an alignment between business and information technology the appropriate paradigm to manage IT in today's organizations. *Management Decision, 39*(10), 797–802. doi:10.1108/EUM0000000006521

Sugahara, S., Daidj, N., & Ushio, S. (2017). *Value Creation in Management Accounting and Strategic Management*. London, UK: ISTE-Wiley. doi:10.1002/9781119419921

Trevor, J. (2018, January 12). Is anyone in your company paying attention to strategic alignment. *Harvard Business Review Digital Articles*.

Vargo, R. F., & Lusch, S. L. (2004). Evolving for a new dominant logic for marketing. *Journal of Marketing, 68*(1), 1–17. doi:10.1509/jmkg.68.1.1.24036

Venkatraman, N., & Henderson, J. C. (1998). Real strategies for virtual organizing. *Sloan Management Review, 40*(1), 33–48.

Von Leipzig, T., Gampa, M., Manza, D., Schöttlea, K., Ohlhausena, P., Oosthuizenb, G., & Palma, D. (2017). Initialising customer-orientated digital transformation in enterprises. *Procedia Manufacturing, 8*, 517–524. doi:10.1016/j.promfg.2017.02.066

Wachal, R. (1971). Humanities and Computers: A Personal View. *The North American Review, 256*(1), 30–33.

Weill, P., & Broadbent, M. (1998). *Leveraging the New Infrastructure*. Boston: Harvard Business School Press.

Wernerfelt, B. (1984). The Resource-Based View of the Firm. *Strategic Management Journal, 5*(2), 171–180. doi:10.1002mj.4250050207

Wernerfelt, B. (1989). From critical resources to corporate strategy. *Journal of General Management*, *14*(3), 4–12. doi:10.1177/030630708901400301

Westerman, G., Bonnet, D., & McAfee, A. (2014). *Leading Digital: Turning Technology into Business Transformation*. Boston: Harvard Business Press.

Westerman, G., Calméjane, C., Bonnet, D., Ferraris, P., & McAfee, A. (2011). *Digital Transformation: A Roadmap for Billion-Dollar Organizations*. Retrieved from https://www.capgemini.com/resource-file-access/resource/pdf/Digital_Transformation__A_Road-Map_for_Billion-Dollar_Organizations.pdf

Yoo, Y., Henfridsson, O., & Lyytinen, K. (2010). Research commentary-The new organizing logic of digital innovation: An agenda for information systems research. *Information Systems Research*, *21*(4), 724–735. doi:10.1287/isre.1100.0322

ADDITIONAL READING

Bain & Company. (2017). Orchestrating a Successful Digital Transformation. Retrieved from http://www.bain.com/publications/articles/orchestrating-a-successful-digital-transformation.aspx

Cap Gemini. (2018). Digital Transformation Review Series. Retrieved from https://www.capgemini.com/consulting/digital-transformation-institute/digital-transformation-review

Daidj, N. (2017). *Uberization (or uberification) of the economy. In Encyclopedia of Information Science and Technology* (4th ed., pp. 2345–2355). PA: IGI Global.

Deloitte & MIT Sloan Management Review. (2015). *Strategy, not Technology, Drives Digital Transformation*. Retrieved from https://www2.deloitte.com/cn/en/pages/technology-media-and-telecommunications/articles/strategy-not-technology-drives-digital-transformation.html

Forbes. (2016). *How Digital Transformation Elevates Human Capital Management: Turning Talent into a Strategic Business Force. Insight Report.* Retrieved from https://www.forbes.com/forbesinsights/digitalhr/index.html

Gartner. (2017). IT Glossary. Retrieved from https://www.gartner.com/it-glossary

Kane, G. C., Palmer, D., Nguyen Phillips, A., Kiron, D., & Buckley, N. (2016). Digitally savvy executives are already aligning their people, processes, and culture to achieve their organizations' long- term digital success. *Sloan Business Review*. Retrieved from https://sloanreview.mit.edu/projects/aligning-for-digital-future/

KEY TERMS AND DEFINITIONS

Competencies: They include knowledge, know how, skills, abilities, and behavior.

Digital Business Strategy: Fusion between IT strategy and business strategy.

Digital Economy: The digital economy marks a historical shift in approach to businesses and potential commercial activities thanks to ICT and to new technologies (artificial intelligence, big data analytics, cloud computing, IOT, machine learning).

Digital Transformation: This new trend (in relation with new technologies) will entirely transform organizations, value capture models, current business models, customer experience and will lead to the implementation of radical changes.

Disruption: Often associated to technology or innovation, this phenomenon can create also major changes in industry frontiers, business processes and business models.

IT Alignment: Fit between organization/strategy and IT/IS capabilities.

Resources: They include two main categories tangible (human, physical, technological etc.) and intangible assets. Data are considered as resources and data analytics as skills (see above).

Strategic Fit: Match or congruence between a firm's organization/strategy and environmental issues (internal and external consistency).

Chapter 7
Big Data Analytics Driven Supply Chain Transformation

Mondher Feki
Lemna Research Center, France

ABSTRACT

Big data has emerged as the new frontier in supply chain management; however, few firms know how to embrace big data and capitalize on its value. The non-stop production of massive amounts of data on various digital platforms has prompted academics and practitioners to focus on the data economy. Companies must rethink how to harness big data and take full advantage of its possibilities. Big data analytics can help them in giving valuable insights. This chapter provides an overview of big data analytics use in the supply chain field and underlines its potential role in the supply chain transformation. The results show that big data analytics techniques can be categorized into three types: descriptive, predictive, and prescriptive. These techniques influence supply chain processes and create business value. This study sets out future research directions.

INTRODUCTION

Big data promises to trigger a revolution in supply chain management (SCM) (Waller & Fawcett, 2013). Fawcett and Waller (2014) argued that big data is one of the forces that will redefine supply chain design. The big-data-driven digital economy facilitates this change by capturing, analyzing and using big data to make evidence-based decisions. Thousands of exabytes of new data are generated each year on a variety of digital platforms such as social media, mobile devices and the Internet of Things. Many companies can capitalize on this big data by managing risks, reducing costs and improving supply chain visibility.

Big data is often characterized by the five Vs: Volume, Velocity, Variety, Veracity, and Value (Fosso Wamba, Akter, Edwards, Chopin, & Gnanzou, 2015). The volume of big data refers to the quantity of data, which is increasing exponentially. Velocity is the speed of data collection, processing and analyzing in real time. Variety refers to the different types of data collected. The data can be structured (e.g., data found in relational databases), semi-structured (e.g., Extensible Markup Language – XML), or unstructured (e.g., images, audio, and video). Veracity represents the reliability of data sources. The

DOI: 10.4018/978-1-5225-7262-6.ch007

variation in the data flow rates reflects the variability of big data, while the myriad of big data sources reflects its complexity. Finally, value represents the process of creating value from big data (Gandomi & Haider, 2015; Hashem et al., 2015; Kshetri, 2014). Fosso Wamba et al. (2015, p. 235) define big data analytics (BDA) as "a holistic approach to manage, process and analyze 5 Vs in order to create actionable insights for sustained value delivery, measuring performance and establishing competitive advantages."

Companies apply BDA in their supply chain to reduce cycle time, react faster to changes, optimize performance and gain insight into the future. A supply chain is defined as "a bidirectional flow of information, products and money between the initial suppliers and final customers through different organizations" (Nurmilaakso, 2008, p. 721); SCM includes planning, implementing and controlling this flow. BDA is expected to transform the supply chain (Fosso Wamba & Akter, 2015). Compared with traditional analytic tools, BDA could help companies to better understand customers' preferences and behavior and launch new products and services that are more customized (Duta & Bose, 2015). Several companies, such as Procter & Gamble, Walmart and Tesco, have benefited from the implementation of supply chain analytics (SCA), which enabled them to improve their operational efficiency and reduce costs (Chae, Olson, & Sheu, 2013). SCA refers to the use of supply chain data and analytical technologies and methods to improve operational performance (Chae et al., 2013). It represents the intersection of three academic disciplines: technologies (tools that support data processing), quantitative approaches (methods for analyzing data) and decision-making (tools used to support the decision-making process). These disciplines share a similar purpose: "...the improvement of business operations and decision making through the utilization of information, quantitative analyses, and/or technologies..." (Mortenson, Doherty, & Robinson, 2015, p. 585).

Three different aspects of analytics can be distinguished: (1) Descriptive analytics uses statistical methods and reports on the past; it is designed to answer the question "What happened?" (2) Predictive analytics uses models based on past data to predict the future and answer the question "What will happen next?" (3) Prescriptive analytics uses models to specify optimal behaviors and actions and answer the question "What should the business do next?" (Davenport, 2013; Lustig, Dietrich, Johnson, & Dziekan, 2010; Mortenson et al., 2015). Despite the benefits of BDA in SCM, companies may find it difficult to adopt the approach if they lack the capacity to make large investments, an analytic culture, executive support, a strong security framework, or analytic capability. In addition, creating business value from big data still represents a challenging and controversial mission as the steep growth curve of performance using analytics is flattening out (Kiron, Prentice, & Ferguson, 2014). Some scholars also describe the hype about big data as a myth, as it does not reflect innovative capability and improved firm performance (Manyika et al., 2011). Ross, Beath, and Quaadgras (2013, p. 90) state that "The biggest reason that investments in big data fail to pay off, though, is that most companies don't do a good job with the information they already have. They don't know how to manage it, analyse it in ways that enhance their understanding, and then make changes in response to new insights."

Motivated by this debate, the main objective of this research is to provide a comprehensive overview on BDA application in SCM and underscore its potential role in supply chain transformation. To this end, the author conducted a literature review to arrive at answers to the following questions:

- What big data analytic techniques are used in the supply chain field?
- How does big data analytics influence supply chain management?
- How can big data analytics create business value in supply chain?

The remainder of this paper is structured as follows. Section 2 presents a sample of business cases. Section 3 describes the big data analytic techniques used in the supply chain field and section 4 summarizes the supply chain analytics approaches. Section 5 and 6 presents respectively the impact of BDA on SCM processes and business value. Finally, section 7 sets out future research directions.

ILLUSTRATIVE EXAMPLES OF BIG DATA ANALYTICS APPLICATIONS

Big data has generated tremendous attention worldwide recently mainly due to its operational and strategic potentials. For example, Anaya et al. (2015) found that by applying data analytics tools into data accumulated by firms in various enterprise systems may lead to the extraction of new insights as well as innovative practices. Similarly, Kohlborn et al. (2014) argued that the firms could use of big data analytics capabilities to generate new insights "into business processes, improve organizational learning, and enable innovative business models". Table 1 shows the business value created by the world's Top 10 most innovative companies in big data (Fast Company, 2014).

Big data promises to generate a revolution in supply chain management (Waller and Fawcett, 2013). A recent study by SCM World states that "64% of supply chain executives consider big data analytics a disruptive and important technology, setting the foundation for long-term change management in their organizations" (Columbus, 2015). Many companies can capitalize on big data analytics by managing risks, reducing costs and improving supply chain visibility and traceability. An Accenture research reveals that big data analytics helping some companies to "improve customer service and demand fulfillment, experience faster and more effective reaction time to supply chain issues, increase supply chain efficiency, and drive greater integration across the supply chain" (Accenture, 2014, p. 11). Table 2 presents the benefits of companies from the implementation of analytics in supply chain (Dell, 2014).

Table 1. Business value of Top 10 most innovative companies in big data

Company	Business Value
1. General Electric	Big data help airlines to predict mechanical malfunctions and reduce flight cancellations.
2. Kaggle	Algorithms developed by its data scientists allow analyzing complex data and making big decisions.
3. Ayasdi	Visual approach of big data reveals new trends.
4. IBM	IBM was launched a new big data technology in the French city of Lyon, for predicting points of congestion to improve traffic flow.
5. Mount Sinai Icahn School of Medicine	It processes gigabytes of health data from more than 25000 patients in order to predict diseases and streamline electronic medical records.
6. The Weather Company	Analyzing millions of local climates to predict habits and behavior patterns of its digital and mobile users in worldwide.
7. Knewton	Through its digital platform, Knewton analyzes the progress of millions of students to create better test questions and personalized course goals.
8. Splunk	As a pure-play leader in the big data space, Splunk provides businesses with hundreds of homegrown apps to sniff out error files and keep things humming.
9. Gnip	Gnip provides and analyzes data from social media websites via API services in order to understand customer profile.
10. Evolv	Evolv provides workforce performance solutions to help employers better understand employees and job candidates by comparing their skills, work experience, and personalities.

Source: Fast Company (2014)

Table 2. Business value from using big data analytics in supply chain

Industry	Company	Business Value From Using Big Data Analytics
Food	McDonald's	"McDonalds tracks vast amounts of data in order to **improve operations** and **boost the customer experience**. The company looks at factors such as the design of the drive-thru, information provided on the menu, wait times, the size of orders and ordering patterns in order to optimize each restaurant to its particular market." (p.12)
	Coca-Cola Co.	"Coca-Cola uses an algorithm to **ensure that its orange juice has a consistent taste throughout the year**. The algorithm incorporates satellite imagery, crop yields, consumer preferences and details about the flavours that make up a particular fruit in order to determine how the juice should be blended." (p.29)
Retail	Nordstrom	"Nordstrom collects data from its website, social media, transactions and customer rewards program in order to **create customized marketing messages** and shopping experiences for each customer, based on the products and channels that customer prefers." (p.17)
	Procter & Gamble	"P&G uses simulation models and predictive analytics in order to **create the best design for its products**. It creates and sorts through thousands of iterations in order to develop the best design for a disposable diaper, and uses predictive analytics to determine how moisture affects the fragrance molecules in dish soap, so the right fragrance comes out at the right time in the dishwashing process." (p.27)
	Amazon	"With more than 1.5 billion items in its catalog, Amazon has a lot of product to keep track of and protect. It uses its cloud system, S3, to **predict which items are most likely to be stolen**, so it can **better secure its warehouses**." (p.32)
Transportation	Union Pacific Railroad	"With predictive analytics and tools such as visual sensors and thermometers, Union Pacific can detect imminent problems with railway tracks in order to **predict potential derailments days** before they would likely occur. So far the sensors have reduced derailments by 75 percent." (p.24)
	Kayak	"Kayak uses big data analytics **to create a predictive model that tells users if the price** for a particular flight will go up or down within the next week. The system uses one billion search queries to find the cheapest flights, as well as popular destinations and the busiest airports. The algorithm is constantly improved by tracking the flights to see if its predictions are correct." (p.38)
Oil	Shell	"Shell uses sensor data to map its oil and gas wells in order to **increase output and boost the efficiency of its operations**. The data received from the sensors is analyzed by artificial intelligence and rendered in 3D and 4D maps." (p.41)

Source: Dell (2014)

BIG DATA ANALYTIC TECHNIQUES IN SUPPLY CHAIN

Supply chain analytics refers to the use of supply chain data and analytical technologies and methods to improve operational performance (Chae et al., 2013). Below, the author presents the advanced analytics techniques founded in literature.

Data mining approaches incorporate statistical and analytical techniques such as classification, regression, clustering and semantic analysis (Z. Y. Chen et al., 2015; Chongwatpol, 2015; Kulkarni et al., 2014; Liao, Chu, & Hsiao, 2012). The inputs to these techniques are large amounts of structured or unstructured data which will be interpreted as useful information in a specific business context by identifying intrinsic models (Chongwatpol, 2015).

Data mining techniques allow users "to find the best predictive model and identify top predictors" (Ballings & Van den Poel, 2015, p. 249). Generally applied in the domains of retail, marketing and customer relationship management, these techniques make it possible to identify customer behavior patterns. For example, data mining allows data captured by RFID and data extracted from point-of-sale systems to be analyzed in order to predict consumer purchasing behavior and detect any changes in it (Chongwatpol, 2015).

Machine learning techniques allow for hierarchical representations to be learned automatically (Chen & Lin, 2014). These techniques include neural networks, Bayesian networks, genetic algorithms or programming, decision trees, support vector machines, and feature selection (Bose & Chen, 2009; Chen et al., 2015; Chongwatpol, 2015).

Advanced machine learning techniques enhance conjoint analysis capabilities and improve prediction performance, making it possible to better identify consumer preferences and relevant attributes (Chen et al., 2015; Maldonado et al., 2015). These techniques, which are frequently used in direct marketing (Chongwatpol, 2015), also allow business managers to make decisions concerning promotional and advertising campaigns and new product design. In addition, they enable managers to assess consumers' willingness to pay and decide on a pricing strategy (Maldonado et al., 2015).

Optimization methods use data with models and algorithms in simulations to solve optimization problems and to help make strategic decisions about operations. For example, in the gas industry, they can improve job scheduling and crew assignment (Angalakudati et al., 2014).

The classification of the examined literature by BDA techniques is summarized in Table 3.

SUPPLY CHAIN ANALYTICS APPROACHES

Big data analytics in supply chain focuses on the use of analytical techniques to drive decisions and actions regarding flows in the supply chain (Souza, 2014). These techniques, found from our literature review, can be categorized into three types: descriptive, predictive, and prescriptive (Souza, 2014; Hahn and Packowski, 2015; Wang, Gunasekaran, Ngai, & Papadopoulos, 2016).

The descriptive analytics bases on the use of statistics techniques to transform big data into meaningful information. Indeed, the GPS (Global Positioning System), RFID (Radio-Frequency Identification) technologies and sensors collect data on a real-time which will be summarized and converted into information relative to location and quantity of goods in supply chain (Groves et al., 2014; Chae, 2015). Thus, the manager can have real-time information, for example, on carrier's location or stock state and consequently, he can make adjustments concerning the delivery schedules or replenishment orders for instance (Souza, 2014). Thereby, the descriptive analytics provide information regarding "what has happened?" and "what is happening at the moment?" for reporting and monitoring purposes (Souza,

Table 3. Literature references on big data analytics techniques in supply chain

Techniques	References
Statistics	Wang et al. (2016), Chae (2015), Hahn and Packowski (2015), Groves et al. (2014), Chae et al. (2014).
Data visualization approaches	Zhong et al. (2016), Duta and Bose (2015), Tan et al. (2015), Groves et al. (2014), Souza (2014)
Data mining	Krumeich et al. (2016), Chae (2015), Zhong et al. (2015), Tan et al. (2015), Hahn and Packowski (2015), Souza (2014), Kahn (2014), Kwon et al. (2014)
Machine learning	Zhong et al. (2016), Zhong et al. (2015), Chae (2015)
Social network analysis	Chae (2015), Tan et al. (2015)
Optimization methods	Wang et al. (2016), Hahn and Packowski (2015), Duta and Bose (2015), Souza (2014), Groves et al. (2014), Hazen et al. (2014), Chae et al. (2014)

2014; Hahn and Packowski, 2015). Statistics techniques can be combined with visualization techniques in order to give an overview of data represented in form of performance scorecard with key performance indicator such as delivery deadline and sales growth (Duta and Bose, 2015).

The predictive analytics concerns the use of techniques such as data mining, machine learning and social network analysis. The data mining is a set of tools, including classification, clustering, association analysis, sentiment analysis and regression (Souza, 2014; Kahn, 2014; Chae, 2015; Zhong et al., 2015), to extract subjective information from big data (e.g., emotion, opinions) (Chae, 2015). The machine learning uses algorithms to discover knowledge and evolve behaviors (Chen and Zhang, 2014; Chae, 2015; Zhong et al., 2015). On the other hand, network analytics explores network-level characteristics (Chae, 2015). Other tools for forecasting are used such as time series methods and market basket analysis (Souza, 2014; Kahn, 2014; Hahn and Packowski, 2015). All these techniques allow to make predictions concerning the future intended from big data. So, predictive analytics in supply chain answer the questions "what will happen?" and "why will it happen?" (Souza, 2014; Hahn and Packowski, 2015).

The prescriptive analytics is based on descriptive and predictive analytics and optimization methods (Souza, 2014). This last tool includes mathematical models and simulation techniques (Groves et al., 2014; Hazen et al., 2014; Chae et al., 2014) to support decision and optimize process on a real-time (Souza, 2014; Duta and Bose, 2015). For example, by using this technique in outbound logistics, every order is analyzed with regard to the availability of stock in order to manage expeditions and determine the appropriate deadlines of delivery on a real-time, what allows to decrease the logistics cost, increase the efficiency and provide a better service to the customer (Duta and Bose, 2015). Accordingly, perspective analytics address the question "what should be happening?" (Souza, 2014; Hahn and Packowski, 2015).

Hahn and Packowski (2015) distinguish four types of use cases of analytics in supply chain management: (1) "monitor and navigate", (2) "sense and respond", (3) "predict and act", and (4) "plan and optimize". The first use case requires descriptive analytics and data visualization techniques. "Plan and optimize" use cases correspond to prescriptive analytics. By contrast, predictive analytics enable "sense and respond" and "predict and act" use cases.

Figure 1 presents three types of big data analytics approaches in supply chain with data examples and references.

BIG DATA ANALYTICS IN SUPPLY CHAIN PROCESSES

Big data can influence business processes in the context of SCM. To study processes concerned by big data-enabled supply chain transformation, the author uses SCOR (Supply Chain Operations Reference) model developed by the Supply Chain Council (www.supply-chain.org) as such a framework of classification. This model divides supply chain processes into six main areas: Plan, Source, Make, Deliver, Return and Enable. Table 4 presents the objectives of the SCORE model (Supply Chain Council, 2008).

With a huge increasing quantity of business data, consumption data and contextual, companies need to use analytics in order to make sense of big data so as to drive decisions and actions. The big data analytics use in supply chain was influenced SCOR areas.

For supply chain planning, companies use predictive approach such as time series, causal forecasting or data mining methods. Usually, they begin with data mining techniques such as clustering or market basket analysis for analyzing purchase models, knowing customers' perceptions with regard to products and services and determining demand factors. These factors will then be analyzed by using causal fore-

Figure 1. Supply chain analytics approaches

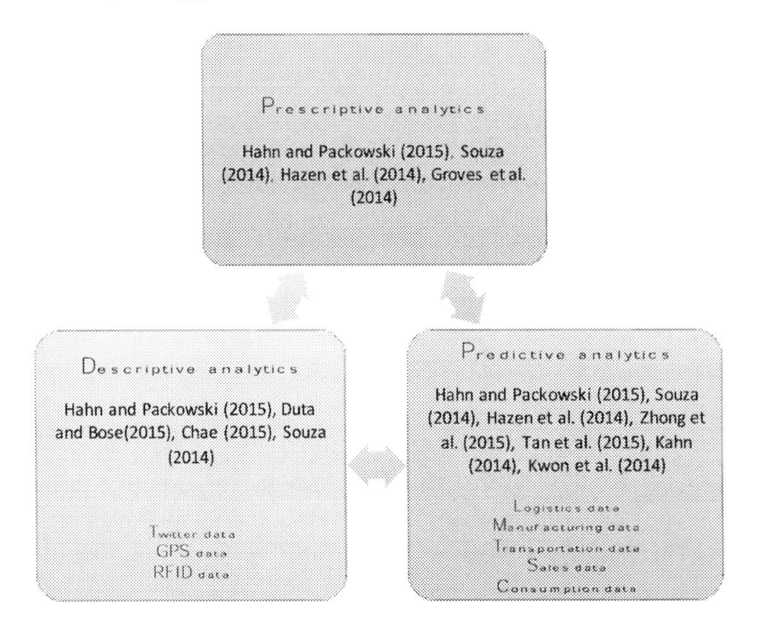

Table 4. Objectives of SCORE model

Processes	Objectives
Source	The ordering, delivery, receipt and transfer of raw material items, subassemblies, products and/or services
Make	The process of adding value to products through mixing, separating, forming, machining, and chemical processes
Deliver	Performing customer-facing order management and order fulfillment activities including outbound logistics
Return	Moving material from customer back through supply chain to address defects in product, ordering, or manufacturing, or to perform upkeep activities.
Plan	The process of determining requirements and corrective actions to achieve supply chain objectives.
Enable	Three distinct types of objectives: • Manage process performance • Manage process control data • Manage process relationships

Source: Supply Chain Council (2008)

casting methods such as the regression to predict product demand. The demand prediction constitutes the main input of planning in supply chain. It is carried out at the strategic, tactical and operational levels to plan operations (procurement, production and distribution) and sales in order to synchronize demand with offer (Souza, 2014; Chae, 2015). Hahn and Packowski (2015) argue that strategic and operational planning is mainly based on prescriptive analytics approaches in particular on the use of optimization methods.

For the sourcing, companies use prescriptive approach by means of the analytic hierarchy process at strategic level to estimate and select the key suppliers and use game theory at the tactical level to define the rules of auction and prescribe contracts (Souza, 2014). Wang et al. (2016) argue that supply chain analytics allow managing the supply risks and optimizing the sourcing decision.

In make process, companies use prescriptive approach at strategic level such as genetic algorithms to determine the capacity of plants. They use predictive approach at tactical level to rationalize the product line, schedule workforce and plan inventory level. They also use algorithms at operational level for manufacturing scheduling and automating replenishment decisions (Condea et al., 2012; Souza, 2014; Tan et al., 2015; Zhang et al., 2017).

In deliver and return processes, companies use predictive approach to plan distribution and transport (Souza, 2014). They also use statics and visualization techniques for analyzing sales performance at various levels such as by zones, regions, or districts by using a real-time metric- based performance measurement system (Duta and Bose, 2015).

Table 5 presents the references of the examined literature on the application of big data analytics in supply chain management processes.

VALUE CREATION FROM BIG DATA ANALYTICS IN SUPPLY CHAIN

In supply chain management, the use of big data analytics enables to know customers' perceptions of offered products and services and discover their unobservable characteristics in order to understand market demands and anticipate future consumer product variety desires. The customer's knowledge enables to develop new products and services more customized and consequently improve their satisfaction (Ng et al., 2015; Chae, 2015).

Furthermore, the demand prediction through big data analytics enables to plan and execute supply chain so as to balance supply and demand and improve supply chain operations (Chae, 2015; Tan et al., 2015). Indeed, the combination of analytics techniques enables to optimize manufacturing processes, shop-floor management and manufacturing logistics (Hahn and Packowski, 2015; Groves et al., 2014) which allows producing new products in a more profitable way (Tan et al., 2015) and reducing logistics cost (Duta and Bose, 2015). Hahn and Packowski (2015) argue that the use sensor-based technology with analytics applications on a real-time involve substantial value potential of 10 to 25% in operating cost reductions.

Table 5. Literature references by SCM process

Process	References
Plan	Krumeich et al. (2016), Wang and Zhang (2016), Chae (2015), Li et al. (2015), Hahn and Packowski (2015), Chae et al. (2014), Souza (2014), Kahn (2014)
Source	Wang et al. (2016), Li et al. (2015), Hahn and Packowski (2015), Souza (2014), Groves et al. (2014), Verdouw et al. (2013), Condea et al. (2012)
Make	Zhang et al. (2017), Wang et al. (2016), Ng et al. (2015), Tan et al. (2015), Zhong et al. (2015), Li et al. (2015), Hahn and Packowski (2015), Kwon et al. (2014), Groves et al. (2014), Hazen et al. (2014), Chae et al. (2014), Souza (2014), Verdouw et al. (2013)
Deliver	Wang et al. (2016), Chae (2015), Duta and Bose (2015), Hahn and Packowski (2015), Li et al. (2015), Groves et al. (2014), Souza (2014), Verdouw et al. (2013)
Return	Li et al. (2015), Souza (2014)

In addition, big data analytics enables the management and control of manufacturing process on a real-time. Indeed, an advanced process management method developed for data mining allowed to follow and control stocks, manufacturing workflows and workers on a real-time, and consequently improving the productivity (Kwon et al., 2014). Therefore, the use of visualization techniques allows making real-time corrective actions. For example, RCL (Ramco Cements Limited) is an Indian company that mainly produces cement in the South of India. It has "5 cement plants, 3 grinding units, 2 packing plants, a dry mix plant and a ready-mix concrete plant spread all over India" and "six captive wind mill sites" (Duta and Bose, 2015, p. 297). Thanks to visualization techniques, RCL was able to adjust its marketing plan in some locations what increased its sales and market share (Duta and Bose, 2015). Indeed, the combination of analytical techniques such as data mining and visualization tools allow to generate relevant and viable information for decision-makers (Tan et al., 2015) in particularity production logistics (Zhong et al., 2015). The use of simulation with statistics and visualization techniques allows analyzing markets, production and sales data on a real-time and computing the key performance indicators relative to supply chain for developing strategic, tactical, and operational decision-making (Groves et al., 2014).

However, Chae et al. (2014) argue that for improving operational performance and increasing big data analytics value, it is necessary that supply chain analytics techniques be combined with SCM initiatives such as "Total Quality Management", "Just in time", and "Statistical Process Control" which can be used to monitor and control data quality in a supply chain (Hazen et al., 2014).

Table 6 presents the references of the examined literature on the business value of big data analytics in supply chain management.

FUTURE RESEARCH DIRECTIONS

This section presents potential future research questions based on our literature review in order to capitalize the research development of BDA applications in the SCM context.

Table 6. Literature references by business value

Business Value	References
Improving decision-making	Tan et al. (2015), Zhong et al. (2015), Duta and Bose (2015), Hahn and Packowski (2015), Groves et al. (2014), Souza (2014), Condea et al. (2012)
Creating new products and services	Addo-Tenkorang and Helo (2016), Chae (2015), Ng et al. (2015), Kahn (2014)
Enabling to discover needs and customization	Addo-Tenkorang and Helo (2016), Tan et al. (2015), Ng et al. (2015), Chae (2015)
Operational excellence	Addo-Tenkorang and Helo (2016), Ng et al. (2015), Duta and Bose (2015), Chae (2015), Li et al. (2015), Chae et al. (2014), Kwon et al. (2014), Condea et al. (2012)
Improving supply chain management	Zhong et al. (2016), Zhong et al. (2015), Hahn and Packowski (2015), Tan et al. (2015), Li et al. (2015), Duta and Bose (2015), Chae et al. (2014), Hazen et al. (2014)

Potential Future Research Questions on Methodology and Theories

The majority of the articles analyzed focused on a mathematical analysis, experiment, literature analysis or research derived from opinions, with no empirical evidence. This finding raises a challenge to use other methodologies such as qualitative research, surveys and quantitative research to study and measure the impact of BDA on operations and supply chain management, as well on supply chain performance and operational performance.

Our review of the articles revealed that several theories have been mobilized to study BDA in the supply chain. However, a number of other potentially useful theories, such as resource-based view theory, contingency theory and systems theory, could also be mobilized. The determination of which theories can usefully be mobilized to study big data in supply chains requires further study. Revisiting resource-based view theory could help determine other resources and capabilities for supply chain analytics.

Potential Future Research Questions on BDA in Supply Chain Processes

Big data can influence business processes in the context of SCM. It would be interesting to study the key processes concerned by big-data-enabled supply chain transformation. Table 7 presents the key processes of the SCORE model (Supply Chain Council, 2008).

Future work can study and identify SCORE areas, which are influenced by BDA

- What is the impact of BDA on each SCORE area?
- Which SCORE area is the most influenced?
- What is the impact of BDA on each of the key processes of the SCORE model?
- What is the impact of BDA on supply chain processes' performance?

Potential Future Research Questions on BDA Techniques in Supply Chains

BDA in supply chains focuses on the use of statistics, data mining, and machine learning. The author notes the emergence of optimization methods. Future research on this type of analytical technique in the supply chain context should be encouraged. However, the author noticed from the analyzed articles that no study had been done on data visualization approaches and social network analysis. Case studies and/or reports on experiments that explore these opportunities are recommended. Such studies can inform the community about these tools and demonstrate their potential benefits.

- What are the opportunities for and limitations on using data visualization approaches in a supply chain context?
- What are the potential benefits of using data visualization approaches in a supply chain context?
- What are the opportunities for and limitations on using social network analysis in a supply chain context?
- What are the potential benefits of using social network analysis in a supply chain context?

Table 7. Key processes of SCORE model

Source	Make	Deliver	Return	Plan	Enable
Schedule product deliveries	Schedule production, request and receive material from source and/or Make processes	Product, service and price quotations	Identification of the need to return a product or asset	Supply chain revenue planning/ forecasting	Manage business rules and monitor adherence
Receive, inspect, and hold materials	Manufacture, assemble/ disassemble and test product, package, hold/release product	Order entry and maintenance	Requesting and issuing return authorization	Materials requirement planning	Measure supply chain performance and determine corrective action
Issue material to Make or Deliver processes	Manage product quality and engineering changes	Order consolidation, picking, packing, labeling and shipping	Inspection and disposition decision-making	Factory, repair, maintenance facilities capacity planning	Manage risk and environmental impact
Supplier/vendor agreements	Manage facilities and equipment, production status workflow and capacity management	Import/export documentation	Transfer/disposition of product or asset	Distribution requirements planning	Manage the supply chain network and facilities
Vendor certification and feedback, sourcing quality	Manage work-in-process inventories	Customer delivery and installation	Manage return transportation capacity	Manage planning parameters	
Manage raw materials inventories		Logistics and freight management	Manage returned material inventories		
Freight, import/export documentation		Manage finished goods inventories			

Source: Supply Chain Council (2008, pp. 21, 25, 32, 37, 41)

Potential Future Research Questions on Big Data Strategies in a Supply Chain Context

The development, implementation and management of big data strategies in the context of a supply chain cannot yet be discussed in any depth. Further research should be done to address those challenges:

- How should a big data strategy be developed in the supply chain context?
- What are the facilitators and inhibitors of BDA implementation for supply chains?
- What are the costs of implementing and driving big data applications and capabilities?

Potential Future Research Questions on the Use and Impact of Big Data in a Supply Chain Context

There is a need to develop more exploratory research and build theories to explain the use of BDA in OSM at the intra- and interorganizational levels and assess its impact on supply chain performance. It would be interesting to identify the contingent factors that can moderate the effect of BDA on supply chain performance. The questions that arise are:

- How is BDA used in operations and supply chain management at the intra- and inter- organizational levels?
- What is its impact on supply chain performance?
- How can firms measure supply chain performance?
- What are the contingent factors that influence this effect?
- What is the impact of BDA on predictive maintenance?
- What is the impact of BDA on after-sales management
- What is the impact of BDA on pricing management?
- What is the impact of BDA on risk management in a supply chain context? How can BDA help with supply chain risk detection and disruption recovery?

Potential Future Research Questions on SCM Practices in a Big Data Context

SCM practices, such as total quality management, just-in-time and statistical process control, allow managers to monitor and correct manufacturing processes and performance (Chae et al., 2013). Chae, Yang, Olson, and Shen (2014) argue that supply chain analytic techniques must be combined with SCM practices to increase the value of BDA. The questions that then arise are:

- How can organizations integrate SCM practices into BDA programs?
- What is the impact of BDA on SCM practices?

Potential Future Research Questions on Big Data in Supply Chain Networks

The articles analyzed in our review tend to concentrate on the study of big data or business analytics in the supply chain of a company by assessing their positive impact on decision-making or forecasting and planning. Our review shows only intra-organizational transformation. However, SCM is concerned with integration of activities both within and between organizations. To our knowledge, no the studies have examined big data or business analytics within the interorganizational framework of a multi-level supply chain. In this context, companies involved in the supply chain network need to share data and interact with each other collaboratively. Thus, it would be interesting to study:

- How can one design a dynamic supply chain network using BDA?
- How should big data applications be aligned between the members of a supply chain network?
- Is potential of BDA different for the various members of a supply chain network?
- What is the differential impact on upstream and downstream members?
- How can BDA support supply chain coordination mechanisms?
- How is data shared between the members of a supply chain network?
- How do members of a supply chain network react to using BDA?

CONCLUSION

This study provides a comprehensive literature review on big data analytics in supply chain management. It describes BDA techniques and shows their influence on SCM processes and value creation. BDA has become necessary given the increasing number of data in SCM. It supports all the supply chain processes. BDA can play a critical role in supply chain management on a strategic, tactical and operational level (Souza, 2014). Wang et al. (2016) found that "Strategic supply chain analytics are important for sourcing, network design, and product design. Tactical and operational supply chain analytics are important for demand planning, procurement, production, inventory, and logistics." Tiwari, Wee and Daryanto (2018, p. 321). According to Akter, Wamba, Gunasekaran, Dubey, and Childe (2016), big data analytics has a big impact to enhance firm performance. Big data analytics can help managers to transform the supply chain management. However, a recent study by Accenture states that "while most companies have high expectations for big data analytics in their supply chain, many have had difficulty adopting it. In fact, 97 percent of executives report having an understanding of how big data analytics can benefit their supply chain, but only 17 percent report having already implemented analytics in one or more supply chain functions" (Accenture, 2014, p. 3). So why don't more firms embrace big data analytics (BDA) in SCM? Firms need to embrace a more sophisticated analytics culture so they can handle, manage, analyze, and interpret big data. There is considerable room for big data research in operations and supply chain by focusing on business process transformation, business strategy alignment, deployment and utilization of new technology etc.

REFERENCES

Accenture. (2014). Big data analytics in supply chain: Hype or here to stay? Retrieved from https://acnprod.accenture.com/_acnmedia/Accenture/Conversion-Assets/DotCom/Documents/Global/PDF/Dualpub_2/Accenture-Global-Operations-Megatrends-Study-Big-Data-Analytics.pdf#zoom=50

Addo-Tenkorang, R., & Helo, P. T. (2016). Big data applications in operations/supply-chain management: A literature review. *Computers & Industrial Engineering*, *101*, 528–543. doi:10.1016/j.cie.2016.09.023

Akter, S., Wamba, S. F., Gunasekaran, A., Dubey, R., & Childe, S. J. (2016). How to improve firm performance using big data analytics capability and business strategy alignment? *International Journal of Production Economics*, *182*, 113–131. doi:10.1016/j.ijpe.2016.08.018

Anaya, L., Dulaimi, M., & Abdallah, S. (2015). An investigation into the role of enterprise information systems in enabling business innovation. *Business Process Management Journal*, *21*(4), 771–790. doi:10.1108/BPMJ-11-2014-0108

Angalakudati, M., Balwani, S., Calzada, J., Chatterjee, B., Perakis, G., Raad, N., & Uichanco, J. (2014). Business analytics for flexible resource allocation under random emergencies. *Management Science*, *60*(6), 1552–1573. doi:10.1287/mnsc.2014.1919

Ballings, M., & Van den Poel, D. (2015). CRM in social media: Predicting increases in Facebook usage frequency. *European Journal of Operational Research*, *244*(1), 248–260. doi:10.1016/j.ejor.2015.01.001

Barton, D., & Court, D. (2012). Making advanced analytics work for you. *Harvard Business Review*, *90*(10), 78–83. PMID:23074867

Bose, I., & Chen, X. (2009). Quantitative models for direct marketing: A review from systems perspective. *European Journal of Operational Research*, *195*(1), 1–16. doi:10.1016/j.ejor.2008.04.006

Chae, B. (2015). Insights from hashtag #supplychain and Twitter Analytics: Considering Twitter and Twitter data for supply chain practice and research. *International Journal of Production Economics*, *165*, 247–259. doi:10.1016/j.ijpe.2014.12.037

Chae, B. K., Olson, D., & Sheu, C. (2013). The impact of supply chain analytics on operational performance: A resource-based view. *International Journal of Production Research*, *52*(16), 4695–4710. doi:10.1080/00207543.2013.861616

Chae, B. K., Yang, C., Olson, D., & Sheu, C. (2014). The impact of advanced analytics and data accuracy on operational performance: A contingent resource based theory (RBT) perspective. *Decision Support Systems*, *59*, 119–126. doi:10.1016/j.dss.2013.10.012

Chen, C. L. P., & Zhang, C. Y. (2014). Data-intensive applications, challenges, techniques and technologies: A survey on big data. *Information Sciences*, *275*, 314–347. doi:10.1016/j.ins.2014.01.015

Chen, X. W., & Lin, X. X. (2014). Big data deep learning: Challenges and perspectives. *IEEE Access: Practical Innovations, Open Solutions*, *2*, 514–526. doi:10.1109/ACCESS.2014.2325029

Chen, Z. Y., Fan, Z. P., & Sun, M. (2015). Behavior-aware user response modeling in social media: Learning from diverse heterogeneous data. *European Journal of Operational Research*, *241*(2), 422–434. doi:10.1016/j.ejor.2014.09.008

Chongwatpol, J. (2015). Integration of RFID and business analytics for trade show exhibitors. *European Journal of Operational Research*, *244*(2), 662–673. doi:10.1016/j.ejor.2015.01.054

Columbus, L. (2015). Ten ways big data is revolutionizing supply chain management. *Forbes*. Retrieved from http://www.forbes.com/sites/louiscolumbus/2015/07/13/ten-ways-big-data-is-revolutionizing-supply-chain-management/

Condea, C., Thiesse, F., & Fleisch, E. (2012). RFID-enabled shelf replenishment with backroom monitoring in retail stores. *Decision Support Systems*, *52*(4), 839–849. doi:10.1016/j.dss.2011.11.018

Danaher, B., Huang, Y., Smith, M. D., & Telang, R. (2014). An empirical analysis of digital music bundling strategies. *Management Science*, *60*(6), 1413–1433. doi:10.1287/mnsc.2014.1958

Davenport, T. H. (2013). Analytics 3.0. *Harvard Business Review*, *91*(12), 64–72.

Dell. (2014). Big Data Use Cases. Retrieved from https://fr.slideshare.net/Dell/big-data-use-cases-36019892

Dobrzykowski, D. D., Leuschner, R., Hong, P. C., & Roh, J. J. (2015). Examining Absorptive Capacity in Supply Chains: Linking Responsive Strategy and Firm Performance. *The Journal of Supply Chain Management*, *51*(4), 3–28. doi:10.1111/jscm.12085

Duta, D., & Bose, I. (2015). Managing a big data project: The case of Ramco Cements Limited. *International Journal of Production Economics*, *165*, 293–306. doi:10.1016/j.ijpe.2014.12.032

Fast Company. (2014). The world's Top 10 most innovative companies in big data. Retrieved from http://www.fastcompany.com/most-innovative-companies/2014/industry/big-data

Fawcett, S. E., & Waller, M. A. (2014). Supply chain game changers—mega, nano, and virtual trends—and forces that impede supply chain design (i.e., building a winning team). *Journal of Business Logistics*, *35*(3), 157–164. doi:10.1111/jbl.12058

Fosso Wamba, S., & Akter, S. (2015). Big data analytics for supply chain management: A literature review and research agenda. In *The 11th International Workshop on Enterprise & Organizational Modeling And Simulation* (EOMAS 2015), Stockholm, Sweden, June 8–9.

Fosso Wamba, S., Akter, S., Edwards, A., Chopin, G., & Gnanzou, D. (2015). How "big data" can make big impact: Findings from a systematic review and a longitudinal case study. *International Journal of Production Economics*, *165*, 234–246. doi:10.1016/j.ijpe.2014.12.031

Gandomi, A., & Haider, M. (2015). Beyond the hype: Big data concepts, methods, and analytics. *International Journal of Information Management*, *35*(2), 137–144. doi:10.1016/j.ijinfomgt.2014.10.007

Groves, W., Collins, J., Gini, M., & Ketter, W. (2014). Agent-assisted supply chain management: Analysis and lessons learned. *Decision Support Systems*, *57*, 274–284. doi:10.1016/j.dss.2013.09.006

Hahn, G. J., & Packowski, J. (2015). A perspective on applications of in-memory analytics in supply chain management. *Decision Support Systems*, *76*, 45–52. doi:10.1016/j.dss.2015.01.003

Hashem, I. A. T., Yaqoob, I., Anuar, N. B., Mokhtar, S., Gani, A., & Khan, S. U. (2015). The rise of "big data" on cloud computing: Review and open research issues. *Information Systems*, *47*, 98–115. doi:10.1016/j.is.2014.07.006

Hazen, H. T., Boone, C. A., Ezell, J. D., & Jones-Farmer, L. A. (2014). Data quality for data science, predictive analytics, and big data in supply chain management: An introduction to the problem and suggestions for research and applications. *International Journal of Production Economics*, *154*, 72–80. doi:10.1016/j.ijpe.2014.04.018

Hofmann, E. (2015). Big data and supply chain decisions: The impact of volume, variety and velocity properties on the bullwhip effect. *International Journal of Production Research*. doi:10.1080/00207543.2015.1061222

Hogarth, R.M., and Soyer, E. (2015). Using Simulated Experience to Make Sense of Big Data. *MIT Sloan Management Review,* (spring), 5-10.

Issacs, L. (2013). Rolling the Dice with Predictive Coding: Leveraging Analytics Technology for Information Governance. *Information & Management*, *47*(1), 22–26.

Kahn, K. B. (2014). Solving the problems of new product Forecasting. *Business Horizons*, *57*(5), 607–615. doi:10.1016/j.bushor.2014.05.003

Kemp, R. (2014). Legal Aspects of Managing Big Data. *Computer Law & Security Review*, *30*(5), 482–491. doi:10.1016/j.clsr.2014.07.006

Kiron, D., Prentice, P. K., & Ferguson, R. B. (2014). The analytics mandate. *MIT Sloan Management Review*, *55*(4), 1–25.

Kohlborn, T., Mueller, O., Poeppelbuss, J., & Roeglinger, M. (2014). Interview with Michael Rosemann on ambidextrous business process management. *Business Process Management Journal*, *20*(4), 634–638. doi:10.1108/BPMJ-02-2014-0012

Krumeich, J., Werth, D., & Loos, P. (2016). Prescriptive control of business processes. *Business & Information Systems Engineering*, *58*(4), 261–280. doi:10.100712599-015-0412-2

Kshetri, N. (2014). Big data's impact on privacy, security and consumer welfare. *Telecommunications Policy*, *38*(11), 1134–1145. doi:10.1016/j.telpol.2014.10.002

Kulkarni, S. S., Apte, U. M., & Evangelopoulos, N. E. (2014). The use of latent semantic analysis in operations management research. *Decision Sciences*, *45*(5), 971–994. doi:10.1111/deci.12095

Kwon, K., Kang, D., Yoon, Y., Sohn, J. S., & Chung, I. J. (2014). A real time process management system using RFID data mining. *Computers in Industry*, *65*(4), 721–732. doi:10.1016/j.compind.2014.02.007

Li, Q., Luo, H., Xie, P. X., Feng, X. Q., & Du, R. Y. (2015). Product whole life-cycle and omni-channels data convergence oriented enterprise networks integration in a sensing environment. *Computers in Industry*, *70*, 23–45. doi:10.1016/j.compind.2015.01.011

Liao, S. H., Chu, P. H., & Hsiao, P. Y. (2012). Data mining techniques and applications – A decade review from 2000 to 2011. *Expert Systems with Applications*, *39*(12), 11303–11311. doi:10.1016/j.eswa.2012.02.063

Lustig, I., Dietrich, B., Johnson, C., & Dziekan, C. (2010). The analytics journey. *Analytics Magazine*, November/December, 11–13.

Maldonado, S., Montoya, R., & Weber, R. (2015). Advanced conjoint analysis using feature selection via support vector machines. *European Journal of Operational Research*, *241*(2), 564–574. doi:10.1016/j.ejor.2014.09.051

Malhotra, M. K., & Kher, H. V. (1996). Institutional research productivity in production and operations management. *Journal of Operations Management*, *14*(1), 55–77. doi:10.1016/0272-6963(95)00037-2

Manyika, J., Chui, M., Brown, B., Bughin, J., Dobbs, R., Roxburgh, C., & Byers, A. H. (2011). *Big data: The next frontier for innovation, competition and productivity*. New York: McKinsey Global Institute.

Mortenson, M. J., Doherty, N. F., & Robinson, S. (2015). Operational research from Taylorism to terabytes: A research agenda for the analytics age. *European Journal of Operational Research*, *241*(3), 583–595. doi:10.1016/j.ejor.2014.08.029

Ng, I., Scharf, K., Pogrebna, G., & Maull, R. (2015). Contextual variety, Internet-of-Things and the choice of tailoring over platform: Mass customisation strategy in supply chain management. *International Journal of Production Economics*, *159*(0), 76–87. doi:10.1016/j.ijpe.2014.09.007

Ngai, E. W. T., & Wat, F. K. T. (2002). A literature review and classification of electronic commerce research. *Information & Management*, *39*(5), 415–429. doi:10.1016/S0378-7206(01)00107-0

Nguyen-Duc, A., Cruzes, D. S., & Conradi, R. (2015). The impact of Global Dispersion on Coordination, Team Performance and Software Quality: A Systematic Literature Review. *Information and Software Technology*, *57*, 277–294. doi:10.1016/j.infsof.2014.06.002

Nurmilaakso, J. M. (2008). Adoption of e-business functions and migration from EDI-based to XML-based e-business frameworks in supply chain integration. *International Journal of Production Economics*, *113*(2), 721–733. doi:10.1016/j.ijpe.2007.11.001

Ranyard, J. C., Fildes, R., & Hu, T. I. (2015). Reassessing the scope of OR practice: The influences of problem structuring methods and the analytics movement. *European Journal of Operational Research*, *245*(1), 1–13. doi:10.1016/j.ejor.2015.01.058

Ross, J. W., Beath, C. M., & Quaadgras, A. (2013). You may not need big data after all. *Harvard Business Review*, *91*(12), 90–98. PMID:23593770

Sampler, J. L., & Earl, M. J. (2014). What's your information footprint? *Sloan Management Review*, *55*(2), 95–96.

Shang, G., Saladin, B., Fry, T., & Donohue, J. (2015). Twenty-six years of operations management research (1985–2010): Authorship patterns and research constituents in eleven top rated journals. *International Journal of Production Research*, *53*(20), 6161–6197. doi:10.1080/00207543.2015.1037935

Sodhi, M. S., & Tang, C. S. (2010). *A Long View of Research and Practice in Operations Research and Management Science*. US: Springer. doi:10.1007/978-1-4419-6810-4

Souza, G. C. (2014). Supply chain analytics. *Business Horizons*, *57*(5), 595–605. doi:10.1016/j.bushor.2014.06.004

Supply Chain Council. (2008). SCOR Framework – Introducing all elements of the supply chain reference model: Standard processes, metrics and best practices. Available at supplychainresearch.com/images/SCOR_Framework_2.1.ppt

Tambe, P. (2014). Big data investment, skills, and firm value. *Management Science*, *60*(6), 1452–1469. doi:10.1287/mnsc.2014.1899

Tan, K. H., Zhan, Y. Z., Ji, G., Ye, F., & Chang, C. (2015). Harvesting big data to enhance supply chain innovation capabilities: An analytic infrastructure based on deduction graph. *International Journal of Production Economics*, *165*, 223–233. doi:10.1016/j.ijpe.2014.12.034

Tiwari, S., Wee, H. M., & Daryanto, Y. (2018). Big data analytics in supply chain management between 2010 and 2016: Insights to industries. *Computers & Industrial Engineering*, *115*, 319–330. doi:10.1016/j.cie.2017.11.017

Verdouw, C. N., Beulens, A. J. M., & van der Vorst, J. G. A. J. (2013). Virtualisation of floricultural supply chains: A review from an Internet of Things perspective. *Computers and Electronics in Agriculture*, *99*, 160–175. doi:10.1016/j.compag.2013.09.006

Waller, M. A., & Fawcett, S. E. (2013). Data science, predictive analytics, and big data: A revolution that will transform supply chain design and management. *Journal of Business Logistics*, *34*(2), 77–84. doi:10.1111/jbl.12010

Wang, G., Gunasekaran, A., Ngai, E. W., & Papadopoulos, T. (2016). Big data analytics in logistics and supply chain management: Certain investigations for research and applications. *International Journal of Production Economics*, *176*, 98–110. doi:10.1016/j.ijpe.2016.03.014

Wang, J., & Zhang, J. (2016). Big data analytics for forecasting cycle time in semiconductor wafer fabrication system. *International Journal of Production Research*, *54*(23), 7231–7244. doi:10.1080/00207543.2016.1174789

White, M. (2012). Digital Workplaces: Vision and Reality. *Business Information Review*, *29*(4), 205–214. doi:10.1177/0266382112470412

Zhang, Y., Ren, S., Liu, Y., & Si, S. (2017). A big data analytics architecture for cleaner manufacturing and maintenance processes of complex products. *Journal of Cleaner Production*, *142*, 626–641. doi:10.1016/j.jclepro.2016.07.123

Zhong, R. Y., Huang, G. Q., Lan, S., Dai, Q. Y., Xu, C., & Zhang, T. (2015). A big data approach for logistics trajectory discovery from RFID-enabled production data. *International Journal of Production Economics*, *165*, 260–272. doi:10.1016/j.ijpe.2015.02.014

Zhong, R. Y., Lan, S., Xu, C., Dai, Q., & Huang, G. Q. (2016). Visualization of RFID-enabled shopfloor logistics big data in cloud manufacturing. *International Journal of Advanced Manufacturing Technology*, *84*(1–4), 5–16. doi:10.100700170-015-7702-1

APPENDIX

Table 8. Summary of highlights and key findings of some reviewed articles

Authors	Highlights and Key Findings
Angalakudati et al. (2014)	• The resource allocation problem of scheduled and unpredictable tasks is addressed in the gas industry. • "The goal is to perform all the standard jobs by their respective deadlines, to address all emergency jobs in a timely manner, and to minimize maintenance crew overtime" (p. 1552). • This study use models and heuristics to develop a decision support tool.
Ballings and Van den Poel (2015)	• This study assesses the feasibility of predicting increases in Facebook usage frequency. • Six classification algorithms are evaluated and the importance of many predictors is assessed. • The top-performing algorithm is Stochastic Adaptive Boosting. • The top predictor is the deviation from regular usage patterns. • Facebook can use this approach to customize its service (advertisements, recommendations).
Barton and Court (2012)	• To benefit from big data, a firm has to build prediction and optimization models and develop the ability to use advanced analytics to improve its performance.
Chae, Olson, and Sheu (2013)	• The impact of supply chain analytics on supply chain planning satisfaction and operational performance is studied. • From the perspective of a resource-based view, supply chain analytics consists of three types of resources: data management resources, IT-enabled planning resources, and performance management resources. • The analysis of data collected from 537 manufacturing plants shows that data management resources are a stronger predictor of performance management resources than IT planning resources. • The deployment of advanced IT-enabled planning resources occurs after acquisition of data management resources. • All three sets of resources improve supply chain planning satisfaction and operational performance. • "Manufacturers with sophisticated planning technologies are likely to take advantage of data-driven processes and quality control practices" (p. 1).
Z. Y. Chen, Fan, and Sun (2015)	• A hierarchical ensemble learning framework is proposed for behavior-aware user response modeling using diverse heterogeneous data. • A data transformation and feature extraction strategy is developed to transform large-scale, multi-relational data into customer-centered high-order tensors. • "An improved hierarchical multiple kernel support vector machine (H-MK-SVM) is developed to integrate the external, tag and keyword, individual behavioral and engagement behavioral data for feature selection from multiple correlated attributes and for ensemble learning in user response modeling" (p. 422).
Chongwatpol (2015)	• An RFID-enabled traceability framework is proposed to improve information visibility at the exhibition industry. • RFID allows exhibitors to track attendees' movements and activities during their entire exhibition visit. • The integration of RFID data and business analytics improves exhibitors' understanding of customers' purchasing behavior. • The key findings provide feedback to business analysts to promote follow-up marketing strategies.
Danaher et al. (2014)	• This study uses a "panel data on digital song and album sales coupled with a quasi-random price experiment to determine own- and cross-price elasticities for songs and albums" (p. 1413). • A "structural model of consumer demand to estimate welfare under various policy relevant counterfactual scenarios" (p. 1413) is developed. • Guidance on optimal pricing and marketing strategies for digital music is provided. • The results show that "tiered pricing coupled with reduced album pricing increases revenue to the labels by 18% relative to uniform pricing policies traditionally preferred by digital marketplaces while also increasing consumer surplus by 23%" (p. 1413).
Davenport (2013)	• The evolution of analytics is presented. • Ten requirements are proposed for capitalizing on analytics 3.0. • The impact of analytics in creating value is discussed.
Hofmann (2015)	• The potential of big data to improve supply chain processes is studied. • The potential of big data characteristics (velocity, volume and variety) to mitigate the bullwhip effect is assessed. • The operationalization of big data in control engineering analyses is presented. • Velocity has the greatest potential to enhance performance.
Kulkarni, Apte, and Evangelopoulos (2014)	• The use of the Latent Semantic Analysis technique is presented. • The implementation of Latent Semantic Analysis is explained and illustrated. • The field study of operations management is presented in the area of big data.
Maldonado, Montoya, and Weber (2015)	• Advanced machine learning techniques enhance conjoint analysis capabilities to better identify consumer preferences. • Feature selection procedure pools information across consumers while identifying individual preferences. • Applications on experimental data and on two empirical studies show that the proposed approaches outperform traditional techniques for conjoint analysis.
Mortenson, Doherty, and Robinson (2015)	• The lack of research into analytics in the operational research field is underscored. • The histories of operational research, analytics and a range of related disciplines are presented and their relationships are discussed. • Avenues for future research are proposed by combining several key themes related to analytics and operational research.
Ranyard, Fildes, and Hu (2015)	• Results of a global survey of operational research practices are presented. • The scope of operational research practices has been extended via problem structuring methods and business analytics. • The gap between academic research and practice is emphasized. • Business analytics and operational research overlap, presenting challenges and opportunities that must be addressed.
Sampler and Earl (2014)	• This study deals with the information assets a company has to have in order to generate economic value.
Tambe (2014)	• "This paper analyzes how labor market factors have shaped early returns on big data investment using a new data source—the LinkedIn skills database" (p. 1452). • Hadoop investments to complement workers' technical skills were associated with 3% faster productivity growth.

Section 3

Entrepreneurship and Innovation in the Digitalization Era

Chapter 8
The Conceptual Framework for The Examination of a Successful Digital Entrepreneurship in 21st Century

Bilal Ahmad Ali Al-khateeb
Al Imam Mohammad Ibn Saud Islamic University, Saudi Arabia

ABSTRACT

There is still lack of a clear conceptual framework to examine a successful digital entrepreneurship within the developing and emerging contexts despite calls by previous scholars. Also, recent studies shown that majority of the digital enterprises studies available today are mostly from the Western world. Thus, there are only few studies on digital enterprises studies emanating from the developing and emerging countries in the Middle-East and Africa continents. Based on the evidence from the literature, this paper provides an overview of digital entrepreneurship, identifies key variables that determine a successful digital entrepreneurship and then provides a conceptual model to guide the understanding of a successful digital entrepreneurship development within the context of developing and emerging economies. The paper offered some implications for digital entrepreneurs, policy makers and some other people in the business of digital entrepreneurship in Saudi Arabia.

INTRODUCTION

There have been several attempts by both practitioners and academics to link entrepreneurship to the digital world. However, prior to this time, both empirical and practical evidence suggest the rise of a new category of entrepreneurship, that is, the concept of digital entrepreneurship which is believed to be an emerging trend.

Digital entrepreneurship affects every task, activity and process within a company (Conigliaro, 2017). Therefore, companies must disrupt their own traditional business model before competitors do. In essence, the time to build a digital company is now. Digital transformation is not only about changing technology but also about changing the way companies do business, and no company is left out of this.

DOI: 10.4018/978-1-5225-7262-6.ch008

The importance of digital entrepreneurship cannot be undermined. For instance, it attracts talent all over the world. It increases employee's productivity, satisfaction, retention, engagement and improves communication within the organization and equally reduces operating costs for the organization. It equally promotes digital marketing by helping companies to collect valuable information about customers and competitors, particularly in real time. In doing so, it becomes an engine for economic development and growth. Thus, digitalization of the economy is the fundamental mechanism of innovation, competition and growth in today's world, which in turn requires a process of adaptation and transformation. According to Nambisan (2016) and Nambisan (2017) digital entrepreneurship helps in transforming the nature of uncertainty inherent in entrepreneurial processes and outcomes, as well as the ways of dealing with such uncertainty.

Despite the many benefits of digital entrepreneurship as highlighted above, and with the recent increase in the number of digital entrepreneurs across the globe, particularly in the Western world, the academic approach to digital entrepreneurship in the 21st century still needs to be explored. Unfortunately, it is yet to be explored and accorded the needed attention and also treated as one of the areas of entrepreneurship study. Additionally, Ngoasong (2017) recently acknowledged that majority of the digital enterprises studies available today are mostly from the Western world. The author argued that the nature of online reach has made this possible for them. The above suggests that there are few studies on digital enterprises studies emanating from the developing and emerging countries, particularly those from the Middle-East and Africa continent.

Practically, digital entrepreneurship seems strong, but the outlook seems theoretically weak in terms of model development. Thus, digital entrepreneurship appears to lack clear models to advance its understanding, particularly in developing countries such as Middle-East and Africa. Also, it appears that the early writers on digital entrepreneurship sound more of practical than academic, and therefore, may not have found a need for models to advance the knowledge in digital entrepreneurship domain. It is on the basis of this that authors such as Archer, Hull, Soukup, Mayer, Athanasiou, Sevdalis & Darzi (2017) and Nambisan (2016) clamored for theoretical frameworks to guide the understanding of digital entrepreneurship and the relevant factors that affect it. The study noted that only limited efforts have been made by previous studies to provide theories and theoretical framework for the examination of digital entrepreneurship. However, some attempts have been made by few to provide theoretical models. On the contrary, these attempts appear to be ineffective as the key variables were not considered, while on the other hand, majority of the frameworks are perceived to be Western based. For example, the theoretical model provided by Ngoasong (2017) seems to have neglected the relevant factors that actually determine successful digital entrepreneurship as the study only focused on digital entrepreneurial competencies, institutional context, local context and technological context, without consideration to variables such as digital knowledge-based, digital business environment, digital skill, digital mindset and digital infrastructure. Also, Giones & Brem (2017) model is only concerned itself with digital technology entrepreneurship. In 2016, Leong, Pan & Liu (2016) advanced another model which they called digital entrepreneurship of born digital and grown digital firms but failed to include the key variables highlighted above. Authors such as Aldrich (2014); Forrester (2017); Nambisan (2016); Ngoasong (2017) claimed that variables such as digital infrastructures, digital knowledge-based, digital business environment, digital skill and digital mindset, cannot be ignored in an attempt to develop successful digital entrepreneurship. In view of this unfolding reality which forms the basis for this study, this paper provides an overview of digital entrepreneurship, identify key variables that determine a successful digital entrepreneurship and then provide a conceptual framework to guide the understanding of successful digital entrepreneurship development in the 21st century.

CONCEPTUALIZATION OF DIGITAL ENTREPRENEURSHIP

Attempts to describe digital entrepreneurship have led to many contributions from different authors and researchers. While some of the descriptions are contextual, others are general in nature. For example, Davidson & Vaast (2010) described digital entrepreneurship as the practice of pursuing "new venture opportunities through new media and internet technologies. They noted that digital entrepreneurship is like, or related to, the traditional entrepreneurship which tends to pursue entrepreneurial opportunities by creating new enterprises or commercializing products and services. On the other hand, digital entrepreneurship is different from traditional entrepreneurship because some or all of the entrepreneurial ventures take place digitally, instead of traditional formats. Therefore, one can describe digital entrepreneurship is that type of entrepreneurship which manifests in existing businesses through the introduction of new digital technologies, or through the novel use of technologies that change business models or revolutionize products or services (Prodanov, 2018; Tumbas, Berente, Seidel & Brocke, 2015). However, Santana (2017) states that the most acceptable definition of digital entrepreneurship across the world should be that which covers all new enterprises and existing companies, that have been "transformed" or converted, and that create economic and social value through digital technologies. These firms are distinguished by high intensity in the use of new digital technologies, (particularly solutions for the social sphere, big data, mobile phone technology and cloud technology), for the purpose of enhancing business operations, inventing new business models, improving business intelligence and dealing with clients and partners. It depends on five pillars namely; digital knowledge based and ICT market; digital business friendly environment, access to finance, digital skills and e-leadership; and an entrepreneurial culture (Santana, 2017). Also, digital entrepreneurship is conceptualized as the embracing of new ventures and the transformation of existing businesses by creating and using novel digital technologies (Zhao & Collier, 2016).

Consequently, these digital businesses possess certain features such as high intensity of new digital technologies utilization (which are mainly social, mobile, analytic and cloud solutions), in order to increase business actions, invent new (digital) business models, sharpen business intelligence, and engage customers and stakeholders through new (digital) channels. It is equally a term used in referring to the methods of creating a new or novel internet enabled/delivered business, product or service. This description is comprised of not just startups-bringing a new digital product or service to market-but also the digital transformation of an existing business activity inside a firm or the public sector. Accordingly, Vasilchenko & Morrish (2011) view digital entrepreneurship as businesses that provide online accounting, software development, social computing and digital platforms for cataloguing, e-commerce (Javalgi, Todd, Johnston & Granot (2012) and multi-media businesses that sell digitized products and services (Hair, Wetsch, Hull, Perotti & Hung, 2012; Onetti, Zucchella, Jones & McDougall-Covin, 2012). Thus, a digital entrepreneur is an individual who creates and delivers key business activities and functions, such as production, marketing, distribution and stakeholder management, using information and communication technologies (ICTs).

ROLES OF DIGITAL ENTREPRENEURS IN 21ST CENTURY

It is actually very difficult to say precisely when digital aspect of entrepreneurship emerged. However, Leadem (2018) observed that an important landmark in digital revolution can be traced down to 1994 when Amazon was founded, while another landmark can be linked to the first iPhone, which was released in 2007. He noted that these remarkable events not only marked the beginning of the life-changing Smartphone frenzy, but gave room for more major digital revolution, including social media and mobile apps, to surface. It is at this point that company like Netflix's streaming was launched in 2007, the birth of Bitcoin in 2009 and the release of Google Drive collaboration software. All these set the stage for a digital revolution which has given birth to digital entrepreneurship, now ruling the world.

As to the role digital entrepreneurship, it would be very difficult to undermine this, particularly in the 21st century. The reason is that digital entrepreneurship is traceable to every aspect of the business sector. For example, digital entrepreneurs help customers and companies to get the most accurate data from the most authoritative sources. Digital entrepreneurs are regarded as professionals who provide context for maps, info cards and specific answers to customers and companies in need of information, in a professional way. With digital entrepreneurship, this information is provided in an intelligent way (Forrester, 2017).

Digital entrepreneurship promotes digital marketing. In the 21st century, businesses cannot afford to ignore digital marketing, because doing so would amount to opening a business, but not telling anyone (Conigliaro, 2017). It also helps businesses to engage users online and help brands in connecting with customers. Digital entrepreneurship, through digital technologies, has broken down communication barriers, thereby improving employee's experience in the workplaces across the world.

Furthermore, digital entrepreneurs also provide digitalization concepts to ensure sustainable growth for both the government and businesses around the world (Denison, 2017). It equally provides the basis for future-oriented jobs, greater resource efficiency, and responsible economic activity. They engage in smart regulation and focus investment in education, research, and entrepreneurship, and these are considered very crucial for the world economy.

According to Nambisan (2016), digital entrepreneurship tends to balance the playing field in some specific sectors such as ICT and entrepreneurship, thereby providing opportunities to work from remote places at different hours, from the home, or on the go. One can also see the role of digital entrepreneurship in the promotion of gender equality, social and economic inclusion, which stimulate local development, and contributing to sustainable development, particularly when new technologies are joined with the availability of open and public data in cases where data for example on climate/weather, crops and soil, or road and traffic conditions, are used to develop services and apps that alleviate local problems, be it the optimization of agricultural production, emergency and disaster relief response, road safety and accident avoidance, or reducing traffic congestion and parking difficulties. Simple price information of products and services can also greatly improve economic opportunities for local suppliers; also, the bringing together or matching of supply and demand for labor, goods and services by digital technologies, tends to improve economic efficiency, productivity and income opportunities.

CHARACTERISTICS OF DIGITAL ENTREPRENEURSHIP

Psychological theory or model of entrepreneurship, which is equally called hereditary or trait theory by McClelland, Lynn & Schrage (1961), emphasized on the need for entrepreneurs to possess certain characteristics or personalities to enable them become successful entrepreneurs. Based on this, Digital Skills Academy (2017) asserts that there are several skills required by the digital entrepreneurs to effectively function in their business area. These qualities are far more different from those of the conventional entrepreneurs, such as economic entrepreneurs, social entrepreneurs, technopreneurs etc. (Lucky, Rahman & Minai, 2013). Because of the nature of the digital task that digital entrepreneurs engage in; they are forced to possess certain characteristics or skills that are more specific to their tasks. Hence, the abilities and characteristics of the digital entrepreneur become their most important assets.

According to Digital Skills Academy (2017), there are ten key unique qualities often used to qualify digital entrepreneurs in the 21st century. They are as follows:

- **Democratization:** It is one of the most interesting effects of the digital revolution in the present business world. New entrepreneurs in the digital world can gradually build businesses from the ground up without capital and business connection like before, when conventional entrepreneurs may not be able to do so without capital and business connections.
- **Visionary:** It is not just having a plan, but a clear vision of what your business will do, how your business will do it, and the business goals and objectives. This keeps the digital entrepreneurs focused and helps them to move on the right track.
- **Communicator:** Digital entrepreneurs get others off the ground by clearly articulating their business ideas, and it is this skill also needed in order for co-workers and clients to work with digital entrepreneurs efficiently. This can be achieved through strong and effective communication skills.
- **Motivated:** Digital entrepreneurs are motivators as they drive and propel others to really do what they are doing; and what they want them to do. They possess the motivational skill to always keep going on the road to success so as to reach the end. Therefore, with the support of clear vision and goals, motivation can be attained.
- **Nous:** This is human intellect that is necessary for understanding what is true and real. It is a know-how, ability, knowledge, skill or acumen to understand how the system and operation works. With this skill, digital entrepreneurs are able to work regardless whether they are functioning in their area of expertise, other related areas or wider structural systems, such as legislative frameworks or the economy, it is important to know how to manage them and how they can affect your business.
- **Opportunist:** A skill to always see or recognize business opportunities within the environment, whenever they arise, is one of the most valuable traits an entrepreneur can have. The skill equally assists digital entrepreneurs to create their own opportunities. For example, it is this skill that assisted Patrick and John Collison to see a gap in the market, that later made them billionaires.
- **Technologist**: It is important, if not compulsory, for any digital entrepreneur to possess this skill. This requires that one must be more than just a tech entrepreneur, because one needs to understand not only the technology he is developing, but the wider technology ecosystem, having the ability to see where technologies will develop and where they will provide opportunities.

- **Explorer:** Digital entrepreneurs should possess the ability to explore avenues as they arise, because it allows the discovery of new areas of possibility. It will also provide a situation where what was initially the core of your business could become just a branch of a much wider and more lucrative business concept.
- **Adaptable:** This is the quality needed to overcome a number of bumps which are inevitable on the way to success. It helps digital entrepreneurs to remain on track especially when the inevitable and the unforeseen knock them off course. Adaptability allows the digital entrepreneurs to deal with any of these inevitable and unforeseen bumps.
- **Tenacity:** Digital entrepreneurs must be persistent and determined in such a way that they should not allow any challenge they face get on top of them. A tenacious entrepreneur is one that is inspired when they come up against a wall, excited by the prospect of finding a way over or around it, rather than being frustrated by it. Tenacious entrepreneurs do not give up because they don't know how to give up (Digital Skills Academy, 2017).
- **Learn From Mistakes:** Mistakes are an inevitable part of running a digital business. Digital entrepreneurs must learn how to deal with both valuable mistakes and costly mistakes. This skill should help digital entrepreneurs to spend their energy in analyzing mistakes, rather than regretting those mistakes. They should learn from the differences between valuable mistakes and costly mistakes.

DETERMINANTS OF SUCCESSFUL DIGITAL ENTREPRENEURSHIP

This paper identified five major determinants that are believed to promote successful digital entrepreneurship. These determinants include digital knowledge-based, digital business environment, digital skill, digital mindset and digital infrastructures. These determinants therefore form the basis for further discussion below.

- **Digital Knowledge-Based:** Forrester (2017) noted that digital knowledge-based management has recently emerged as one of the digital entrepreneurs with the most important responsibilities. Digital knowledge - based becomes imperative due to the fact that it is almost appearing impracticable to manage or upgrade all of this information content manually. The concept is not a passive responsibility but rather encompasses far more than just local listings management. Digital knowledge base consists of facts about your organization's people, products, and locations. It also includes the public information concerning your business, like your name, address, and phone number (NAP data), which then improves content, such as photos, descriptions, and menu items-information that is dynamic and changes often. Digital knowledge management vendors provide businesses with solutions to effectively manage all of the digital information about their business. If an entrepreneur is serious about making sure customers get accurate information about his/her brand, when and where they need it, then such an entrepreneur is actively managing his/her digital knowledge. Thus, digital knowledge management is concerned with the use of data a company has, to determine what data it needs to solve problems and drive strategy. Therefore, an effective knowledge management solution should support internal social networks that people can use to connect with each other. If business improvement is the ultimate goal of digital transformation, effective knowledge management is what' is going to pave the way there.

Digital entrepreneurs are people who have the ability to track down all the authoritative sources of knowledge about a brand, people, products, events and locations from within your organization. This could be an easy job at a small company, but it could become a huge undertaking for large corporations. It would likely require conversations with departments such as Marketing, IT, Legal, Facilities, Store Operations and others. It is no small feat to identify and find all the public facts about a business that is relevant to customers. Digital knowledge management saves time which is worth the investment. At the same time, it equally puts the company's brand in a position that would address new information needs and opportunities, as they arise. An organization such as Yext is already into digital knowledge based. For example, Yext, through its digital knowledge based, discovered in 2006 alone that its customers generated 334 million phone calls and 622 million requests for driving directions from third-party resources. Customers are not just looking for up driving directions by chance, but it is believed that they have strong intent in searching for how to get to your business; they have strong intent to go, and you want them to be able to find you. Thus, correct information everywhere is therefore needed to drive customers to your business, and ultimately improves your bottom line. From the above, it becomes crystal clear that digital entrepreneurship is greatly affected by the quality of digital knowledge based available to them. The better the quality of the digital knowledge based, the higher the success of the digital entrepreneurship.

- **Digital Business Environment:** From entrepreneurship perspectives, environment refers to external and internal conditions that affect the business. It is a condition, situation, or factors in the surrounding that affect or influence entrepreneurs. Entrepreneurs interact with the environment all the time. Therefore, digital business environment is best described as those technological factors and supports that tend to influence and support digital entrepreneurship. Environment changes from time to time; therefore, environment is full of uncertainties. However, it provides opportunities as well as threats to an entrepreneur's business. So, digital entrepreneurs must always analyze and understand the environment he/she is operating in, and by analyzing and understanding the environment, an entrepreneur will also be able to know his/her business's strengths and weaknesses. Technological factor is one of the aspects of the business environment. It is actually required to increase process efficiency, minimize cost, invent new designs, innovate and effectively use resources. For example, with ICT, digital entrepreneurs can have access to new knowledge, new processes of production, new and faster information flow and new strategy and better decision making.
- **Digital Skill:** Generally, skill is described as the capacity or ability to do something. It is the focus for analytical research (Green, 2011). It is seen as the ability one possesses or acquires mostly from training which enables him to perform his task. Therefore, skill is ability, and ability is a skill. Skill is the expertise in doing something or expertise needed to perform a task. According to European Union (EU) Skills Panorama 2014 (2015), entrepreneurship is an individual's ability to turn ideas into action. It includes creativity, innovation and risk- taking, as well as the ability to plan and manage projects in order to achieve the stated objectives. It is seen as vital to promoting innovation, competitiveness and economic growth. Fostering entrepreneurial spirit supports the creation of new firms and business growth. Basically, the OECD did identify three fundamental skills entrepreneurs needed for effective functioning. These skills include:
    - **Technical:** Communication, environment monitoring, problem solving, technology implementation and use, interpersonal and organizational skills.

- **Business Management:** Planning and goal setting, decision making, human resources management, marketing, finance, accounting, customer relations, quality control, negotiation, business launch, growth management and compliance with regulations skills.
- **Personal Entrepreneurial:** Self-control and discipline, risk management, innovation, persistence, leadership, change management, network building, and strategic thinking.

From the above, it is very clear that the type of entrepreneurial skills needed by the digital entrepreneurs is that of technical skill, which they required to perform the core operations in digital organization. It entails the use of technological tools, techniques and procedures, to function in their digital field. For example, internet skill, computer repair skill, innovative and creative skills etc. The quality of these entrepreneurial skills possessed by the digital entrepreneurs goes a long way to determine the type of success the entrepreneur would achieve.

- **Digital Mindset:** Entrepreneurial mindset refers to a specific state of mind, which orientates human conduct towards entrepreneurial activities and outcomes. Individuals with entrepreneurial mindsets are often drawn to opportunities, innovation and new value creation. Practically, entrepreneurs are all different with different mindsets that contribute to their success. These mindsets have to do with their individual personalities or characteristics, in relation to the field in which they have chosen to work. As pointed out by Hessinger (2018), entrepreneurs have about ten mindsets that actually make then successful in the areas of their endeavours. For example, they have the mindset on reaching out to customers first, finding a new market for an existing product, using networking to build their business, giving without expecting a return, keep control of their vision, understand the power of brand, focus their energy on what is good for their businesses, always maintain quality control, set their product apart and take ownership. Therefore, digital entrepreneurs should strive for those entrepreneurial mindsets that could make them successful in their digital field of business.
- **Digital Infrastructure:** the study by Nambisan (2016) pointed out that one factor that affects digital entrepreneurship in the 21st century is the digital infrastructures. This suggests that digital infrastructures are strongly associated with successful digital entrepreneurship. The author described digital entrepreneurship as digital technology tools and systems (e.g., cloud computing, data analytics, online communities, social media, 3D printing, digital makerspaces, etc.) that offer communication, collaboration, and/or computing capabilities to support innovation and entrepreneurship. Aldrich (2014) and Delacroix, Parguel & Benoit-Moreau (2018) asserts that digital infrastructures such as cloud computing, data analytics, online communities, social media, 3D printing, and digital makerspaces have significantly affected entrepreneurship in the form of democratization of entrepreneurship. Accordingly, Hatch (2013) affirmed that cloud computing, digital makerspaces, and data analytics have made it possible for new ventures to cost- effectively construct and test novel concepts, involving a larger set of potential customers, thereby showing the capability to support end-to-end entrepreneurial activities. A good infrastructure for incubators and corporate accelerators, funding programs for start-ups, and targeted investment in research and development, are the prerequisites for pioneering new developments (Denison, 2017). Finally, good digital infrastructures would bring about successful entrepreneurship.

Empirical Review

Authors have argued that digital entrepreneurship is driven or determined by certain key variables. However, these variables are yet to be properly identified. It is one of the main objectives of this present study to properly and effectively identify these variables, and then link them with digital entrepreneurship, while explaining how these variables affect successful digital entrepreneurship. With respect to this, Forrester (2017); Nambisan (2016); Ngoasong (2017) have made frantic attempts to determine what kind of variables predict a successful digital entrepreneurship. For example, Nambisan (2016) claimed that local context in the form of physical infrastructure affect digital entrepreneurship. However, the study failed to be specific about digital infrastructures. Also, Aldrich (2014) asserts that digital infrastructures such as cloud computing, data analytics, online communities, social media, 3D printing, and digital makerspaces have significantly affected entrepreneurship in the form of democratization of entrepreneurship. In another related development, Hatch (2013) affirmed that cloud computing, digital makerspaces, and data analytics have made it possible for new ventures to cost effectively construct and test novel concepts involving a larger set of potential customers, thereby showing the capability to support end-to-end entrepreneurial activities.

Ngoasong (2017) emphasized on entrepreneurial digital competencies. He argued that small digital businesses require digital competencies to be successful. Thus, digital competencies are referred to as digital skills, and are believed to be strongly related to a successful digital entrepreneurship. Finally, Conigliaro (2017), in his presentation, reported that digital knowledge base, digital business environment, digital skill and digital mindset, are the core pillars that can foster successful digital entrepreneurship. This has been highlighted by the digital entrepreneurship scoreboard in 2015. Hence, this paper proposed a strong relationship between digital knowledge-based, digital business environment, digital skill, digital mindset, digital infrastructure and successful digital entrepreneurship.

MODELS OF DIGITAL ENTREPRENEURSHIP

Ngoasong Model (2017)

In order to explain digital entrepreneurship, Ngoasong (2017) developed a model called entrepreneurial digital competencies. The major aim of the model is to demonstrate the importance of digital competence in the development of digital entrepreneurship. The model depicts the link between context, competencies and outcomes. The research framework suggests that the influence of context on digital entrepreneurship is mediated by entrepreneurial digital competencies (EDCs). Also, the framework is applied to examine how digital entrepreneurs use EDCs to either discover or create entrepreneurial opportunities and make entry choices and post-entry decisions, to overcome context-specific challenges, using Cameroon as an empirical setting. However, this model failed to take cognizance of key factors such as digital knowledge-based, digital business environment, digital skill, digital mindset and digital infrastructure in the development of successful digital entrepreneurship. It should be stressed that these factors are very crucial and must not be excluded in any entrepreneurship study of this nature. For example, Obeng (2010) observed that the need to create an enabling environment that promotes and encourages entrepreneurial activities is a crucial one. While Lucky (2013) argued that studies in entrepreneurship development that failed to consider environment, should be regarded as incomplete.

Figure 1. Ngoasong, (2017)

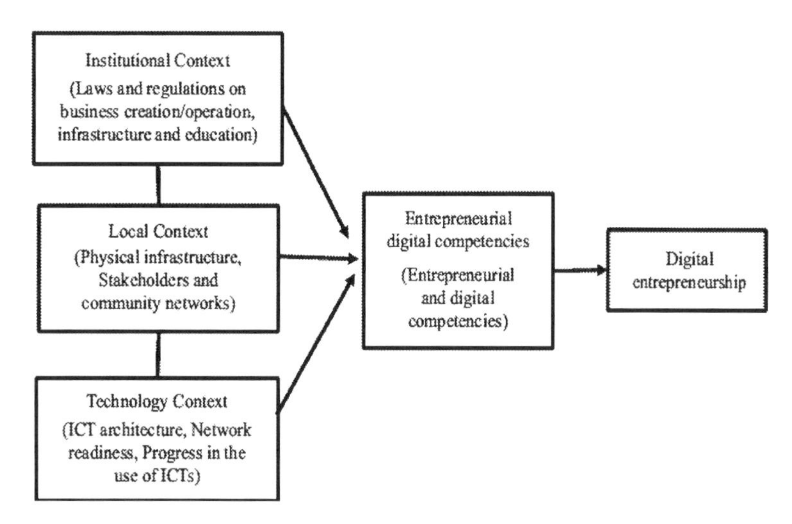

Leong, Pan and Liu Model (2016)

To also understand the concept of digital entrepreneurship, Leong, Pan & Liu (2016) developed a digital entrepreneurship model called digital entrepreneurship of born digital and grown digital firms. The model tends to differentiate between born digital firm and grown digital firm. They claimed that born digital firms are forward generative, while grown digital firms are past-future tension. At the middle of their relationship are digital options exploration and digital opportunity exploitation which affect each other accordingly. There are equally potential-instantiation effectuating and tension-reconciliation effectuating which revolve round to affect both born digital firm and grown digital firm accordingly. Effectuation describes a process of entrepreneurial actions based on the logic of entrepreneurial expertise and a dynamic and interactive process of creating new artifacts in the world (Sarasvathy 2008). However, Leong, Pan & Liu (2016) appear to be silent about the key factors such as digital knowledge-based, digital business environment, digital skill, digital mindset and digital infrastructure in the development of a successful digital entrepreneurship.

Giones and Brem Model (2017)

Giones & Brem (2017) developed a model of digital entrepreneurship which they described as a conceptual representation of a new type of technology entrepreneurship: digital technology entrepreneurship. The author's model describes the change in the meaning of "technology" as a continuum between the extremes represented by the commercialization of the latest scientific breakthroughs (e.g., a new material like graphene) and the latest application for smartphones (e.g., a new food delivery app). The conceptual model demonstrates the overlap between technology entrepreneurship and digital entrepreneurship. The overlap is caused by digital entrepreneurship. They argued that the concept of digital technology entrepreneurship is a combination of elements of technology and digital entrepreneurship. It is on this basis that the authors described technology entrepreneurship to include specific aspects related to this specific form of entrepreneurship: digital technology entrepreneurship is focused on the identification

Figure 2. Digital Entrepreneurship of born digital and grown digital firms

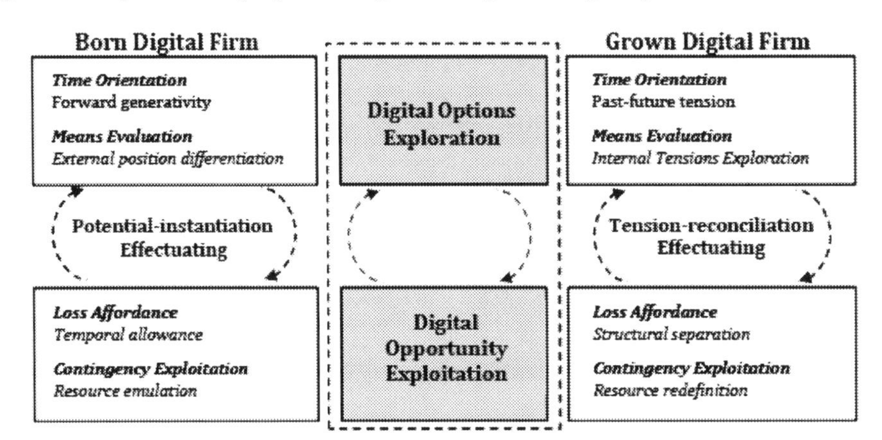

and exploitation of opportunities based on scientific or technological knowledge, through the creation of digital artifacts. Digital technology entrepreneurs build firms based on technologies on one hand, and on services on the other hand. Just like the previous model, this model was silent about the key factors such as digital knowledge-based, digital business environment, digital skill, digital mindset and digital infrastructure, in the development of successful digital entrepreneurship.

THEORETICAL FRAMEWORKS

One of the most important and relevant theories that can guide the understanding of digital entrepreneurship is knowledge-based theory. This theory is one of the theories associated with the Resource-Based View of the firm (RBV) initially promoted by Penrose (1959). Knowledge-based theory became popular due to the belief that resource-based view of the firm (RBV) does not go far enough to explain and differentiate knowledge. For instance, Resource-Based View of the firm (RBV) failed to distinguish between different types of knowledge-based capabilities; rather it only recognizes the important role of knowledge in firms. Accordingly, RBV only specifically treats knowledge as a generic resource, rather

Figure 3. Conceptual representation of a new type of technology entrepreneurship-Digital technology entrepreneurship

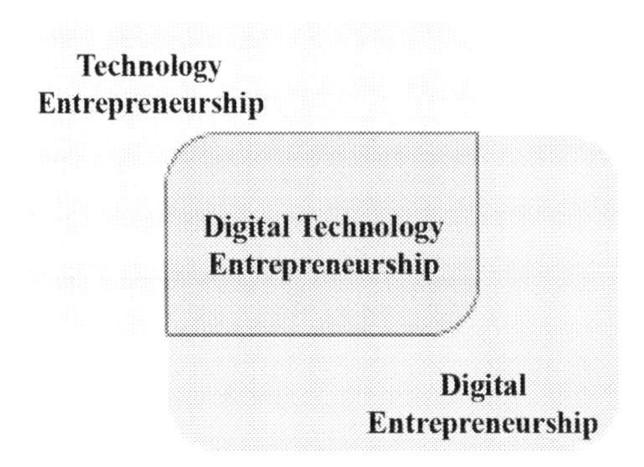

than having special characteristics. Unfortunately, this is not enough to explain the understanding of digital entrepreneurship. It is believed that Knowledge-Based View (KBV) of the firm is a recent extension of the RBV of the firm. It is accepted to be adequate to the present economic context of digital world. According to Curado (2006), KBV of the firm has attracted great interest, as it reflects that academia recognizes the fundamental economic changes resulting from cumulative and availability of knowledge in the past two decades.

In order to effectively guide the understanding of digital entrepreneurship, an emerging theory such as Knowledge-Based theory is required. Digital entrepreneurship is part of the Information and Communication Technology (ICT) that is shaping the world. Information technologies plays a crucial role in the knowledge-based view of the firm, in that information systems can be used to synthesize, enhance, and expedite large-scale intra- and inter-firm knowledge management (Alavi & Leidner 2001). Also, it is believed that heterogeneous knowledge bases and capabilities among firms are the major determinants of sustained competitive advantage and superior corporate performance, instead of RBV. The sustainability of digital entrepreneurship is determined by the amount of knowledge the digital entrepreneurs possess. Apart from that, disseminating the information equally depends on the knowledge of the person. Knowledge-Based theory is a theory that is relevant in linking knowledge to digitalization.

KBV posits that knowledge is the most strategically significant resource of the firm and that knowledge is embedded and carried through multiple entities, including organizational culture and identity, policies, routines, documents, systems, and employees. In the digital entrepreneurship field, knowledge is the most important and relevant resource as well as the only service they offer to their clients. Therefore, knowledge would help the digital entrepreneurs to grow, expand and sustain their business in the 21st century.

Digital Entrepreneurship and Developing/Emerging Economies

The conceptual framework was developed based on the peculiar nature of the developing countries which include Saudi Arabia. According to Ngoasong (2017), advancements in digitization, and the development of ICTs is creating opportunities for new type of entrepreneurial activities in emerging economies. Unfortunately, there are very few international business studies with theoretical models to understand and guide the nature of digital entrepreneurship in developing/emerging economies. Thus, the proposed conceptual model of this present study becomes very relevant in this respect.

Developing/emerging economies are often characterized by slow economic growth, low human capital index, low standard of living, low or middle per capita income, less developed industrial base, lack of rapid development in terms of industrialization and infrastructures etc. Saudi Arabia may not be excluded from this. There are countries that have not achieved a significant degree of industrialization relative to their populations, and have, in most cases, a medium to low standard of living. There is a strong association between low income and high population growth. Developing countries differ from the developed countries, because they still rely primarily on agriculture, have made less impressive gains in infrastructure and industrial growth, and are experiencing low or middle incomes and slow economic growth (Reynolds, 2017).

One greater focus of digital entrepreneurship is the creation of new enterprises. It presents and provides huge opportunities for potential young entrepreneurs through the creation of new ventures. The new enterprises created by digital entrepreneurship covers all new enterprises and existing firms that have been "transformed" or converted, and that create economic and social value through digital technologies. The industry has a very huge capacity to absorb a large and huge population of young people who are ready to become entrepreneurs. This is in line with the crucial and pressing need of many developing countries; which is to provide employment for their many unemployed youths, who have among them those that are graduates from their various tertiary institutions (Lucky, 2015). These young entrepreneurs would in time create new businesses that would hire employees. They create jobs and these economic opportunities uplift and support communities by increasing the quality of life and overall standard of living (Sappin, 2016). Achieving this objective requires that these emerging/developing economies take into account the variables presented by the conceptual model in figure 1. It implies that variables such as digital knowledge based, digital business environment, digital skill, digital mindset and digital infrastructures should be considered for successful digital entrepreneurship in emerging/developing economies.

Secondly, the model provided in this paper would boost development by promoting commerce and regional economic integration. You may ask how? It is very possible because of technology which digital entrepreneurship depends on. Digital entrepreneurship is the creation of new venture opportunities presented by new media and internet technologies; with digital entrepreneurship, small entrepreneur-led businesses in developing countries would have opportunities to expand into regional and global markets. According to (Sappin, 2016), whenever these new digital businesses export goods and services to other regions or countries, they contribute directly to a region's productivity and earnings. There is no doubt, this increase in revenue strengthens an economy and promotes the overall welfare of its population. It has been observed that economies that trade with one another are economically better off. Also, digital entrepreneurship is an easy way to engage in regional and international trade, which in turn promotes investment in regional transportation and infrastructure, and then strengthens economies. Countries such as the United States, China etc. are greatly benefiting from this. Foreign trade, according to some estimates, is responsible for over 90 percent of the world economic growth.

Finally, the conceptual framework emphasizes on factors such as digital knowledge base, digital business environment, digital skill, digital mindset and digital infrastructure. Unfortunately, many developing countries lack all these factors needed for the development of successful digital entrepreneurship. There is no doubt, digital entrepreneurs bring new ideas to life in a digital way; however, their flourishing condition depends on key factors aforementioned above. For example, digital entrepreneurs need education and training to enable them possess the required digital skills for the jobs. The government can play a major role in this aspect, by providing them with training and by so doing, they can raise the human capital index of their countries.

CONCLUSION

This paper examined digital entrepreneurship with particular attention to developing/emerging economies. It identified and discussed the key variables needed to develop and promote successful digital entrepreneurship. Therefore, variables such as digital knowledge - based, digital business environment, digital skill, digital mindset and digital infrastructures, were identified and the extent and how they contribute to the development of a successful digital entrepreneurship in developing/emerging economies was equally discussed. Evidence from the previous studies (Giones & Brem, 2017; Leong, Pan & Liu, 2016; Ngoasong, 2017; Sappin, 2016) helped to establish a strong correlation between identified factors and their contributions to the successful development of a digital entrepreneurship particularly in developing/emerging economies.

Figure 4. Developing a conceptual framework to examine and guide the understanding of a successful digital entrepreneurship development in the 21st century.

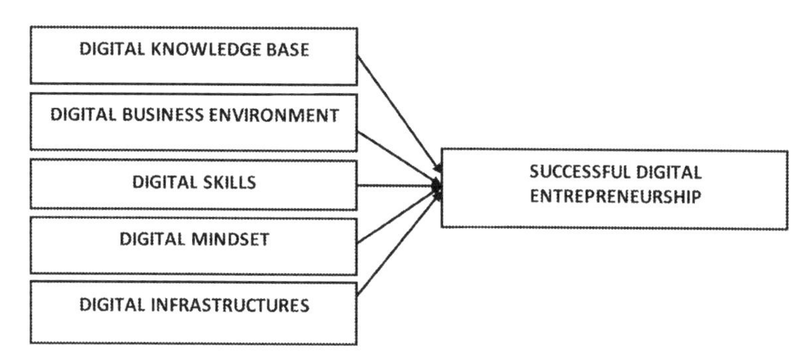

From the ongoing, this paper provided two major contributions to practice and research in the area of digital entrepreneurship. First and foremost, this paper documented the relevance of digital entrepreneurship as an area of inquiry within entrepreneurship and elucidated the potential implications of such research for innovators and entrepreneurs in industries across the digital spectrum.

Secondly, the paper provided a conceptual framework for better understanding and the relationships that exist between different dimensions and components that determine successful digital entrepreneurship in this 21st century. These variables arguably greatly influence successful digital entrepreneurship. In so doing, practitioners and academia will not neglect the relevant factors or variables in their study of digital entrepreneurship. Conclusively, digital entrepreneurship, as an emerging area of entrepreneurship discipline, is creating opportunities for new types of entrepreneurial activities in developing/emerging economies; it however lacks a theoretical model to guide its understanding in the developing/emerging economies. This is therefore being taken care of in this present paper.

REFERENCES

Alavi, M., & Leidner, D. E. (2001). Review: Knowledge management and knowledge management systems. *Management Information Systems Quarterly*, *25*(1), 107–136. doi:10.2307/3250961

Aldrich, H. (2014). The democratization of entrepreneurship? Hackers, makerspaces, and crowdfunding. In Annual Meeting of the Academy of Management, Philadelphia, PA, October 27.

Archer, S., Hull, L., Soukup, T., Mayer, E., Athanasiou, T., Sevdalis, N., & Darzi, A. (2017). Development of a theoretical framework of factors affecting patient safety incident reporting: a theoretical review of the literature. *BMJ open*, *7*(12). Digital entrepreneurship. Retrieved from https://www.slideshare.net/GiovanniConigliaro/digital-entrepreneurship-67720006

Davidson, E., & Vaast, E. 2010.Digital Entrepreneurship and its Sociometrical Enactment. *Paper presented at 43rd Hawaii International Conference on System Sciences (HICSS)*, January 5-8.

Delacroix, E., Parguel, B., & Benoit-Moreau, F. (2018). Digital subsistence entrepreneurs on Facebook. *Technological Forecasting and Social Change*. doi:10.1016/j.techfore.2018.06.018

Denison, E. (2017). Driving prosperity in the digital era. The role of digital talent, innovation and entrepreneurship. Retrieved from https://www2.deloitte.com/de/de/pages/about-deloitte/articles/prosperity-digital-era.html

Digital Skills Academy. (2017). Ten Traits of Successful Digital Entrepreneurs. Retrieved from https://digitalskillsacademy.com/blog/10-traits-of-successful-digital-entrepreneurs

Entrepreneur.com. (October 20, 2016). 7 Ways Entrepreneurs Drive Economic Development. Retrieved from https://www.entrepreneur.com/article/283616

Forrester, D. (2017). Digital knowledge manager: 5 Skills You Need to Succeed at the Newest Marketing Role. Retrieved from https://www.entrepreneur.com/article/299178

Giones, F., & Brem, A. 2017. Digital Technology Entrepreneurship: A Definition and Research Agenda. *Technology Innovation Management Review, 7*(5), 44–51. Retrieved from http://timreview.ca/article/1076

Hair, N., Wetsch, L. R., Hull, C. E., Perotti, V., & Hung, Y.-T. C. (2012). Market Orientation in Digital Entrepreneurship: Advantages and Challenges in A Web 2.0 Networked World. *International Journal of Innovation and Technology Management*, *9*(6), 1250045. doi:10.1142/S0219877012500459

Hatch, M. (2013). *The maker movement manifesto: Rules for innovation in the new world of crafters, hackers, and tinkerers*. New York: McGraw-Hill.

Javalgi, R. G., Todd, P. R., Johnston, W. J., & Granot, E. (2012). Entrepreneurship, muddling through, and Indian Internet-enabled SMEs. *Journal of Business Research*, *65*(6), 740–744. doi:10.1016/j.jbusres.2010.12.010

Leadem, R. (2018). The History of Digital Content (Infographic). How did we get to where we are today? https://www.entrepreneur.com/article/309740

Leong, C., Pan, S. L., & Liu, J. (2016). Digital Entrepreneurship of Born Digital and Grown Digital Firms: Comparing the Effectuation Process of Yihaodian and Suning Research-in-Progress. In *Thirty Seventh International Conference on Information Systems*, Dublin, Ireland.

Lucky, E. O.-I. (2013). Exploring the ineffectiveness of government policy on entrepreneurship in Nigeria. *International Journal of Entrepreneurship and Small Business*, *19*(4), 471–487. doi:10.1504/IJESB.2013.055487

Lucky, E. O.-I., Rahman, H. A., & Minai, M. S. (2013). *A conceptual framework for a successful co-operative entrepreneurship development. In Handbook of entrepreneurship and co-operative development. Co-operative and Entrepreneurship Development institute (CEDI), Universiti Utara Malaysia (UUM)*. Malaysia: Sintok.

Nambisan, S. (2016) Digital Entrepreneurship: Toward a Digital Technology Perspective of Entrepreneurship. SAGE Publications Inc. doi:. doi:10.1111/etap.12254

Nambisan, S. (2017). Digital entrepreneurship: Toward a digital technology perspective of entrepreneurship. *Entrepreneurship Theory and Practice*, *41*(6), 1029–1055. doi:10.1111/etap.12254

Ngoasong, M. Z. (2017). Digital entrepreneurship in a resource-scarce context: A focus on entrepreneurial digital competencies. *Journal of Small Business and Enterprise Development*. doi:10.1108/JSBED-01-2017-0014

O'Sullivan, A., & Sheffrin, S. M. (2003). *Economics: Principles in action*. Upper Saddle River, NJ: Pearson Prentice Hall.

Onetti, A., Zucchella, A., Jones, M., & McDougall-Covin, P. (2012). Internationalization, innovation and entrepreneurship: Business models for new technology-based firms. *The Journal of Management and Governance*, *16*(3), 337–368. doi:10.100710997-010-9154-1

Prodanov, H. (2018). Social Entrepreneurship and Digital Technologies. *Economic Alternatives*, (1), 123-138.

Reynolds, J. (September 26, 2017). Difference Between Developing Countries & Emerging Countries. Retrieved from https://bizfluent.com/info-10002682-difference-between-developing-countries-emerging-countries.html

Santana, M. (2017). *Digital entrepreneurship: expanding the economic frontier in the Mediterranean*. European institute of the Mediterranean.

Sarasvathy, S. D. (2008). *Effectuation: Elements of entrepreneurial expertise*. Cheltenham, U.K.: Edward Elgar Publishing. doi:10.4337/9781848440197

Tumbas, S., Berente, N., Seidel, S., & Brocke, V. J. (2015). The 'digital façade' of rapidly growing entrepreneurial organizations. In *International Conference on Information Systems*, Fort Worth, TX.

Vasilchenko, E., & Morrish, S. (2011). The role of entrepreneurial networks in the exploration and exploitation of internationalization opportunities by information and communication technology firms. *Journal of International Marketing*, *19*(4), 88–105. doi:10.1509/jim.10.0134

Zhao, F., & Collier, A. (2016). Digital entrepreneurship: Research and practice. In *9th Annual conference of the EuroMed academy of business*, Warsaw, Poland, September 14–16.

KEY TERMS AND DEFINITIONS

Digital: A new industrial revolution and the good thing is that digital entrepreneurship is its primary component.

Digital Entrepreneur: Provides the basis for future-oriented jobs, greater resource efficiency, and responsible economic activity.

Digital Entrepreneurship: The creation of new venture from opportunities presented by new media and internet technologies.

Chapter 9
Digitalization and Growth of Small Businesses

Indira Ananth
LIBA, India

Dananjayan Madhava Priya
LIBA, India

ABSTRACT

On November 8, 2016, Government of India declared demonetization of all Rs. 500/- and Rs.1000/- currency notes towards a cashless society and create a digital India. The point of sale (PoS) and prepaid instruments are the most popular systems currently installed by merchants and service providers for receiving payments from customers. The primary focus of the study is to understand the adaptability, affordability, acceptability, and sustainability of the payments system as seen from the point of view of small merchants. A total of 221 responses were collected in Chennai. Results show that cash remains the most preferred mode for business. It is required for the working capital, payment of employee remuneration, wages, and others. With regards to the use of payment systems such as POS and prepaid instruments, awareness needs to be created of the benefits in having non-cash transactions. Improving credit worthiness and eligibility to receive loans from banks is one such benefit which would convince the merchants. However, too many systems could confuse the merchants and customers.

INTRODUCTION

The sudden announcement of demonetisation of higher denomination notes on 8th November 2016 in India led to a fast-tracking of the use of non-cash payment methods for all size of transactions. The move affected small businesses, consumers and producers alike. The disruption was severe for many because of high usage of currency in day to day transactions. Overnight, the situation forced people to opt for payment systems like prepaid instruments, mobile banking and cards for their daily transactions. The main non-cash systems used by merchants were Point of Sale (PoS) terminals and Prepaid Payment Instruments (PPIs). The usage in terms of volume and value of transactions of these increased tremendously for a short time. However, as soon as the currency level improved the usage of these payment systems

DOI: 10.4018/978-1-5225-7262-6.ch009

started decreasing gradually. Could this mean that there is a preference for cash payments? What does this bode for non-cash payment measures?

With the coming of computerization, there have been studies to understand the how and why of its adoption among small businesses for their office operations. Henry M. Levin and Russell W. Rumberger (1986) found that small businesses regarded very highly education and training requirements for the adoption and use of computers. The study clearly emphasized reading, comprehension skills, and reasoning skills, rather than mathematics skills, formal computer training, or prior computer experience.

Small merchants are the decision making authority about the investment and strategy of information systems(Thong and Yap, 1995, Chau, 1995 and Ekanem, 2005). The technology adoption is determined by factors like sector, complexity, size, status, assertiveness, rationality and interaction with organisational strategy(Julien and Raymond, 1994). It was seen that the software selection criteria among small businesses varied between the owners and the managers (Chau, 1995; Raghavan V., Wani M., Abraham D.M. (2018). Owners focussed more on technical aspects while managers looked at non-technical aspects. Further it was seen that the extent of information systems adoption is mainly determined by organizational characteristics. The environmental characteristic of competition had no direct effect on small business adoption of information systems(Thong, 1999).

Milind K. Sharma (2009) surveyed the receptivity of India's small and medium-sized enterprises to the adoption of information systems utilising a portion of Rogers' model of innovation diffusion as the framework, and treating information systems as a form of new innovations. It was seen that the factors affecting the willingness to adopt information systems by SMEs significantly were relative advantage, compatibility and trialability.

Business strategies seemed to have changed for small businesses with the large-scale digitization in the new economy with specific themes – virtualization, molecularisation and disintermediation (Terence Tse and Khaled Soufani, 2003). Even so the adoption of new technologies for payments remained modest (Niina Mallat and Virpi Krishtiina Tuunainen, 2008; Francisco Liébana-Cabanillas, Iviane Ramos de Luna and Francisco Montoro-Ríos, 2017). The main adoption drivers were related to the means of increasing sales or reducing the costs of payment processing, whereas the barriers to adoption included complexity of the systems, unfavorable revenue sharing models, lack of critical mass, and lack of standardization.

The extant literature shows that despite the potential and wide availability, newer technologies are not yet widely used by small merchants. The Point of Sale (PoS) is the focus of this paper as they are predominantly used in transactions involving sale of goods and services. These are the most popular systems installed by merchants and service providers for receiving payments from customers. The paper tries to understand the adaptability, affordability, acceptability and sustainability of the PoS systems available to small merchants and the difficulties faced by them in using them. The paper is based on the survey done in the business district of Chennai, capital city of Tamil Nadu, a state in South India.

Accordingly this paper is structured as follows: Section 1 gives a brief description of the payment systems available in India; Section 2 discusses the research method used to understand the use of PoS among the small businesses; Section 3 is the data analysis and finally the conclusions.

PAYMENT AND SETTLEMENT SYSTEMS IN INDIA

Payment and settlement systems play an important role in improving overall economic efficiency. They consist of all the diverse arrangements used to systematically transfer money-currency, paper instruments such as cheques, and various other electronic channels accepted by the Reserve Bank of India. The payment system available in a country should be easy to use, be efficient and cost effective such that it helps in increasing the level of goods and services traded in the country. The level of confidence the public have in a payment system reflects the level to which it is embraced by the country.

In India, the Reserve Bank of India is responsible for managing the payment and settlement system. The payment and settlement system in India can be classified under three heads namely,

Paper-Based Payments

The paper-based payment and settlement system are the traditional system which is well established as an important system. The paper-based payments system in India includes cheques, drafts and other such similar paper based systems. It accounted for nearly 60% of the total non-cash payment system in terms of volume during the financial year 2010-11. But it has reduced drastically and accounts for only 6% of the non-cash transactions, because of the availability of many more easy options of payment systems.

Electronic Payments

The advancements in computer and communication technology have enabled the introduction of electronic payment systems as they were cost effective, easy to use and fast. They account for around 22% of all non-cash transaction in terms of volume currently. The various electronic payment systems include Electronic Clearing Services (ECS), Electronic Funds Transfer (EFT), National Electronic Funds Transfer (NEFT) system, Real Time Gross Settlement (RTGS) system and Clearing Corporation of India Limited (CCIL).

Other Payment Systems

In terms of volume they account for 70% of all non-cash payment in India currently. These are very popular and used by people in everyday life for purchasing goods and services. The convenience, ease of use and acceptance among traders has made these payment systems an integral part of everyday transactions. These payment systems include pre-paid payment systems, mobile banking systems, ATMs, Online transactions and Point of Sale (POS) terminals.

Point of Sale (PoS) refers to the terminal used to pay for goods and services. It includes both the physical use of cards as well the use of cards for online transactions. The number of Point of Sale (PoS) terminals in India has been increasing because of the ease of use, customer preference. Hence it has become a 'must have' for all merchants.

The number of point of sale terminals in India currently is around 25 lakhs as shown in the Figure 1. During October 2016, there were only 15 lakh POS machines in India. But the necessity created by demonetization and the initiatives of government like removing all taxes and making the import of POS easier resulted in additional 10 lakh machines being installed in the last 5 months of the FY 2016-17.

The POS machine can be categorized into 3 types:

1. **Mobile POS:** It requires a smartphone with an internet connection. No extra is charged for device cost. Monthly maintenance cost starts at Rs.150/-. It accepts payments from all credit cards, debit cards, mobile wallets and UPI enabled apps. It has minimal cost for hardware maintenance.

Figure 1. The actual number of POS in India year wise
Source: Created for the study

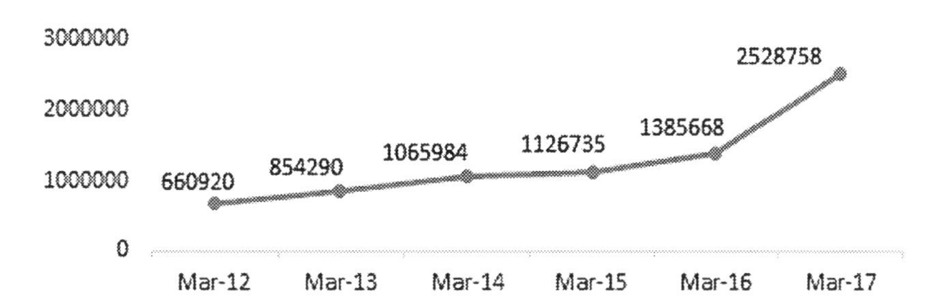

2. **Handheld POS:** It requires a sum card with data plan. The device cost is as low as Rs.6000/-. Monthly maintenance cost starts at Rs.700/-. It accepts payments for all credit cards and debit cards. High cost for maintenance and repair of hardware, software and security upgrades.
3. **Fixed Line POS:** It requires a fixed phone line. The device cost is Rs.8000 to 10,000/-. Monthly maintenance cost starts at Rs.700/-. It accepts payments for all credit cards and debit cards. High cost for maintenance and repair of hardware, software and security upgrades.

For acquiring a POS system, the merchants need to approach a bank which offers the service. Each bank has some pre-requisite conditions that need to be satisfied by the merchant. Most of these conditions are similar for all the banks.

- The merchant should have an active business. Some banks would require that the business is running for a specific number of days or the turnover per month is above a specific limit.
- The merchant should submit Identity proof, Address proof, Business establishment proof as part of the Know Your Customer (KYC) compliance.
- All the merchants who require POS to be installed need to have current account with the bank.
- The merchant can opt for any of the three types of POS.
- In case of fixed line POS/Public Switched Telephone Network(PSTN) the landline connection should be available with the merchant else the merchant should opt for the GPRS based POS.
- The merchant and bank should come to an agreement regarding the Merchant Discount Rate (MDR) and settlement period. Usually it is T+1 day.
- The POS would be installed only after the signing of Merchant Establishment Agreement between the bank and the merchant.

The mobile POS system is offered apart from banks by payment solution companies such as mswipe, Paynear, ezetap and others which are currently dominating this space. The mPOS can accept payment by both card as well as mobile wallets.

These companies have tie ups with banks for the as the transaction process can take place only through acquiring banks. In this system, the mobile payment application would be installed on the mobile which would be linked to the POS. All the transactions are recorded online via the application. Hence a smart phone with internet connectivity is mandatory. The volume due to aggregation of transaction allows the companies to get a better Merchant Discount Rate from the banks. Some companies have tie-ups with many banks. This allows them to have more ON-US transactions and thus transaction charges are reduced.

For security, all POS machine should adhere to Payment Card Industry-Data Security Standards (DCI-DSS) and Payment Application-Data Security Standards (PA-DSS).

A look at the numbers in Table 1 shows how the technology enabled transactions are growing. ATMs and PoS devices are the most commonly accessed touchpoints. It can be seen that the infrastructure for online PoS and the number of transactions of PoS has grown over the years.

However, given the scale and diversity of India, the acceptance infrastructure is still largely under-developed with 15 ATMs per 100,000 adults and 1.2 million PoS terminals for an estimated 14 million merchants. By 2020 around 40% of merchants are expected to have electronic payment acceptance devices. The issuance gap that exists in PoS terminals is quite high. This would ultimately lead to acceptance problems. One way to tackle this would be to take advantage of the existing infrastructure of smartphones and work toward a virtual mobile point of sale (mPoS) solution that converts a merchant's smartphone into a virtual PoS device.

Card Payment System

In India there are of two types of cards based on their characteristics namely

Table 1. Technology-enabled touchpoints and transactions over the years

Detail	2012	2013	2014	2015	2016	Growth
Infrastructure						
Onsite ATMs	47,545	55,760	83,379	89,061	101,950	114%
Offsite ATMs	48,141	58,254	76,676	92,337	97,149	102%
Online PoS	**647,869**	**840,983**	**1,050,323**	**1,126,389**	**1,385,342**	**114%**
Offline PoS	13,051	13,307	15,661	346	326	–98%
Total touchpoints	756,606	968,304	1,226,039	1,308,133	1,584,767	109%
Credit Cards						
Outstanding credit cards	17,653,818	19,538,329	19,181,567	21,110,653	24,505,219	39%
Transactions at ATMs	202,106	225,770	296,548	437,278	612,531	203%
Transactions at PoS	**28,744,710**	**35,616,482**	**46,105,415**	**56,906,942**	**72,220,394**	**151%**
Amnts ` million at ATM	1,209	1,493	1,662	2,344	2,803	132%
Amnts ` million at PoS	**88,374**	**111,217**	**145,487**	**178,988**	**226,943**	**157%**
Debit Cards						
Outstanding debit cards	278,282,839	331,196,720	394,421,738	553,451,553	661,824,092	138%
Transactions at ATMs	471,031,623	482,004,645	571,497,661	624,205,135	731,722,405	55%
Transactions at PoS	**30,668,922**	**45,376,619**	**56,981,333**	**76,105,726**	**112,868,336**	**268%**
Amnts ` million at ATM	1,317,168	1,556,406	1,796,099	1,987,480	2,245,822	71%
Amnts ` million at PoS	**46,534**	**66,873**	**85,771**	**108,283**	**134,632**	**189%**

Source: ATM/PoS/Card Statistics, https://rbi.org.in/scripts/ATMView.aspx?atmid=61, accessed on July 14, 2017.

- **Debit Cards:** They do not offer any credit to the customer. The value of the card is limited by the balance amount present in the associated bank account.
- **Credit Cards:** They offer credit to the customer. The value of the card is limited by the credit limit that is set for each card.

Table 2 shows the growth of credit and debit cards in India. The cards have become a popular payments and settlement system in India and are spread widely throughout India.

Debit cards are the most popular card in India. Over the last decade, the number of debit cards has been constantly increasing. At present, there are 28 debit cards in circulation for every credit card (28:1).

The advantage of cards is that they can be utilized for withdrawing cash from the Automated teller machine (ATM) as well and they are very easy to use. Hence the usage of cards in terms of volume and value has increased tremendously over the years. The number of ATMs and PoS terminals have also increased and made card system an essential part of the economy.

The usage of cards by volume has increased from around 15 million in March of 2005 to 1087 million in March of 2017. In terms of value the card usage has increased from Rs.29 billion to Rs.2947 billion when compared between March 2005 and March 2017. Over the past 12 years it has increased almost by 100 times.

Card Usage at Point of Sale (PoS)

Cards can be categorized by usage into 2 categories

- Usage at Point of Sale (PoS)
- Usage at ATM.

Table 2. Number of debit and credit card circulation in India over the years (2007-17)

	Credit Card		Debit Card	
	Volume (Million)	YoY Growth %	Volume (Million)	YoY Growth %
Mar-17	29.84	22%	854.87	29%
Mar-16	24.51	16%	661.54	20%
Mar-15	21.11	10%	553.45	40%
Mar-14	19.18	-2%	394.42	19%
Mar-13	19.55	11%	331.20	19%
Mar-12	17.65	-2%	278.28	22%
Mar-11	18.04	-2%	227.84	25%
Mar-10	18.33	-26%	181.97	32%
Mar-09	24.70	-10%	137.43	34%
Mar-08	27.55	19%	102.44	37%
Mar-07	23.12		74.98	

Source: www.rbi.org.in

The focus of our study is the usage of cards at Point of sale terminals as these reflect the cards usage in making purchases.

The usage of cards at PoS has increased manifold over the last decade. The Reserve Bank of India has been improving the payments and settlement system in terms of cost to customer, security to encourage their use. The advancements in technology have been leveraged to delivery benefits to the customers.

Even though the number of debit cards is around 28 times the number of credit cards, the usage of cards in terms of value is almost similar. Till the financial year 2015-16 the usage of credit cards in terms of value was more than that of debit cards. The value versus volume usage of cards shows that the transaction value for each credit card usage is more than that of debit card.

As per March 2017 data each credit card transaction was worth Rs. 3096, whereas each debit card transaction was worth Rs.1316 as shown in the Table 3. This shows that the debit cards are used even for low value transactions.

However, in terms of volume the usage of debit cards is more than credit cards. As per the latest data the usage of debit cards is almost 2.5 times the credit cards in terms of volume.

PREPAID PAYMENT INSTRUMENTS

In this system of payment, one can purchase goods and services based on the value present in the instruments. This value is the amount paid by the holder to be stored in the instrument. The payments can be done by means of cash, debiting bank account or credit cards. The instruments can be in the form of paper voucher, card, mobile wallets.

As per Reserve Bank of India Guidelines for Prepaid Instruments, Prepaid Instruments are categorized into four categories

Table 3. Value of transaction using point of sale (PoS)

	Value in Rs. Per Credit Card Transaction at PoS	Value in Rs. Per Debit Card Transaction at PoS
Mar-05	2,095	1,291
Mar-06	2,113	1,289
Mar-07	2,589	1,608
Mar-08	2,518	1,363
Mar-09	2,389	1,340
Mar-10	2,766	1,554
Mar-11	2,976	1,571
Mar-12	3,075	1,517
Mar-13	3,123	1,474
Mar-14	3,156	1,505
Mar-15	3,145	1,423
Mar-16	3,142	1,193
Mar-17	3,096	1,316

Source: www.rbi.org.in

Figure 2. Volume of transaction using debit and credit cards in point of sale (PoS)

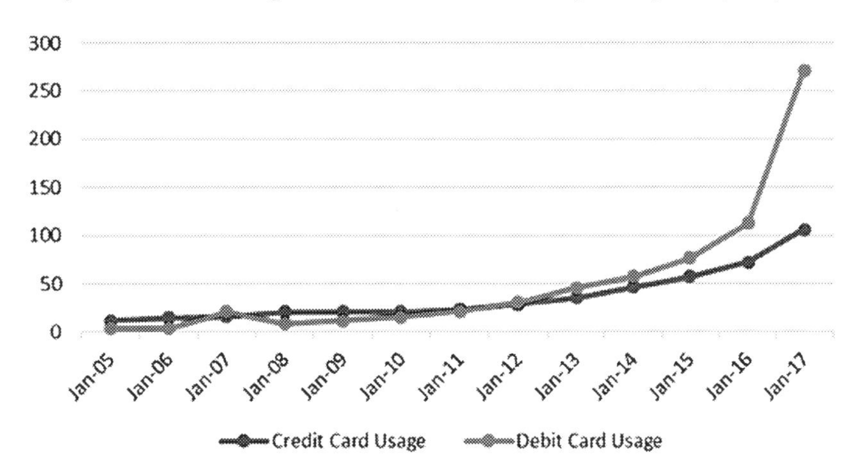

- **Open PPIs:** These are payment instruments which can be used for purchase of goods and services and they also permit cash withdrawal at ATMs. Vodafone M-pesa is an example of open PPI.
- **Semi-Open PPIs:** These are payment instruments which can be used for purchase of goods and services at any card accepting merchant locations (Point of sale terminals). These instruments do not permit cash withdrawal or redemption by the holder. Gift card and food cards issued by banks are semi-open PPIs.
- **Semi-Closed PPIs:** These are payment instruments which are redeemable at a group of clearly identified merchant locations/ establishments which contract specifically with the issuer to accept the payment instrument. These instruments do not permit cash withdrawal or redemption by the holder. Oxigen, Paytm, Mobikwick are examples of semi-closed PPIs.
- **Closed PPIs:** These are payment instruments generally issued by business establishments for use at their respective establishment only. These instruments do not permit cash withdrawal or redemption. Gift cards issued by branded retail stores, Flipkart wallet are examples of closed PPIs.

Many entities were issuing prepaid instruments in the country previously. Among these only banks were seeking Reserve Bank of India's authorizations. However, with the passage of Payments and Settlement systems, Act 2007 all entities which were issuing pre-paid cards and those planning to issue the cards had to get Reserve Bank of India's approval. The usage of PPIs in India by volume and value has shown manifold increase during the last financial year.

As on 6th June 2017 there were 55 entities approved by Reserve Bank of India to operate PPIs.

In terms of volume there has been a 374% in usage of PPIs in the last year and in terms of value there has been an increase of around 86% as shown in Table 4. This shows that the value per transaction of PPIs has come down. During March of 2016 the value per transaction was Rs.790 and this value has come down to Rs. 312 by March 2017. This shows that consumers have started using PPIs for even smaller transactions.

Table 4. Prepaid payment instruments volume and value

Month and Year	PPIs Volume (Million)	PPIs Value (Rupees Billion)
Mar-17	342.09	106.77
Mar-16	72.05	57.16
Mar-15	54.10	29.80
Mar-14	16.57	9.80
Mar-13	10.35	5.61

Source: www.rbi.org.in

Transaction Limits and Other Notifications

RBI has issued the limits on value that can be loaded on PPIs and on transactions that can be carried out using PPIs would have within these limits. These limits vary for each type of PPIs. The following limitations apply only for semi-closed wallets.

1. An amount up to Rs.10,000 can be loaded in mobile wallets by accepting minimum details from the customers. The amount outstanding and total value of reloads in any month also should not breach this limit.
2. An amount between Rs.10,001 and 50,000 can be loaded in the account if valid documents accepted under Prevention of Money Laundering (PML) . These PPIs are non-reloadable.
3. An amount upto Rs. 1,00,000 if full Know Your Customer (KYC). It is reloadable in nature.

During demonetization the limit for customers was increase to Rs.20,000 on a temporary basis.

The PPI issuer is mandated to set up adequate data security and information infrastructure. These steps are aimed at prevention and detection of frauds.

The issuers should disclose all charges, terms and conditions clearly to the customers. The language should be English, Hindi and a local language.

All these steps are taken by RBI to regulate the payment system and protect the interest of the customers. These steps help in increasing the customer's confidence in the system.

OBJECTIVE OF THE PAPER

This paper looks at the different types of payment system available for small merchants and the difficulties faced by them in the payment systems and its effect on the business. The objective is to understand the adaptability, affordability, acceptability and sustainability of the non-cash payment systems available at small merchants. Specifically, the paper seeks to look at the following questions:

- Is there a preference among merchants for cash?
- If so does this preference of the merchant for cash the reason for public preference to go for cash transactions?
- Are there any factors which affect the merchant's decision to whether install a payment system and continue using them?

RESEARCH METHODOLOGY

The study focuses on understanding small businesses, which is done by looking at small merchants shops' which were chosen as they would help in understanding the penetration of the payment system. From the product life-cycle approach it can be seen that any new technology or innovation gets adopted in stages. The new technology is first adopted by the innovators and early adopters. In the Indian context, the organized retail segment accounting for 10-20% of the overall retail market would be the first-movers. Small merchants would typically fall into the category of late majority when it comes to adoption of technology. The reason being they do not have the scale or volume to go for newer innovations or technologies without any tangible proof of benefit or necessity. Hence if a payment system is used by small merchants extensively it shows that the system has penetrated successfully.

The small merchants surveyed were categorized into groups to check if there is a variation among them with respect to the objectives of the study.

The small merchants were as follows:

1. Clothing, Textile, Shoes and other such similar shops
2. Departmental Stores
3. Hardware, Electrical other such similar shops
4. Pharmacy
5. Shops present at shopping malls

A total of 221 responses were collected among the 5 categories. The factors of adaptability, affordability, sustainability of the payments system as seen from the point of view of small merchants were studied. This study was based on both primary and secondary data. A well-structured questionnaire was developed and tested in order to confirm the relevancy, clarity and applicability. Population was infinite and hence, the sample has selected as 221 by convenience sampling techniques among the prominent business areas in the capital city of Chennai, state of Tamil Nadu one of the largest urban agglomerations in India. The entire selected sample were personally contacted and interviewed.

Hypothesis of the Study

Based upon the objectives, the following hypotheses was tested:

Hypothesis 1: Organisational characteristics of the small firms are negatively related to their adoption of PoS system.
Hypothesis 2: Adaptability is positively related to the adoption of PoS system.
Hypothesis 3: Affordability is positively related to the adoption of PoS system.

Hypothesis 4: Acceptability is positively related to the adoption of PoS system.
Hypothesis 5: Sustainability is positively related to the adoption of PoS system.
Hypothesis 6: Perception of small merchants towards PoS is positively related to the adoption of PoS system.

DATA ANALYSIS AND DISCUSSION

Profile of Small Merchants

The sample had 221 small merchants, out of which 57.9 percent of the shops were located on the main road, 23.5 percent on the side roads and 18.6 percent in the malls. Out of 221 respondents 88 were Textile shops, 49 Departmental Stores, 49 Hardware stores and 35 Pharmacies respectively.

Around 42.1 percent of the store owners were in the age group of 30-40 years, 30.3 percent around 40-50 years and 17.2 and 10.4 percent of the owners were aged around 50-60 and 20-30 respectively as shown in Table 5.

Types of Payment System Used by Small Merchants

There were three types of payments system currently in use by merchants. They were card payments, mwallets and mobile banking. Overall there were 215 merchants having card payments system, 76 merchants having mwallets and 27 merchants having mobile banking system.

Table 5. Profile of small merchants

Location of Shop	Frequency	Percent
Main Road	128	57.9
Side Road	52	23.5
Mall	41	18.6
Total	221	100.0
Type of Shop	**Frequency**	**Percent**
Textile shops	88	39.8
Departmental Stores	49	22.2
Hardware Stores	49	22.2
Pharmacy	35	15.8
Total	221	100.0
Age of Owner	**Frequency**	**Percent**
20-30	23	10.4
30-40	93	42.1
40-50	67	30.3
50-60	38	17.2
Total	221	100.0

Figure 3. Types of payment system used by small merchants

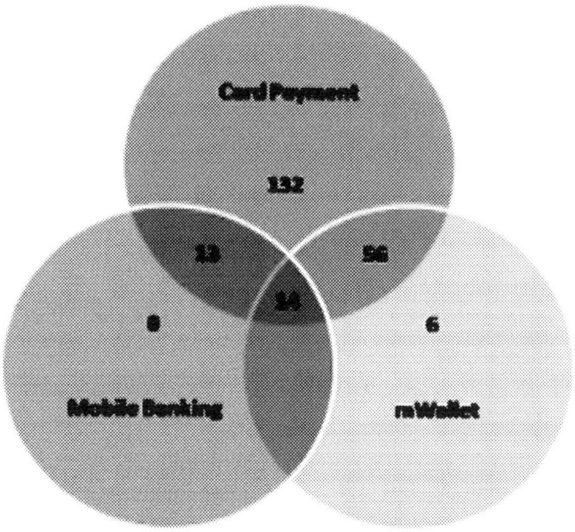

In percentage terms 97% of all merchants surveyed had card payment option, whereas for mwallets and mobile banking the percentage of merchants offering the system was only 34% and 12.2% respectively.

Figure 3 shows the combination of systems used by the vendors, it was found that the card payments and the mwallet was the most preferred combination when compared to others.

Adaptability, Affordability and Sustainability

The mwallets and mobile banking usage at small merchants was very negligible. Hence the adaptability, affordability and sustainability was studied only for card payments system.

Figure 4 shows the various parameters of card usage perception. The Adaptability of card system was studied on parameters like the ease of use, support from vendor, ease of learning to use the system and ease of integration. The score of these parameters showed that the adaptability score of the card system was good, though support from vendor score could be improved.

Figure 4. Card usage perception scores

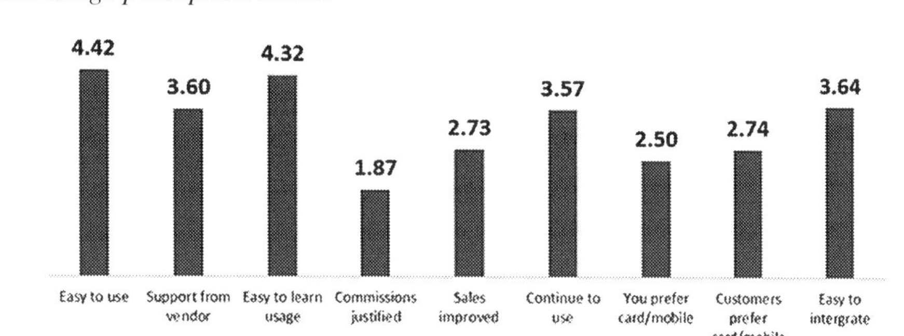

The Affordability of the card system can be gauged from commission justified and sales improved. The low score in the commission question showed that the merchants felt that the bank charges were too much for the transactions. The increasing sales in value and volume could improve the bottom line of the merchants and make the cost of the system affordable. According to the merchant's perception the sales did not drastically improve because of the installation of the card system.

The merchants would continue to use the system as it had become an essential part of customer's payment method. Most of the customers now had either a debit card or credit card. The issuance of Rupay card to Pradhan Mantri Jan Dhan Yojana accounts holders has further increased the presence of cards in India. This is reflected in the high score.

However, would the public be using the payment using plastic money extensively? The preference of merchants and customers as seen from the merchant's point of view does not paint a positive picture. The low score shows the lack of preference for non-cash payment systems.

Independent samples Kruskal-Wallis test (conducted at 0.05 significance level) was conducted to check if there were any variations among the parameters with respect to the type of merchants.

Since the significance level scores were all above 0.05 as shown in Table 6, null hypothesis is accepted. The distribution is same across different types of merchants. All the merchants had a similar perception when it came to adaptability of POS system.

The above Table 7 shows the significance level scores and the null hypothesis is accepted. There is no difference between different types of merchants when it comes to affordability of the small merchants towards the PoS system. However, the impact of installing POS system on sales shows a difference among different merchant types.

With respect to sustainability parameters there seems to be differences in the perception of the different types of merchants as shown in the table 8. The null hypothesis is rejected in all the cases and can be inferred that the small merchants are not able to sustain the Point of Sales system in the process of digitalization.

Table 6. Adaptability among merchants

	The POS System is Easy to Use	**There Is Enough Support From Vendor**	**The POS System Usage Is Easy to Teach**	**The POS Is Easy to Integrate With**
Chi-Square	5.423	4.066	1.598	1.797
df	3	3	3	3
Asymp. Sig.	**.143**	**.254**	**.660**	**.615**

Table 7. Affordability among merchants

	The Commission Payed Is Justified	**Sales Has Improved After Installation**
Chi-Square	5.017	27.912
df	3	3
Asymp. Sig.	**.171**	**.000**

Table 8. Sustainability among merchants

	I Will Continue to Use the System	The Merchant Prefers Card/ Mobile	The Customer Prefers Card/ Mobile
Chi-Square	13.478	10.853	41.468
Df	3.000	3.000	3.000
Asymp. Sig.	**0.004**	**0.013**	**0.000**

Table 9. Adaptability between shops at malls and others

	The POS System Is Easy To Use	There Is Enough Support From Vendor	The POS System Usage Is Easy to Teach	The POS Is Easy to Integrate With
Mann-Whitney U	862.000	618.000	846.000	884.000
Wilcoxon W	2240.000	1996.000	2224.000	2262.000
Z	-0.741	-3.056	-0.964	-0.496
Asymp. Sig. (2-tailed)	**0.459**	**0.002**	**0.335**	**0.620**

All the adaptability parameters are same across malls and roads except the question regarding the support from vendor. Only the significance level for question 'on support' is lesser than 0.05 and hence null hypothesis can be rejected. Shops in malls seem to get greater support from the banks as shown in Table 9.

On the face of it, with regards to affordability namely commission and sales improvement seem to be same across both shops at malls and roads. However, the significance level of sales improved is just above the critical point (Table 10). If the significance level is increased, then the null hypothesis needs to be rejected in sales improved parameter also.

On sustainability parameters, the distribution of small merchant's continuation of the system varies across shops at malls and roads (Table 11). The preference of merchants and customers as seen from merchant's point of view is similar.

To understand the variations in distribution, the means of these parameters which differed between malls and roads was compared. This helps in understanding which parameters score higher in which location (Table12). The support from banks and sales seems to have a better score for shops located in malls. Despite this the score relating to 'continue to use the system' fares poorly at malls when compared to shops at roads.

Table 10. Affordability between shops at malls and others

	The Commission Paid Is Justified	Sales Has Improved After Installation
Mann-Whitney U	911.500	733.000
Wilcoxon W	1577.500	2111.000
Z	-0.221	-1.951
Asymp. Sig. (2-tailed)	**0.825**	**0.051**

Table 11. Sustainability between shops at malls and others

	I Will Continue to Use the System	The Merchant Prefers Card/ Mobile	The Customer Prefers Card/ Mobile
Mann-Whitney U	722.000	820.500	714.500
Wilcoxon W	1388.000	1948.500	1344.500
Z	-2.085	-0.021	-1.294
Asymp. Sig. (2-tailed)	**0.037**	**0.983**	**0.196**

Table 12. Comparison of the means between the shops at malls and roads

		Mean Rank	Sum of Ranks
There is enough support from vendor	Shops at Road	38.38	1996.00
	Shops at Mall	53.33	1920.00
Sales has improved after installation	Shops at Road	40.60	2111.00
	Shops at Mall	50.14	1805.00
I will continue to use the system	Shops at Road	48.62	2528.00
	Shops at Mall	38.56	1388.00

In order to find out the non-cash sales across different types of shops, one-way ANOVA test was done to compare the percentage of non-cash sales before and after demonetization with respect to the type of shop.

The distribution of percentage of non-cash sales varied between different type of shops before demonetization. The distribution percentage of non-cash sales is different at shops. The level of significance is less than 0.05 in both the cases (Table 13).

The above Table 14 shows the means of the different types of shops that shows the sales in non-cash forms is maximum at Clothing stores and minimum at Hardware stores.

Small Merchants perception towards benefits related to non-cash transactions were calculated (Table 15). Non-cash transactions would help the customer in getting credit from formal sources as it shows the credit worthiness of the merchant. But the awareness among merchants in this aspect is not at the desirable level.

Table 13. One-way ANOVA test for non-cash sales across different types of shops

		Sum of Squares	df	Mean Square	F	Sig.
% of non-cash transactions before demonitisation	Between Groups	30745.30	3.00	10248.43	86.10	0.00
	Within Groups	25828.68	217.00	119.03		
	Total	56573.98	220.00			
% of non-cash transactions now	Between Groups	41142.04	3.00	13714.01	91.49	0.00
	Within Groups	32526.96	217.00	149.89		
	Total	73669.00	220.00			

Table 14. Means calculation for non-cash transactions across different types of shops

		Mean	Std. Deviation	Std. Error
% of non-cash transactions before demonitisation	Cloth, Textile and Shoes	40.68	10.62	1.13
	Departmental Stores	20.51	14.08	2.01
	Hardware Stores	11.53	9.14	1.31
	Pharmacy	23.29	8.57	1.45
% of non-cash transactions now	Cloth, Textile and Shoes	49.94	11.58	1.23
	Departmental Stores	25.92	15.67	2.24
	Hardware Stores	16.43	9.57	1.37
	Pharmacy	30.43	11.72	1.98

Table 15. Perception towards benefits related to non-cash transaction

Type of Shop	Noncash Transactions Will Help in Getting Credit			% of Merchants Who Think Non-Cash Transaction Will Help Them in Getting Credit
	Yes	No	Not Sure/Blank	
Cloth, Textile	44	16	28	50.0%
Departmental Stores	11	21	17	22.4%
Hardware Stores	12	22	15	24.5%
Pharmacy	15	6	14	42.9%

The category of shops in which percentage of bank loans are more seem to agree that having non-cash transactions will help them in getting credit from formal source (Table 16).

Demonetization and Impact on Non-Cash Transactions

The effect of demonetization on non-cash transactions usage is analysed by comparing the difference in percentage of sales before and after demonetization with the annual average increase in non-cash transactions. The annual average increase in non-cash transactions was calculated by comparing non-cash transactions of a month with the same month of previous year. The average was calculated from the year 2011 to 2016. The value of transaction per unit of POS machine was used for the calculation.

Table 16. Sources of loans

Type of Shop	What Type of Loan Have You Taken				% of Merchants With Bank Loans
	Bank	Hand Loan	Others	No Loan	
Cloth, Textile	34	16	10	28	38.6%
Departmental Stores	14	17	0	18	28.6%
Hardware Stores	13	21	4	11	26.5%
Pharmacy	12	6	0	17	34.3%

The average annual increase in value of transactions per POS machine was found to be 7.8%. one sample T-test was conducted based on this parameter (Table 17).

The increase in POS transaction has not increased drastically post demonetization. The increase in POS transactions is similar to the annual average increase. The percentage increase in March 2016 when compared to March 2015 is similar to March 2017 when compared with March 2016 (Table 18).

CONCLUSION

This paper looked at a specific issue of adoption of Point-of-Scale (PoS) system among small merchants after demonetisation in India. The small merchants were affected and suffered acutely due to shortage of cash. It seemed that PoS could be a possible answer.

From the analysis it is found that although organisational characteristics were important, the more significant variable identified was the issue of affordability and sustainability.

Although the initiatives regarding PoS were good and important, they appeared to have very little impact, because of the lack of knowledge about its importance among the small merchants. The reasons were different depending on types of businesses, location, age group of owners.

It was seen that only the small merchants who knew about the technology and who realised how it helped them in reducing the cash transactions and brought in accountability were using the PoS. Others had not even got into the stage of using PoS. It shows that a strong knowledge about the PoS system and its effects is important to change this trend.

Finally, the government has been encouraging use of a variety of payments system post demonetization. However too many systems could confuse the merchants and customers. The focus needs to be just on a few.

Table 17. One sample T-test for average annual increase in transaction value per PoS

	Test Value = 7.8					
	t	df	Sig. (2-tailed)	Mean Difference	95% Confidence Interval of the Difference	
					Lower	Upper
Diff in non-cash transactions before and after demonetisation	-1.31	220.00	0.19	-0.70	-1.75	0.35

Table 18. PoS transaction post demonetization

	Number of POS (in Actuals)	Volume (Million)	Value (Rupees Billion)	Volume per POS	Value in Rs. per POS	% Change in Volume	% Change in Value
Mar-17	2528758	377.40	685.86	149.24	271,223.79	11.7%	3.9%
Mar-16	1385668	185.09	361.57	133.57	260,939.05	13.1%	2.3%
Mar-15	1126735	133.01	287.27	118.05	254,958.94		

REFERENCES

Chau, P. Y. K. (1995). Factors Used in the Selection of Packaged Software in Small Businesses: Views of Owners and Managers. *Information & Management, 29*(2), 71–78. doi:10.1016/0378-7206(95)00016-P

Ekanem, I. (2005). 'Bootstrapping': The Investment Decision-Making Process in Small Firms. *The British Accounting Review, 37*(3), 299–318. doi:10.1016/j.bar.2005.04.004

Julien, P. A., & Raymond, L. (1994). Factors of New Technology Adoption in The Retail Sector. *Entrepreneurship Theory and Practice, 18*(4), 79–90. doi:10.1177/104225879401800405

Kadamudimatha. (2016). Digital Wallet: The Next Way of Growth. *International Journal of Commerce and Management Research, 2*(12).

Levin, H. M., & Rumberger, R. W. (1986). Education and Training Needs for Using Computers in Small Businesses. *Educational Evaluation and Policy Analysis, 8*(4), 423–434. doi:10.3102/01623737008004423

Liébana-Cabanillas, F., Ramos de Luna, I., & Montoro-Ríos, F. (2017). Intention to use new mobile payment systems: A comparative analysis of SMS and NFC payments. *Economic Research-Ekonomska Istraživanja, 30*(1), 892–910. doi:10.1080/1331677X.2017.1305784

Mallat & Tuunainen. (2008). Exploring Merchant Adoption of Mobile Payment Systems: An Empirical Study. *e-Service Journal, 6*(2), 24-57. Doi:10.2979/esj.2008.6.2.24

Raghavan, V., Wani, M., & Abraham, D. M. (2018). Exploring E-Business in Indian SMEs: Adoption, Trends and the Way Forward. In *Emerging Markets from a Multidisciplinary Perspective. Advances in Theory and Practice of Emerging Markets*. Cham: Springer. doi:10.1007/978-3-319-75013-2_9

Sharma, M. K. (2009). Receptivity of India's small and medium-sized enterprises to information system adoption. *Enterprise Information Systems, 3*(1), 95–115. doi:10.1080/17517570802317901

Thong, J. Y. L. (1999, Spring). An Integrated Model of Information Systems Adoption in Small Businesses. *Journal of Management Information Systems, 15*(4), 187–214. doi:10.1080/07421222.1999.11518227

Thong, J. Y. L., & Yap, C. S. (1995). CEO Characteristics, Organizational Characteristics and Information Technology Adoption in Small Businesses. *Omega, 23*(4), 429–442. doi:10.1016/0305-0483(95)00017-I

Tse, T., & Soufani, K. (2003). Business strategies for small firms in the new economy. *Journal of Small Business and Enterprise Development, 10*(3), 306–320. doi:10.1108/14626000310489781

Chapter 10
Mapping Innovation in the Digital Transformation Era:
The Role of Technology Convergence

Elona Marku
University of Cagliari, Italy

Manuel Castriotta
University of Cagliari, Italy

Maria Chiara Di Guardo
University of Cagliari, Italy

Michela Loi
University of Cagliari, Italy

ABSTRACT

Digital transformation is imperative for gaining and sustaining a firm's competitive advantage. Hence, understanding the dynamics of technology evolution becomes salient for both scholars and practitioners. This chapter aims to provide a complementary perspective to the field of innovation by mapping and visualizing the patterns of digital transformation at the industry level with a particular focus on the role of technology convergence. The authors tracked 20 years of the technology of the U.S. communications industry in order to investigate how digital transformation has shaped the industry technological structure, which are the technological gaps and potential future technology trends. The results show a deep transformation of the industry with many interconnections between technology domains and a high degree of overlap between technology areas.

DOI: 10.4018/978-1-5225-7262-6.ch010

INTRODUCTION

In the past few decades, high-technology industries have experienced an important evolution connected with the advancement of digital innovations, regulatory changes, as well as new consumer preferences (Curran & Leker, 2011). At the same time, scholars have been drawing increasing attention to the phenomenon of digital transformation, defining it as crucial for firm growth and success (Yoo, Boland, Lyytinen, & Majchrzak, 2012; Hess, Matt, Benlian, & Wiesböck, 2016; Soule, Puram, Westerman, & Bonnet, 2016; Schweer & Sahl, 2017; Singh & Hess, 2017). Indeed, the potential benefits resulting from this digitization trend—including big data, the Internet of things, mobile computing and cloud computing—are enormous (Fitzgerald et al., 2014; Peppard & Ward, 2016). Digitalization has reshaped the way we work and live by introducing new and more disruptive technologies. This rapid transformation continually leaves traces in the technological space reshaping the industry's technological structure. In order to capture these changes across time, the present study aims at mapping and visualizing technology and innovation in the digital transformation era. More specifically, the authors attempt to explore the technology trajectory and patterns at the industry level in order to unveil the dynamics of the evolution of digitalization, to identify its most active technological areas and focal streams.

Digital transformation is a very broad phenomenon that covers several aspects. Zhu, Kramer, and Xu (2006) have examined the digital transformation process in different countries showing that it is influenced by contextual factors including the technological, organizational, and environmental contexts have focused on the technological, organizational, and environmental factors. Other studies have revisited the innovation appropriability dynamics (Teece, 2018; Helfat & Raubitchek, 2018), while other works have examined digital transformation at individual level (i.e. Stolterman & Fors, 2004; Lanzolla & Giudici, 2017) or at the industry level (*i.e.* Agarwal, Gao, DesRoches, & Jha, 2010). However, studies that investigate the digital transformation process within the technology convergence and industry evolution paradigms are still absent.

Our goal in this chapter is to explore one aspect of digital transformation related to how it has affected the technology innovations generated by firms and how it can be explained through the technological convergence paradigm. Consistent with our goal, we analyzed the communication services industry as its digital transformation is evident, it has been a key driver of the worldwide digitization and as such it is positioned at the forefront of the deep transformation that comprehends new information technologies such as broadband networks, mobile communications, and the Internet (Maitland, Bauer, & Westerveld, 2002; Andal-Ancion *et al.,* 2003).

Moreover, the technology literature has widely shown the crucial role of patents as a meaningful instrument to measure the innovation performance (Trajtenberg, 1990; Ahuja & Katila, 2001; Hagedoorn & Cloodt, 2003; Di Guardo & Harrigan, 2016), to capture the multifaceted dimensions of technology (*i.e.* Hall, Jaffe, & Trajtenberg, 2001; Harrigan, Di Guardo, Marku, & Velez, 2017), to track the knowledge flows and spillovers (*i.e.* Jaffe A., 1986), and to monitor convergence and emerging technologies in the digital transformation era (Tijssen,1992; Engelsman & van Raan, 1994; Archibugi & Pianta, 1996; Curran & Leker, 2011; De Rassenfosse *et al.*, 2013; Lee, Park, & Kang, 2018).

Based on patent co-classification analysis (Engelsman & van Raan, 1994), this study examines the digital transformation of the U.S. communications industry in a 20-year time interval that goes from 1992 to 2011. Patent co-classification analysis is widely acknowledged as a valid alternative to the most widespread patent co-citation analysis (Tijssen, 1992; Leydersdorf, 2008; Luan, Liu, & Wang, 2013; Marku, Castriotta, & Di Guardo, 2018; Marku & Zaitsava, 2018). Furthermore, the present work introduces a new approach for mapping and visualizing an industry's profile and evolution by focusing on fine-grained patent data. Departing far from extant literature, we map the co-occurrence of the technology classification codes using a novel and validated tool in management studies: the VosViewer software, whose algorithm for computing the similarity measure allows the overcoming of some of the artifacts produced by the multidimensional scaling (van Eck, Waltman, van den Berg, & Kaymak, 2006; van Eck & Waltman, 2007; Waaijer, van Bochove, & van Eck, 2010; van Eck, Waltman, Dekker, & van den Berg, 2010; Zupic & Čater 2015). In addition, following the latest trends in technology studies and avoiding the limitations of the International Patent Classification (IPC) system (Luan, Liu, & Wang, 2013), we use the Derwent World Patent Index (DWPI) classification codes (Calcagno, 2008; Luan, Liu, & Wang, 2013; Luan, Hou, Wang, & Wang, 2014). The peculiarity of the DWPI system consists of the assignment of one or several manual codes to a single patent document, aimed at covering all the relevant aspects of the invention. In this way, we can capture the smallest technology elements possessed by firms.

The remainder of this chapter is structured as follows. First, the authors discuss the relevant characteristics of digital transformation, technology convergence, and patent analysis. Second, the method based on patent co-classification and a brief explanation of the VosViewer algorithm are introduced. Third, descriptive results are presented and last, discussion, concluding remarks and implications are provided.

LITERATURE BACKGROUND

Digital Transformation

Digital transformation is imperative for firm growth and success, and even for firm survival (Andal-Ancion, Cartwright, & Yip, 2003; Fitzgerald *et al.,* 2014; Abolhassan, 2016; Riedl *et al.,* 2017; Reis *et al.,* 2018). Although terms like "digitization", "digitalization", or "digital transformation" are increasingly appearing in management studies, no clear definition has been provided yet (Agarwal et al., 2010; Andal-Ancion *et al.,* 2003; Yoo *et al.,* 2012; Hansen & Sia, 2015; Hess *et al.,* 2016; Lanzolla & Giudici, 2017; Teece, 2018; Helfat & Raubitchek, 2018). In this chapter, the authors conceive "digitization" as the creation of a digital version of physical objects such as paper documents, photographs, etc., "digitalization" encompasses the adoption and the leverage of digital technologies and digital data, meanwhile, "digital transformation" refers to the transformation from partial digitized businesses to fully digitized ones, it is an is an umbrella concept that includes artificial intelligence, robotics, machine learning, Internet of things, etc. (Riedl *et al.*, 2017).

Although, existing literature has pointed out that the digital transformation represents an autonomous field (Mckelvey, Tanriverdi, & Yoo, 2016; Urbach, Drews, & Ross, 2017), the concept of digital transformation is very broad and covers several aspects. Most studies have focused on the analysis of the phenomenon at firm level, for example, Andal-Ancion, Cartwright, and George (2003) investigated the digital transformation of traditional businesses pointing out that the key to success consists in knowing how and when to apply new technologies. Other studies have examined the digital transformation process in different countries showing that it is influenced by contextual factors including the technological, organizational, and environmental contexts (Zhu, Kramer, & Xu, 2006). Moreover, other works have analyzed digital transformation at the individual level (i.e. Stolterman & Fors, 2004; Lanzolla & Giudici, 2017) as well as industry level (*i.e.* Agarwal *et al.,* 2010; Angst et al., 2017; Wang et al., 2018).

In this chapter, we examine the aspect of digital transformation related to technology innovation at the industry level. We refer to the concept of innovation in terms of inventions since they represent essential outputs of the R&D and inventive activities (Valentini & Di Guardo, 2012). It is extensively acknowledged that digital transformation has changed our work, our home, and our life, producing in this way far-reaching effects on our entire society, including firms and their inventive activities (Brynjolfsson & McAfee, 2014; Fitzgerald et al., 2014). Thus, digital technologies with their capability to enable pervasive connectivity, immediacy of interaction as well as wide access to data have changed radically the nature of innovations produced by firms (Yoo *et al.,* 2012; Matt, Hess, & Benlian, 2015; Lanzolla & Giudici, 2017). Digital transformation, empowered by the combination of different digital innovations, provides the building blocks for the development of new products and services (Hinings, Gegenhuber, & Greenwood, 2018; Teece, 2018). For this reason, no industry is immune to digital transformation as it continually modifies entire ecosystems (Hess *et al.,* 2016); almost all firms have started a process of exploration of new digital technologies in order to exploit their benefits and potentiality (Matt *et al.,* 2015; Hess et al., 2016).

In particular, the way the digital transformation has affected the technology innovations is evident in the case of the communication services industry. The latter has been a key driver of the worldwide digitization and as such it is positioned at the forefront of the deep transformation that comprehends new information technologies such as broadband networks, mobile communications, and the Internet (Maitland, Bauer, & Westerveld, 2002; Andal Ancion *et al.,* 2003). Firms operating in this industry including hardware, semiconductors, network-based, and new information and communications technologies (ICT) have been growing up simultaneously creating a unique ecosystem progressively permeated with digital technology (Suh & Lee, 2017). Very often, the integration of digital technologies goes beyond the firm's boundaries redefining the rules of the game of the whole industry (Downes & Nunes, 2013; Matt *et al.,* 2015).

In the last decades, new technologies introduced in this industry have been disruptive leading to a paradigm shift that has changed not only the boundaries but especially the core technologies of the communications industry (Li & Whalley, 2002; Peppard & Rylander, 2006; Kim, Cho, & Kim, 2015; Lee, Kang, & Shin, 2015). The crucial role of the Internet and data communication services has moved the industry towards greater convergence and complexity (Maitland *et al.,* 2002). Indeed, everyday products such as smartphones or TVs have embedded software digital capabilities which include intelligent machines with sensors, network, and processors (Yoo *et al.,* 2012). Finally, this rapid transformation process with radical changes and technology convergence has continually left traces in the technological space, redesigning the industry's technological structure and landscape.

Technological Convergence

The concept of *technological convergence* was introduced by Rosenberg (1963) who defined it as "the process by which two hitherto different industrial sectors come to share a common knowledge and technological base". Other scholars have defined it as the blurring of boundaries between disjoint fields of science, technology, markets, or industries (Curran & Leker, 2011; Curran, 2013; Kim *et al.,* 2015). In this vein, technological convergence is a process where technologies move from their prior positioning to new and common place. Consistent with this stream of research, we conceive technological convergence as the merging or overlapping of different fields of technology as a result of scientific and technological progress. More specifically, in this chapter technological convergence is defined as a combination of at least two existing technologies into hybrid technologies (Kodama, 1995; Curran et al., 2010; Kim *et al.,* 2018).

Technological convergence has become a key feature of all high-tech industries as they have rapidly evolved to meet market demands (Iwai, 2000; Yoo *et al.*, 2012; Lee *et al.,* 2015) moving the boundaries of existing technologies and increasing their overlapped technological area (Gambardella & Torrisi, 1998; Athreye & Keeble, 2000; Borés, Saurina, & Torres, 2003; Kim *et al.,* 2015; Aharonson & Schilling, 2016). For example, the '90s were characterized by a great technological convergence in many fields: the power of computers boosted the information technology and at the same time the communications technology rapidly evolved leading to a convergence in the ICT; personal digital assistants evolved into the omnipresent Blackberry, iPhone, MP3; the digitalization involved an integration between analog and digital technologies, voice and data, leading to a convergence of telecommunication and broadcasting (Maitland *et al.*, 2002; Lee et al., 2010; Suh & Lee, 2017).

At the same pace, the innovation literature has analyzed this phenomenon (Lee et al., 2010; Curran & Leker, 2011; Jeong et al., 2015; Kim, Kim, & Lee, 2018). Caviggioli (2016) examined the main drivers of technological convergence, using patents granted by the European Patent Office (EPO) from 1991 to 2007 he found that merges are more frequent if the focal technology fields are closely related. Curran and colleagues (2010) focused on the specific phases of the convergence process such as science, technology, market, and industry. Moreover, the study of Curran and Leker (2011) highlighted how these phases are not static and convergence may occur without including all these steps, a new product, service, or business model can cause convergence in an industry. Although, technological convergence is important because it can lead to a convergence of whole industries (D'Aveni, Dagnino, & Smith, 2010; Hacklin, Marxt, & Fahrni, 2010; Zhang & Li, 2010; Hacklin & Wallin, 2013), *how* technological convergence reshapes the industry technological structure and landscape is still unexplored.

Indeed, the gaining of awareness about the features of converging technologies in a wide scenario becomes essential since technological convergence is a key driver of innovation (Curran & Leker, 2011; Karvonen & Kässi, 2013; Lee et al., 2015). Innovations that come up from converging fields seem to be more novel and breakthrough (Schumpeter,1939; Fleming, 2001; Hacklin, 2007; No & Park, 2010; Nemet & Johnson, 2012; Karvonen & Kässi, 2013). Last, sectors that face technological convergence appear to have higher innovation performance and technological impact on industry evolution, making the phenomenon even more interesting to disentangle (No & Park, 2010; Kim & Kim, 2012; Curran, 2013; Hacklinetal, 2013; Caviggioli, 2016).

Patent Analysis

Digital transformation and technology convergence have intensified the technological competition among firms leading to a higher need to secure technologies via the patenting activity (Lee *et al.,* 2016; Lee & Kim, 2017). Patents grant to their owners an exclusive monopoly power over the use of an invention and as such they play a pivotal role of patents to protect a firm's R&D output (Oh, Cho, & Kim, 2014). The technology literature broadly acknowledges their importance of assessing firm innovation performance (Trajtenberg, 1990; Ahuja & Katila, 2001; Hagedoorn & Cloodt, 2003), seizing the different dimensions of technology (*i.e.* Hall *et al.*, 2001; Harrigan *et al.,* 2017), tracking the knowledge flows and spillovers (*i.e.* Jaffe A., 1986), and also monitoring technology convergence and industry evolution (Tijssen,1992; Engelsman & van Raan, 1994; Archibugi & Pianta, 1996; Curran & Leker, 2011; De Rassenfosse *et al.*, 2013; Karvonen & Kässi, 2013; Suh & Sohn, 2015; Han & Sohn, 2016; Lee, Park, & Kang, 2018).

Moreover, in comparison with other information sources, patents can signal the latest technological change (Soete & Wyatt, 1983; Chen, 2011).

Specifically, patent analysis represents a common method to transform patent data into useful information (Tseng, Hsieh, Peng, & Chu, 2011; Niemann, Moehrle, & Frischkorn, 2017). In order to better understand the dynamics of technology evolution and convergence in different time frames, prior research has applied occurrence and co-occurrence network analysis using technology classification codes, assignee, applicants, inventors, citations, and co-citations analysis (Curran et al., 2010; Curran & Leker, 2011; Karvonen & Kässi, 2013; Jeong, Kim, & Choi, 2015; Castriotta & Di Guardo, 2016; : Loi, Castriotta, & Di Guardo, 2016; Di Guardo, Galvagno, & Cabiddu, 2012). Patent citation analysis is more suitable for the identification of technology flows among different elements as well as the investigating of convergence mechanisms using those knowledge flows (Kim & Lee, 2017). Instead, patent co-occurrence analysis through a science mapping approach allows a more detailed visualization of the phenomena occurring in contexts with technology convergence.

Patent co-classification analysis is widely acknowledged as a valid alternative to the most widespread patent co-citation analysis (Tijssen, 1992; Leydersdorf, 2008; Luan, Liu, & Wang, 2013; Marku, Castriotta, & Di Guardo, 2018). It draws upon the assumption that patents classification systems assign one or several manual codes to a single patent document aimed at covering all the relevant aspects of the invention. In this regard, the DWPI patent co-classification codes analysis is a more appropriate tool for mapping and visualizing an industry's profile and evolution by focusing on fine grained patent data (Calcagno, 2008; Luan *et al*, 2013; 2014).

Operationally, two codes that co-occur within the same patent are an indication of their technological relatedness (Park & Yoon, 2014). Then, the presence of many co-occurrences around the same patents or firms' portfolios contributes to forming a technological structure map of an entire industry or sector (Park & Yoon, 2014). In the context of co-occurrence patent studies, Curran and Leker (2011) applied patent co-classification to monitor technology convergence while Karvonen and Kässi (2011) used large-scale patent data to anticipate the early stages of technology convergence. Lee, Kang, & Shin (2015) by adopting a patent maps approach focused on the identification of potential technology opportunities in a wider scenario like an industry. In summary, this study adopts a Derwent codes patent co-classification analysis in order to detect the patterns of digital transformation and technology convergence.

METHODOLOGY

Sample and Data

Technology convergence has involved many ICT fields including telecommunications, broadcasting, information technologies, and entertainment (Borés *et al*., 2003; Han & Sohn, 2016). In particular, the technology convergence faced by the communications industry has been fueled by the introduction into the market and subsequent adoption of several radical digital technologies that have transformed the industry structure into a complex network created by the combination of wireless technologies, the Internet, and data communications (Maitland *et al*., 2002; Suh & Lee, 2017). Mainly for this reason, in this chapter, we examine the U.S. communications industry because of its high technological dynamism and complexity (Di Guardo, Harrigan, & Marku, 2018; Harrigan *et al*., 2017). Indeed, the U.S. Telecom Act of 1996 deregulated the provision of specialized communications services and the technology convergence began. Communications services technologies evolved rapidly as it became possible to digitize and transmit voice, data and video over one network and many firms made acquisitions in the late 1990s to supplement their internal technology gaps. Entry into the various service specialties brought diffusion of communications technologies that were used elsewhere. The result of this cross-pollination was a huge disruption in industry structure.

Using the COMPUSTAT database (Standard & Poor's, 2013), firms were selected according to the following SIC (Standard Industrial Classification) codes: 4812 (Radiotelephone communications), 4813 (Telephone communications, except radiotelephone), 4822 (Telegraph and other message communications), 4841 (Cable and other pay television services), and 4899 (Communication services not elsewhere classified). Patent information was retrieved from the DWPI database (2013) using a 20-year time interval that goes from 1992 to 2011; in addition to the whole dataset, four 5-year time-frame datasets were created. In order to capture the earliest stage of technology convergence, the earliest priority year was considered. This selection procedure led to a final sample that consists of 120,476 U.S. patents granted to 180 firms.

Moreover, different from prior research that mainly uses IPC (International Patent Classification) codes or U.S. classification codes, this work adopts the Derwent classification system. Although, the Derwent manual codes——similarly to the IPC codes or the U.S. classification codes——have a logical hierarchy assigned to the primary inventive features of a basic patent, their peculiarity consists of the assignment of one or several manual codes to a single patent document aimed at covering all the relevant aspects of the invention (Calcagno, 2008; Harrigan & Di Guardo, 2017; Harrigan, Di Guardo, & Marku, 2018). From an operational point of view, co-classifications can be represented by a frequency matrix of co-occurrences in which the patents fitting with our sample are paired (Engelsman & van Raan, 1994, Curran & Leker, 2011; Karvonen & Kässi, 2013).

We retrieved the co-classification codes frequencies for each patent in our sample and compiled them into a raw matrix. In our case, the co-occurrence matrix is composed of 284 Derwent codes. This 284 X 284 asymmetric matrix is the source from which we ran multivariate science mapping analysis (Van Eck *et al.*, 2010). Further, these fine-grained data allow to account for a multiplicity of technological domains and to better monitor the technology convergence and digital transformation of the industry.

Multivariate Analysis

By relying on previous patent and bibliometric studies (Di Stefano, Gambardella, & Verona, 2012), and in coherence with our research goals of exploring technology convergence and evolution of the U.S. communications industry, we apply two patent co-occurrence techniques to perform our co-classification analysis: cluster analysis and multidimensional scaling (MDS). Furthermore, a network visualization is applied to better identify the connections among the different technology classification codes and clusters (if any). Cluster analysis is a common technique well suited to find subgroups in a field that gathers the similarity among groups' constituting objects (McCain, 1990). As regards to the MDS, it attempts to find a structure in a set of proximity measures between objects (Kruskal, 1977) by producing a map in a low dimensional space (usually two) that optimizes distances between those objects and that reflects a similarity measure. Accordingly, similar items will appear closer in the map (Leydesdorff & Vaughan, 2006).

Visualization Software: VosViewer

The increase of the amount of patent documents has fostered the development of sophisticated patent analysis instruments (Abbas, Zhang, & Khan, 2014). In this study, we introduce a novel instrument to track technology convergence by using the VosViewer software. The latter is a validated tool in management studies whose algorithm for computing the similarity measure allows the overcoming of some of the artifacts produced by the multidimensional scaling (van Eck *et al.* 2006; van Eck & Waltman, 2007; Waaijer, van Bochove, & van Eck, 2010; van Eck *et al.,* 2010; Zupic & Čater 2015). The VosViewer methodology is developed by van Eck & Waltman (2007) and it is based on the association strength s_{ij} between items (concepts) that can be synthetized as follows:

$$s_{i,j} = \frac{c_{i,j}}{w_i w_j} \quad (1)$$

where c_{ij} refers to the number of co-occurrences of items *i* and *j*, while w_i and w_j concern to either the total number of occurrences of items *i* and *j* or to the total number of co-occurrences of these items. It could be considered a weighted MDS that presents the advantage to overcome two artifacts of the latter: the first regards the MDS tendency to map the most important items in the center while the less important on the periphery, and the second concerns the use of a circular structure (van Eck, *et al.,* 2010). This situation occurs because when the MDS approach is applied to similarity data, it tries to assign the same distance to each pair of items, including those with similarity equal to zero. In our case, there are many classification codes that are never combined together in a patented invention, their frequency and similarity measure is equal to zero. According to van Eck et al., (2010), in the case of similarity data that consists largely of zeros, it is not possible to build a map with an exact same distance between each pair, hence, the MDS provides an approximation of this distance. In contrast, the VosViewer approach, being a weighted MDS, it does not give equal weight to all pairs of items, it gives more weight to more similar pairs of items, low weight to pairs of items with low similarity. As similarity data are typically dominated by low values, the VosViewer pairs of items with low similarity will have little weight and consequently will receive a little effect on the map.

RESULTS

Evidence of the digital transformation experienced by firms operating in the U.S. communications industry clearly emerges by looking at Table 1 which summarizes the top-20 technologies in the four 5-year time intervals. Not surprisingly, this list includes technology groups that strongly characterize the industry, namely, "W" (Communications), "T" (Computing and control), "L" (Refractories, ceramics, cement and electro(in)organics), "P" (General), "U" (Semiconductors and electronic circuitry), "V" (Electronic components), "S" (Instrumentation, measuring and testing), "A" (Polymers plastics), and "X" (Electric power engineering). The first rows of Table 1 highlight those technologies that can be defined as core, we find "T01 E" - Digital computers (data processing), "W01 E" - Telephone and data transmission systems (wireless), "W02 E" - Broadcasting, radio and line transmission systems, and "W04 E" - Audio/visual recording and systems.

In particular, it is interesting to observe the paths of two dominant technologies, Digital computers (T01 E) and Telephone (W01 E). In the first and second time frame the percentage of patents that contains a technology classified as "W01 E" was 6.49% and 9.72%, respectively. This frequency was higher than the technology "T01 E" that was embodied in the 5.02% (1992-1996) and 9.45% (1997-2001) of the patented inventions. This result can be explained by the predominance of the telephone technologies during 1992-2001, however, starting from the third time frame (2002-2006), the situation is overturned as "T01 E" boosted and scored 13.14% (2002-2006) increasing even more in the last time frame (2007-2011) with 17.44%. Although "W01 E" shows a positive trend, its speed was lower than "T01 E". The digital transformation process has started.

The last column of Table 1 depicts the technological trend of the different technologies, most of them seem to have a negative trend; digitalization fosters the development of technologies like digital computers while it hinders other technologies such as data recording (tape/filament). Among technologies that decreased their technological presence in the industry, we identify: Broadcasting, radio and line transmission systems (W02 E); Electro-(in)organic (L03 C); Optics (P81 N); Fibre-optics and light

Table 1. Frequency percentage of the top-20 technologies in each 5-year time frame

Technology		1992-1996	1997-2001	2002-2006	2007-2011	Trend
T01 E	Digital computers (data processing)	5.02%	9.45%	13.14%	17.44%	↑
W01 E	Telephone, data transmission syst. (wireless)	6.49%	9.72%	10.31%	13.95%	↑
W02 E	Broadcasting, radio and line transmission syst.	6.96%	7.95%	6.14%	5.56%	↓
W04 E	Audio/visual recording and systems	6.65%	4.62%	4.80%	7.11%	↑
L03 C	Electro-(in)organic (electric discharge lamps)	3.94%	3.25%	2.73%	2.45%	↓
P81 N	Optics	3.37%	2.57%	2.42%	2.39%	↓
T04 E	Computer peripheral equip. (graph reading)	1.72%	1.75%	2.73%	3.73%	↑
U11 E	Semiconductor materials and processing	2.66%	2.34%	1.92%	2.19%	↑
W03 E	Tv and broadcast radio receivers	1.44%	1.45%	1.85%	3.71%	↑
U14 E	Memories, film and hybrid circuits	1.68%	1.54%	1.80%	3.01%	↑
V07 E	Fibre-optics and light control	2.13%	2.13%	1.68%	1.57%	↓
U12 E	Discrete devices	1.60%	1.73%	2.02%	2.08%	↑
P85 N	Education, cryptography, adverts	1.85%	1.93%	1.71%	1.42%	↓
S03 E	Scientific instrumentation	1.51%	1.52%	2.33%	1.24%	↓
W06 E	Aviation, marine and radar systems	1.62%	1.31%	1.42%	1.78%	↑
T03 E	Data recording (tape(/filament) transport)	2.78%	1.61%	-	-	↓
V04 E	Printed circuits and connectors	1.30%	1.43%	-	1.57%	↑
U21 E	Logic circuits, electronic switching and coding	1.46%	1.45%	1.38%	-	↓
W05 E	Alarms, signalling, telemetry and telecontrol	-	-	1.57%	1.49%	↓
P86 N	Musical instruments, acoustics	1.39%	1.44%	-	-	↓
A85 C	Electrical applications	-	1.17%	-	1.33%	↑
S05 E	Electrical medical equipment	-	-	1.85%	-	↓
V06 E	Electromechanical transducers	-	-	1.59%	-	↓
U13 E	Integrated circuits	1.39%	-	-	-	↓
S02 E	Engineering instrumentation	-	-	1.29%	-	↓
X26 E	Lighting (portable lights)	-	-	-	1.14%	↑
A89 C	Photographic, laboratory equipment, optical	-	-	-	1.10%	↑

control (V07 E); Scientific instrumentation (S03 E); etc. Technologies with a positive trend comprehend: Computer peripheral equipment (T04 E); Semiconductor materials and processing (U11 E); Receivers of Tv and broadcast radio (W03 E); Memories (U14 E); etc. Another interesting result is connected with the appearance and especially the disappearance of technologies from the top-20 ranking technologies. While Data recording technologies that use tape or filament (T03 E) and Musical instruments (P86 N) were shifted out from this ranking, other technologies reinforced their position, this is the case of Portable lights (X26 E) and Photographic laboratory equipment (A89 C).

The same data can be seen from another perspective, by mapping and visualizing innovation. Figure 1 illustrates the most active technological areas in the 20-year span. The core industry is held by the communications technologies "W01 E" and "W02 E" that are positioned very close to each other, and the digital technology "T01 E" which seems to have a similar distance from "W01 E" and "W04 E". The other technologies are shaped like clouds around the core presenting different densities.

Moreover, Figure 2 shows four maps (one map per time window) that can be useful to understand the dynamics of technology convergence in the industry. In general, technology classification codes present our nodes in the map, thick swaths evidence a high frequency of two technology codes classified together in the same patent document. Nodes are also gathered into different clusters according to their degree of similarity. Specifically, the first quadrant of Figure 2 (1992- 1996) shows an industry characterized by clusters with well-defined boundaries with few links between them. The core technologies identified by the big nodes are commonly combined with each other with few hubs (T01 E and W04 E). Moreover, the industry appears to gain more dynamicity in the second time frame, technologies are positioned closer to each other and they are more intertwined. This signals a cross-pollination of the field. This process continues during the years that go from 2002 to 2006. Finally, the most meaningful picture of the industry is enclosed in the map that refers to the last time span (2007-2011). Technology boundaries are blurred, the map does not present distinct clusters, it is a massive cloud with a specific trajectory pulled by the core technologies (T01 E and W01E), showing in this way a high convergence among different technologies.

Figure 1. Density map of the 20-year time span

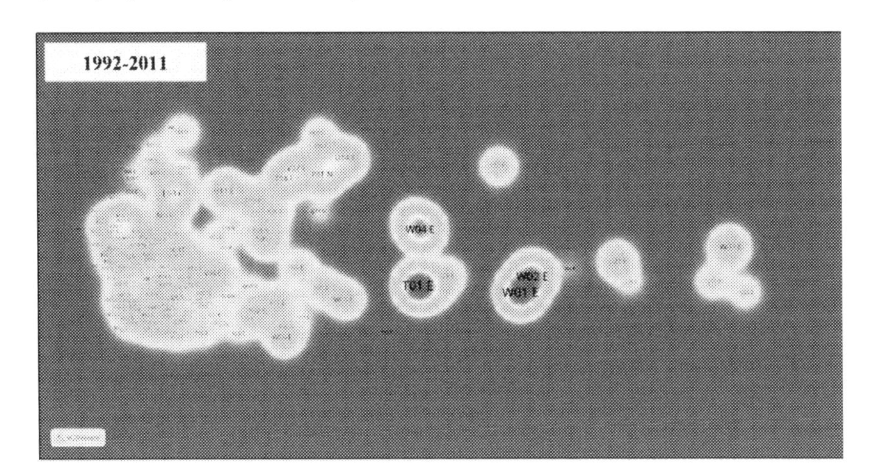

**For a more accurate representation see the electronic version.*

Figure 2. Visualization of technology convergence in four 5-year time frames

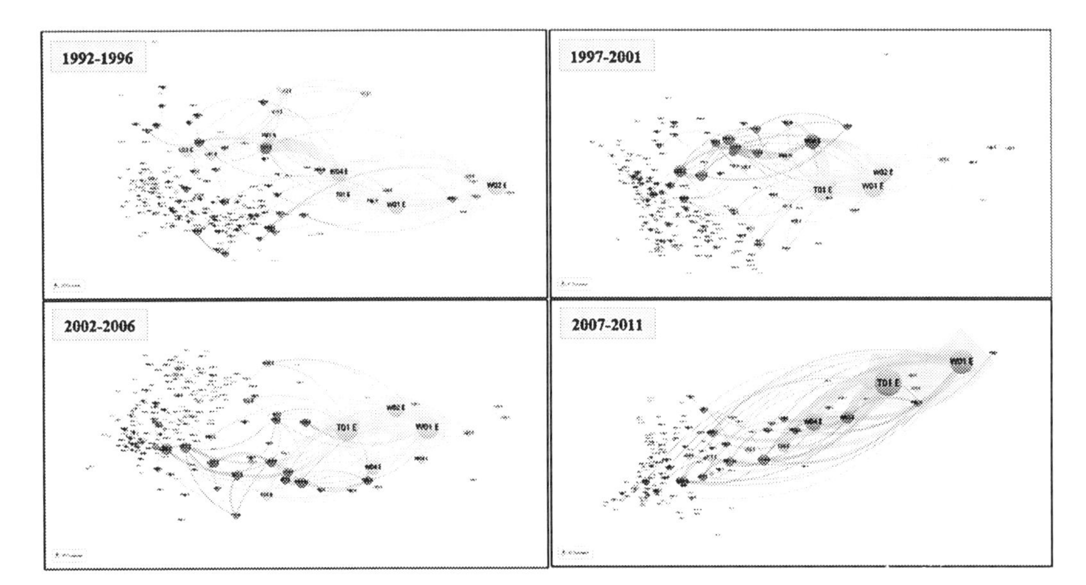

DISCUSSION AND CONCLUSION

The present study explored the technology trajectory and emerging patterns at the industry level in order to unveil the dynamics of the evolution of digitalization, to identify its most active technological areas and focal streams. Our analysis exploited the information included in patent documents which are commonly used as a proxy of technology innovation and their classification provides a useful means to identify technological domains and their overlap across time. Using a co-classification methodology on 20-year window data, it was possible to identify the different phases in which the digital transformation affected the technology innovations in the specific case of the U.S. communication services industry. The mapping and visualization of the industry's technological structure in different landscapes show a deep transformation of the industry, from a polycentric structure (in the timespan that goes from 1992 to 1996) to a diffused overlap between technology fields (in the timeframe 2007-2011). This is consistent with the technological convergence paradigm, two or more technologies move together in the technological space, overlapping or merging with each other while generating new innovations.

It is interesting to observe how the entry and further development of a disruptive innovation (this is the case of digital computers and data processing technologies) lead to a redefinition of all the existing links in terms of strength, new links formed or even canceled. In general, being aware of the strength of links between the technology portfolio possessed by the firm and the industry's technology core can be fundamental for the sustaining of competitive advantage or identifying new opportunities. This exploratory study increases the awareness of scholars by detecting and visualizing the technology management phenomenon as it provides a descriptive overview of the positioning of the technologies involved in the industry's inventive activity, as well as the links between them.

The contributions of the present study are threefold. It contributes to the digital transformation literature by exploring how it shapes the technology structure of an industry. Second, it sheds new light on the dynamics of technology convergence in the process of digital transformation, contributing to the technology innovation literature. A methodological contribution was provided by introducing a novel (although validated) tool for mapping the co-occurrence of the technology classification codes, the VosViewer software.

Managerial contributions can be also highlighted. This paper examined 20 years technology of one of the industries in which digital transformation was more evident, our findings suggest a potential scenario to managers of industries that now are facing digital transformation. If an industry's picture is similar to the first 5-year window in the case analyzed in this chapter, a suggestion for managers could be the development of a higher degree of heterogeneity of the resources involved in the R&D and inventive activities as well as the enhancement of firm's capabilities to recombine new knowledge. Moreover, the industry analyzed is characterized by a fast pace of technology, having an idea of how much time does it take for a technology structure to change its connotations can be useful for managers in their decision-making process.

In addition to the above-mentioned contributions, the present study presents several limitations. First, in this chapter, we analyzed the effect of digital transformation at a sector level on the technological structure from 1992-2011, future research might focus on what happened from 2011 to 2018 in terms of technological change as well as changes in the industry structure. Second, we examined the technology structure of a single industry, future research might focus on two more industries in order to provide a broader overview of industry convergence. Third, although the industry under examination has a high intensity of patent activities, not all inventions are patented; one common reason is the firms' reliance on secrecy.

REFERENCES

Abbas, A., Zhang, L., & Khan, S. U. (2014). A literature review on the state-of-the-art in patent analysis. *World Patent Information*, *37*, 3–13. doi:10.1016/j.wpi.2013.12.006

Abolhassan, F. (2016). *The Drivers of Digital Transformation*. Springer International Publishing.

Agarwal, R., Gao, G., DesRoches, C., & Jha, A. K. (2010). Research commentary—The digital transformation of healthcare: Current status and the road ahead. *Information Systems Research*, *21*(4), 796–809. doi:10.1287/isre.1100.0327

Aharonson, B. S., & Schilling, M. A. (2016). Mapping the technological landscape: Measuring technology distance, technological footprints, and technology evolution. *Research Policy*, *45*(1), 81–96. doi:10.1016/j.respol.2015.08.001

Ahlgren, P., Jarneving, B., & Rousseau, R. (2003). Requirements for a cocitation similarity measure, with special reference to Pearson's correlation coefficient. *Journal of the Association for Information Science and Technology*, *54*(6), 550–560.

Ahuja, G., & Katila, R. (2001). Technological acquisitions and the innovation performance of acquiring firms: A longitudinal study. *Strategic Management Journal*, *22*(3), 197–220. doi:10.1002mj.157

Andal-Ancion, A., Cartwright, P. A., & Yip, G. S. (2003). The digital transformation of traditional business. *MIT Sloan Management Review*, *44*(4), 34–41.

Angst, C. M., Wowak, K. D., Handley, S. M., & Kelley, K. (2017). Antecedents of Information Systems Sourcing Strategies in US Hospitals: A Longitudinal Study. *Management Information Systems Quarterly*, *41*(4), 1129–1152. doi:10.25300/MISQ/2017/41.4.06

Archibugi, D., & Planta, M. (1996). Measuring technological change through patents and innovation surveys. *Technovation*, *16*(9), 451519–468. doi:10.1016/0166-4972(96)00031-4

Athreye, S., & Keeble, D. (2000). Technological convergence, globalisation and ownership in the UK computer industry. *Technovation*, *20*(5), 227–245. doi:10.1016/S0166-4972(99)00135-2

Borés, C., Saurina, C., & Torres, R. (2003). Technological convergence: A strategic perspective. *Technovation*, *23*(1), 1–13. doi:10.1016/S0166-4972(01)00094-3

Breschi, S., Lissoni, F., & Malerba, F. (2003). Knowledge-relatedness in firm technological diversification. *Research Policy*, *32*(1), 69–87. doi:10.1016/S0048-7333(02)00004-5

Brynjolfsson, E., & McAfee, A. (2014). *The second machine age: Work, progress, and prosperity in a time of brilliant technologies*. WW Norton & Company.

Calcagno, M. (2008). An investigation into analyzing patents by chemical structure using Thomson's Derwent World Patent Index codes. *World Patent Information*, *30*(3), 188–198. doi:10.1016/j.wpi.2007.10.007

Castriotta, M., & Di Guardo, M. C. (2016). Disentangling the automotive technology structure: A patent co-citation analysis. *Scientometrics*, *107*(2), 819–837. doi:10.100711192-016-1862-0

Caviggioli, F. (2016). Technology fusion: Identification and analysis of the drivers of technology convergence using patent data. *Technovation*, *55*, 22–32. doi:10.1016/j.technovation.2016.04.003

Chen, Y. S. (2011). Using patent analysis to explore corporate growth. *Scientometrics*, *88*(2), 433–448. doi:10.100711192-011-0396-8

Curran, C. S. (2013). *The Anticipation of Converging Industries*. London: Springer. doi:10.1007/978-1-4471-5170-8

Curran, C. S., Bröring, S., & Leker, J. (2010). Anticipating converging industries using publicly available data. *Technological Forecasting and Social Change*, *77*(3), 385–395. doi:10.1016/j.techfore.2009.10.002

Curran, C. S., & Leker, J. (2011). Patent indicators for monitoring convergence–examples from NFF and ICT. *Technological Forecasting and Social Change*, *78*(2), 256–273. doi:10.1016/j.techfore.2010.06.021

De Rassenfosse, G., Dernis, H., Guellec, D., Picci, L., & de la Potterie, B. V. P. (2013). The worldwide count of priority patents: A new indicator of inventive activity. *Research Policy*, *42*(3), 720–737. doi:10.1016/j.respol.2012.11.002

Di Guardo, M. C., Galvagno, M., & Cabiddu, F. (2012). Analysing the intellectual structure of e-service research. [IJESMA]. *International Journal of E-Services and Mobile Applications*, *4*(2), 19–36. doi:10.4018/jesma.2012040102

Di Guardo, M. C., & Harrigan, K. R. (2016). Shaping the path to inventive activity: The role of past experience in R&D alliances. *The Journal of Technology Transfer*, *41*(2), 250–269. doi:10.100710961-015-9409-8

Di Guardo, M. C., Harrigan, K. R., & Marku, E. (2018). M&A and diversification strategies: What effect on quality of inventive activity? *The Journal of Management and Governance*, 1–24. doi:10.100710997-018-9437-5

Di Stefano, G., Gambardella, A., & Verona, G. (2012). Technology push and demand pull perspectives in innovation studies: Current findings and future research directions. *Research Policy*, *41*(8), 1283–1295. doi:10.1016/j.respol.2012.03.021

Downes, L., & Nunes, P. F. (2013). Big-bang disruption. *Harvard Business Review*, *91*(3), 44–56.

Engelsman, E. C., & van Raan, A. F. (1994). A patent-based cartography of technology. *Research Policy*, *23*(1), 1–26. doi:10.1016/0048-7333(94)90024-8

Fitzgerald, M., Kruschwitz, N., Bonnet, D., & Welch, M. (2014). Embracing digital technology: A new strategic imperative. *MIT Sloan Management Review*, *55*(2), 1.

Fleming, L. (2001). Recombinant uncertainty in technological search. *Management Science*, *47*(1), 117–132. doi:10.1287/mnsc.47.1.117.10671

Gambardella, A., & Torrisi, S. (1998). Does technological convergence imply convergence in markets? Evidence from the electronics industry. *Research Policy*, *27*(5), 445–463. doi:10.1016/S0048-7333(98)00062-6

Hacklin, F. (2007). *Management of convergence in innovation: strategies and capabilities for value creation beyond blurring industry boundaries*. Springer Science & Business Media.

Hacklin, F., & Wallin, M. W. (2013). Convergence and interdisciplinarity in innovation management: A review, critique, and future directions. *Service Industries Journal*, *33*(7-8), 774–788. doi:10.1080/02642069.2013.740471

Hagedoorn, J., & Cloodt, M. (2003). Measuring innovative performance: Is there an advantage in using multiple indicators? *Research Policy*, *32*(8), 1365–1379. doi:10.1016/S0048-7333(02)00137-3

Hall, B. H., Jaffe, A. B., & Trajtenberg, M. (2001). *The NBER patent citation data file: Lessons, insights and methodological tools (No. w8498)*. National Bureau of Economic Research. doi:10.3386/w8498

Han, E. J., & Sohn, S. Y. (2016). Technological convergence in standards for information and communication technologies. *Technological Forecasting and Social Change*, *106*, 1–10. doi:10.1016/j.techfore.2016.02.003

Hansen, R., & Sia, S. K. (2015). Hummel's Digital Transformation Toward Omnichannel Retailing: Key Lessons Learned. *MIS Quarterly Executive*, *14*(2).

Harrigan, K. R., & Di Guardo, M. C. (2017). Sustainability of patent-based competitive advantage in the US communications services industry. *The Journal of Technology Transfer*, *42*(6), 1334–1361. doi:10.100710961-016-9515-2

Harrigan, K. R., Di Guardo, M. C., & Marku, E. (2018). Patent value and the Tobin's q ratio in media services. *The Journal of Technology Transfer*, *43*(1), 1–19. doi:10.100710961-017-9564-1

Harrigan, K. R., Di Guardo, M. C., Marku, E., & Velez, B. N. (2017). Using a distance measure to operationalise patent originality. *Technology Analysis and Strategic Management*, *29*(9), 988–1001. doi: 10.1080/09537325.2016.1260106

Helfat, C. E., & Raubitschek, R. S. (2018). Dynamic and integrative capabilities for profiting from innovation in digital platform-based ecosystems. *Research Policy*, *47*(8), 1391–1399. doi:10.1016/j.respol.2018.01.019

Hess, T., Matt, C., Benlian, A., & Wiesböck, F. (2016). Options for Formulating a Digital Transformation Strategy. *MIS Quarterly Executive*, *15*(2).

Hinings, B., Gegenhuber, T., & Greenwood, R. (2018). Digital innovation and transformation: An institutional perspective. *Information and Organization*, *28*(1), 52–61. doi:10.1016/j.infoandorg.2018.02.004

Iwai, K. (2000). A contribution to the evolutionary theory of innovation, imitation and growth. *Journal of Economic Behavior & Organization*, *43*(2), 167–198. doi:10.1016/S0167-2681(00)00115-3

Jaffe, A. B. (1986). *Technological opportunity and spillovers of R&D: evidence from firms' patents, profits and market value*. Academic Press.

Jeong, S., Kim, J. C., & Choi, J. Y. (2015). Technology convergence: What developmental stage are we in? *Scientometrics*, *104*(3), 841–871. doi:10.100711192-015-1606-6

Karvonen, M., & Kässi, T. (2013). Patent citations as a tool for analysing the early stages of convergence. *Technological Forecasting and Social Change*, *80*(6), 1094–1107. doi:10.1016/j.techfore.2012.05.006

Kim, J., & Lee, S. (2017). Forecasting and identifying multi-technology convergence based on patent data: The case of IT and BT industries in 2020. *Scientometrics*, *111*(1), 47–65. doi:10.100711192-017-2275-4

Kim, N., Lee, H., Kim, W., Lee, H., & Suh, J. H. (2015). Dynamic patterns of industry convergence: Evidence from a large amount of unstructured data. *Research Policy*, *44*(9), 1734–1748. doi:10.1016/j.respol.2015.02.001

Kruskal, J. (1977). The relationship between multidimensional scaling and clustering. In Classification and clustering (pp. 17-44). Academic Press.

Lanzolla, G., & Giudici, A. (2017). Pioneering strategies in the digital world. Insights from the Axel Springer case. *Business History*, *59*(5), 744–777. doi:10.1080/00076791.2016.1269752

Lavie, D., & Rosenkopf, L. (2006). Balancing exploration and exploitation in alliance formation. *Academy of Management Journal*, *49*(4), 797–818. doi:10.5465/amj.2006.22083085

Lee, C., Kang, B., & Shin, J. (2015). Novelty-focused patent mapping for technology opportunity analysis. *Technological Forecasting and Social Change*, *90*, 355–365. doi:10.1016/j.techfore.2014.05.010

Lee, C., Park, G., & Kang, J. (2018). The impact of convergence between science and technology on innovation. *The Journal of Technology Transfer*, *43*(2), 522–544. doi:10.100710961-016-9480-9

Lee, S., & Kim, W. (2017). The knowledge network dynamics in a mobile ecosystem: A patent citation analysis. *Scientometrics*, *111*(2), 717–742. doi:10.100711192-017-2270-9

Lee, S., Kim, W., Lee, H., & Jeon, J. (2016). Identifying the structure of knowledge networks in the US mobile ecosystems: Patent citation analysis. *Technology Analysis and Strategic Management*, *28*(4), 411–434. doi:10.1080/09537325.2015.1096336

Lee, S. M., Olson, D. L., & Trimi, S. (2010). The impact of convergence on organizational innovation. *Organizational Dynamics*, *39*(3), 218–225. doi:10.1016/j.orgdyn.2010.03.004

Leydesdorff, L. (2008). Patent classifications as indicators of intellectual organization. *Journal of the Association for Information Science and Technology*, *59*(10), 1582–1597.

Leydesdorff, L., & Vaughan, L. (2006). Co-occurrence matrices and their applications in information science: Extending ACA to the Web environment. *Journal of the Association for Information Science and Technology*, *57*(12), 1616–1628.

Li, F., & Whalley, J. (2002). Deconstruction of the telecommunications industry: From value chains to value networks. *Telecommunications Policy*, *26*(9-10), 451–472. doi:10.1016/S0308-5961(02)00056-3

Loi, M., Castriotta, M., & Di Guardo, M. C. (2016). The theoretical foundations of entrepreneurship education: How co-citations are shaping the field. *International Small Business Journal*, *34*(7), 948–971. doi:10.1177/0266242615602322

Luan, C., Hou, H., Wang, Y., & Wang, X. (2014). Are significant inventions more diversified? *Scientometrics*, *100*(2), 459–470. doi:10.100711192-014-1303-x

Luan, C., Liu, Z., & Wang, X. (2013). Divergence and convergence: Technology-relatedness evolution in solar energy industry. *Scientometrics*, *97*(2), 461–475. doi:10.100711192-013-1057-x

Maitland, C. F., Bauer, J. M., & Westerveld, R. (2002). The European market for mobile data: Evolving value chains and industry structures. *Telecommunications Policy*, *26*(9-10), 485–504. doi:10.1016/S0308-5961(02)00028-9

Marku, E., Castriotta, M., & Di Guardo, M. C. (2017). Disentangling the Intellectual Structure of Innovation and M&A Literature. *Technological Innovation Networks: Collaboration and Partnership*, 47.

Marku, E., & Zaitsava, M. (2018). Smart Grid Domain: Technology Structure and Innovation Trends. International Journal of Economics. *Business and Management Research*, *2*(4), 390–403.

Matt, C., Hess, T., & Benlian, A. (2015). Digital transformation strategies. *Business & Information Systems Engineering*, *57*(5), 339–343. doi:10.100712599-015-0401-5

McCain, K. W. (1990). Mapping authors in intellectual space: A technical overview. *Journal of the American Society for Information Science*, *41*(6), 433–443. doi:10.1002/(SICI)1097-4571(199009)41:6<433::AID-ASI11>3.0.CO;2-Q

McKelvey, B., Tanriverdi, H., & Yoo, Y. (2016). Complexity and Information Systems. Research in the Emerging Digital World. *Management Information Systems Quarterly*.

Mêgnigbêto, E. (2017). Controversies arising from which similarity measures can be used in co-citation analysis. *Malaysian Journal of Library and Information Science*, *18*(2), 25–31.

Nicita, A., Ramello, G. B., & Scherer, F. M. (2005). Intellectual property rights and the organization of industries: New perspectives in law and economics. *International Journal of the Economics of Business*, *12*(3), 289–296. doi:10.1080/13571510500299029

Niemann, H., Moehrle, M. G., & Frischkorn, J. (2017). Use of a new patent text-mining and visualization method for identifying patenting patterns over time: Concept, method and test application. *Technological Forecasting and Social Change*, *115*, 210–220. doi:10.1016/j.techfore.2016.10.004

Oh, C., Cho, Y., & Kim, W. (2015). The effect of a firm's strategic innovation decisions on its market performance. *Technology Analysis and Strategic Management*, *27*(1), 39–53. doi:10.1080/09537325.2014.945413

Park, H., & Yoon, J. (2014). Assessing coreness and intermediarity of technology sectors using patent co-classification analysis: The case of Korean national R&D. *Scientometrics*, *98*(2), 853–850. doi:10.100711192-013-1109-2

Peppard, J., & Rylander, A. (2006). From value chain to value network: Insights for mobile operators. *European Management Journal*, *24*(2-3), 128–141. doi:10.1016/j.emj.2006.03.003

Peppard, J., & Ward, J. (2016). *The strategic management of information systems: Building a digital strategy*. John Wiley & Sons.

Reis, J., Amorim, M., Melão, N., & Matos, P. (2018, March). Digital Transformation: A Literature Review and Guidelines for Future Research. In *World Conference on Information Systems and Technologies* (pp. 411-421). Springer.

Riedl, R., Benlian, A., Hess, T., Stelzer, D., & Sikora, H. (2017). On the relationship between information management and digitalization. *Business & Information Systems Engineering*, *59*(6), 475–482. doi:10.100712599-017-0498-9

Rosenberg, N. (1963). Technological change in the machine tool industry, 1840–1910. *The Journal of Economic History*, *23*(04), 414–443. doi:10.1017/S0022050700109155

Schweer, D., & Sahl, J. C. (2017). The Digital Transformation of Industry–The Benefit for Germany. In *The Drivers of Digital Transformation* (pp. 23–31). Cham: Springer. doi:10.1007/978-3-319-31824-0_3

Singh, A., & Hess, T. (2017). How Chief Digital Officers Promote the Digital Transformation of their Companies. *MIS Quarterly Executive*, *16*(1).

Soete, L. G., & Wyatt, S. M. (1983). The use of foreign patenting as an internationally comparable science and technology output indicator. *Scientometrics*, *5*(1), 31–54. doi:10.1007/BF02097176

Soule, D. L., Puram, A. D., Westerman, G. F., & Bonnet, D. (2016). *Becoming a Digital Organization: The Journey to Digital Dexterity*. Academic Press.

Stolterman, E., & Fors, A. C. (2004). Information technology and the good life. In *Information Systems Research* (pp. 687–692). Boston, MA: Springer. doi:10.1007/1-4020-8095-6_45

Suh, J., & Sohn, S. Y. (2015). Analyzing technological convergence trends in a business ecosystem. *Industrial Management & Data Systems*, *115*(4), 718–739. doi:10.1108/IMDS-10-2014-0310

Suh, Y., & Lee, H. (2017). Developing ecological index for identifying roles of ICT industries in mobile ecosystems: The inter-industry analysis approach. *Telematics and Informatics*, *34*(1), 425–437. doi:10.1016/j.tele.2016.06.007

Sung, K., Kim, T., & Kong, H. K. (2010). Microscopic approach to evaluating technological convergence using patent citation analysis. In T.-h. Kim, J. Ma, W. c. Fang, B. Park, B.-H. Kang, & D. Ślęzak (Eds.), *U-and E-Service, Science and Technology* (pp. 188–194). Berlin: Springer. doi:10.1007/978-3-642-17644-9_21

Teece, D. J. (2018). Profiting from innovation in the digital economy: Enabling technologies, standards, and licensing models in the wireless world. *Research Policy*, *47*(8), 1367–1387. doi:10.1016/j.respol.2017.01.015

Tijssen, R. J. (1992). A quantitative assessment of interdisciplinary structures in science and technology: Co-classification analysis of energy research. *Research Policy*, *21*(1), 27–44. doi:10.1016/0048-7333(92)90025-Y

Trajtenberg, M. (1990). A penny for your quotes: Patent citations and the value of innovations. *The Rand Journal of Economics*, *21*(1), 172–187. doi:10.2307/2555502

Tseng, F. M., Hsieh, C. H., Peng, Y. N., & Chu, Y. W. (2011). Using patent data to analyze trends and the technological strategies of the amorphous silicon thin-film solar cell industry. *Technological Forecasting and Social Change*, *78*(2), 332–345. doi:10.1016/j.techfore.2010.10.010

Urbach, N., Drews, P., & Ross, J. (2017). Digital business transformation and the changing role of the IT Function. *Comments on the special issue. MIS Quarterly Executive*, *16*(2).

Valentini, G., & Di Guardo, M. C. (2012). M&A and the profile of inventive activity. *Strategic Organization*, *10*(4), 384–405. doi:10.1177/1476127012457980

van Eck, N. J., & Waltman, L. (2007). Bibliometric mapping of the computational intelligence field. *International Journal of Uncertainty, Fuzziness and Knowledge-based Systems*, *15*(05), 625–645. doi:10.1142/S0218488507004911

van Eck, N. J., Waltman, L., Dekker, R., & van den Berg, J. (2010). A comparison of two techniques for bibliometric mapping: Multidimensional scaling and VOS. *Journal of the American Society for Information Science and Technology*, *61*(12), 2405–2416. doi:10.1002/asi.21421

van Eck, N. J., Waltman, L., van den Berg, J., & Kaymak, U. (2006). Visualizing the computational intelligence field [Application Notes]. *IEEE Computational Intelligence Magazine*, *1*(4), 6–10.

Waaijer, C. J., van Bochove, C. A., & van Eck, N. J. (2010). Journal Editorials give indication of driving science issues. *Nature*, *463*(7278), 157–157. doi:10.1038/463157a PMID:20075899

Wang, Y., Kung, L., & Byrd, T. A. (2018). Big data analytics: Understanding its capabilities and potential benefits for healthcare organizations. *Technological Forecasting and Social Change*, *126*, 3–13. doi:10.1016/j.techfore.2015.12.019

Yoo, Y., Boland, R. J. Jr, Lyytinen, K., & Majchrzak, A. (2012). Organizing for innovation in the digitized world. *Organization Science*, *23*(5), 1398–1408. doi:10.1287/orsc.1120.0771

Zhu, K., Kraemer, K. L., & Xu, S. (2006). The process of innovation assimilation by firms in different countries: A technology diffusion perspective on e-business. *Management Science*, *52*(10), 1557–1576. doi:10.1287/mnsc.1050.0487

Zupic, I., & Čater, T. (2015). Bibliometric methods in management and organization. *Organizational Research Methods*, *18*(3), 429–472. doi:10.1177/1094428114562629

Chapter 11
Entrepreneurship and Innovation in the Digitalization Era:
Exploring Uncharted Territories

Wassim Aloulou
Al Imam Mohammad Ibn Saud Islamic University, Saudi Arabia & University of Sfax, Tunisia

ABSTRACT

Digital transformation is not a new phenomenon. Neither is digital entrepreneurship. But during the last decade, these phenomena are taking another dimension with the emergence of new digital-to-disruptive technologies that need to be mastered by individuals, groups, firms, organizations, and governments. Based on key concepts such as digitalization, entrepreneurship, and innovation, this chapter contributes to the literature on digital entrepreneurship and innovation by adopting an ecosystem approach. Then, this chapter provides an overview of the digital entrepreneurship and innovation ecosystem and its main components. Within this new philosophy of digital entrepreneuring, the chapter presents new trendy phenomena as precursors and enablers to boost digital entrepreneurial ventures and certain uncharted territories that need to be explored. At the end, the chapter advances new directions for future research in digital entrepreneurship and innovation. It concludes with the idea of democratization gained for entrepreneurship, innovation, and digitalization in this era.

INTRODUCTION

We stand on the brink of a technological revolution that will fundamentally alter the way we live, work, and relate to one another. Klaus Schwab, Founder and Executive Chairman, World Economic Forum.

Influenced by the dramatic and unprecedented evolution of digital-to-disruptive technologies (big data, IOT, AI, AR, VR, Blockchain, drones, 3D/4D printing, robotics and machine learning, autonomous cars…), we witness a technological shifting and deeper transformation of businesses in several activities,

DOI: 10.4018/978-1-5225-7262-6.ch011

value chains and industries. The technology world has evolved since the emergence of such technologies and many of them have been widespread commercial deployment by individuals, firms and governments.

We live in a world where everyone is connected. Digital is everywhere and the digital economy becomes a golden opportunity for entrepreneurs and managers and their firms. In fact, in the digital economy and with the ever-expanding and ever-deepening reach of the Internet, new technologies from internet, mobile, and media have generated opportunities for entrepreneurs and innovative managers alike. Entering an era of the fourth industrial revolution, it is now the time of brilliant and disruptive technologies to gain global competitiveness, prosperity and sustainability (Brynjolfsson & McAfee, 2014; Curley & Salmelin, 2017; Schwab, 2017).

In his book, Kelly (2016) identified 12 inevitable technological forces that will shape the future and represent the momentum of an ongoing technological shift toward digital technologies, not just North America, but to the entire world. According to him, much of what will happen in the next decades is inevitable. The future will bring with it even more screens, tracking, and lack of privacy.

Being outlined in his book, these twelve trends will forever change the ways in which we work, learn and communicate. In other words, increased the fast-moving system of technology amplifies the following forces: Becoming, Cognifying, Flowing, Screening, Accessing, Sharing, Filtering, Remixing, Interacting, Tracking, Questioning, and then Beginning. In fact, *sharing* and collaborating at mass-scale both encourages increased *flowing* in real-time for everything and depends upon it. *Cognifying* by making everything much smarter using cheap powerful Artificial Intelligence requires *tracking* of everything for the benefit of citizens and consumers (diet, fitness, sleep patterns, moods, blood factors, genes, location, and so on). *Screening* or turning all surfaces into screens is inseparable from *interacting* and being immersed in a different world (Virtual Reality). According to Kelly (2016), the verbs themselves can be *remixed*, and all of these actions are variations on the process of *becoming within* a series of endless upgrades. Thus, a*ccessing* means having access to services at all times without owning them and a *remixing* of products in all possible ways to harness intense personalization through *filtering* and anticipating customers' desires. It's the *beginning of a* planetary system constructed to connect all humans and machines into a global matrix.

The technological shifting toward digital-to-disruptive technologies has impacted the ways of creating, doing and innovating businesses but also the ways of transforming established ones. According to the European Commission (2017), digital technologies diffusion has been growing rapidly over the years and is expected to continue to expand around 8% of GDP in 2015 to around 25% by 2030. Accordingly, calling firms to optimize digital investments to realize higher productivity and growth, Knickrehm et al., (2016) showed how the smarter use of digital skills, technologies and other assets could boost productivity and generate US$2 trillion of additional economic output by 2020.

Since entrepreneurship and innovation have become the watchwords for so many individuals and organizations seeking for prosperity (Acs et al., 2017c), digital entrepreneurship and innovation has been viewed also as a critical pillar for economic growth, job creation and competitiveness by many countries (European Commission, 2015). To boost them successfully, Nepelski et al., (2017) focused on 7 issues published within the European Innovation Policies for the Digital Shift (EURIPIDIS) project and related to capacity building; entrepreneurial culture; ecosystem and collaborative interactions between various players under an easy technological interoperability; adequate funding for the scaling-up of digital businesses; and a balance between provision of incentives to create new products and the stimulation of knowledge dissemination.

This brief description does not mean that digital transformation is a new phenomenon (Lanzolla & Anderson, 2008), neither digital entrepreneurship (Rosenbaum & Cronin, 1993; Hull et al., 2007). But during the last decade, the phenomenon is taking another dimension with the emergence of new digital and disruptive technologies (Rogers, 2016). It is taking its road all over the world, in different countries (e.g., European Commission, 2015; Meltzer & Pérez, 2016) and several sectors of manufacturing, automotive and creative industries (Li, 2018; Li et al., 2017; Liere-Netheler et al., 2018; Piccinini et al., 2015). Such phenomenon of transformation needs to be mastered in order to benefit the whole ecosystem and drive global prosperity (Caudron & Van Peteghem, 2014; Hanna, 2016).

New, established and incumbent businesses have to be prepared to cope with digital transformation and disruption by mastering disruptive business models in the way of creating, proposing and capturing value for customers (Rogers, 2016). A huge number of opportunities are offered by the technological disruption.

Having this in mind, firms, organizations and governments have to invest heavily in digital talent to develop skills and fill in the digital gap. They also have to invest in different digital infrastructures for boosting entrepreneurship and innovation and digital life in general (Denison, 2017; Sussan et al., 2016).

The research in digital entrepreneurship and innovation has emerged in the last decade and it is much closer to the information systems' concepts of artifacts, platforms, and information infrastructure (Giones & Brem, 2017). However, entrepreneurship research has ignored both the role that digital technologies play in entrepreneurship and the role that users and agents play in digital entrepreneurship within the digital entrepreneurial ecosystem (Sussan & Acs, 2017). Several mainstreams of research on digital entrepreneurship and innovation have been recently identified by the pioneering works of Kraus et al. (2018) and Zhao & Collier (2016) among others.

This chapter is organized as follows: in section "Background", we present the main ideas related to the concepts of digitization, digitalization and digital transformation and disruption. Then, the concepts of digital-to-disruptive technologies; digital entrepreneurship and innovation and digital entrepreneurship ecosystem. Certain labels and uncharted territories of research related to our main concepts are explored. In section "Future research directions", we present some urgent calls from confirmed researchers to deepen the study of the phenomenon of digital entrepreneurship and innovation in order to be understood by the research community. In the end, we conclude the chapter with the idea of democratization gained for entrepreneurship, innovation and digitalization in this era.

BACKGROUND

The chapter explores the main background and ideas related to key concepts of "digital", "entrepreneurship" and "innovation".

Digitization, Digitalization, Digital Transformation and Digital Disruption: The Differences?

The terms "digitization" and "digitalization" will both be used but they are not interchangeable. In this chapter we intend to distinguish between them according to the recent literature in the field. Digitization is creating a second economy that's vast, automatic, and invisible – thereby bringing the biggest change since the Industrial Revolution (Arthur, 2011). For businesses the only reasonable reaction to persistent digitization is digitalization (Schreckling & Steiger, 2017). And the digital transformation is a process of change allowed to grow and taken place by the digitized data used (Gobble, 2018).

Digital transformation is affecting all sectors of society, in particular the economy. It opens up new networking possibilities and cooperation between different actors, who, for example, exchange data and thus initiate processes.

There is a huge pressure, now, for actors to begin such transformation that needs to be mastered (Caudron & Van Peteghem, 2014; Hanna, 2016b; Schallmo & Williams, 2018).

Digitization

Schreckling, & Steiger (2017) defined digitization as the process of changing from analog to digital form, is inevitable, irreversible, tremendously fast, and ubiquitous.

They enumerated four main drivers of digitization including the digital technology breakthroughs; changes in people's behavior, attitudes and expectations; comparatively low barriers to entry (allowing anyone with an internet connection and a great idea to become an entrepreneur, even with limited capital); and the availability of huge amounts of venture capital (available from investors who are looking for profitable opportunities outside the old economy).

Be that as it may, the benefits of digitization are many and include, inter alia, the easier access to digitized information anytime and anywhere helping companies to make faster business decisions; better real-time tools for communication improving consumer satisfaction; the cutting down of costs, the increase of the value efficiency and profitability. Briefly, digitalization drives new or disruptive business models or fundamental business strategies (Gobble, 2018; Li, 2018)

Digitalization

Digitization is the technical process, whereas digitalization is a socio-technological process of applying digitization techniques to broader social and institutional contexts that render digital technologies infrastructure (Autio, 2017; Autio et al., 2017).

A digital business consultancy, I-SCOOP (2016), offers a concise definition of digitalization by meaning "*the use of digital technologies and of data (digitized and natively digital) in order to create revenue, improve business, replace/transform business processes (not simply digitizing them) and create an environment for digital business, whereby digital information is at the core.*"

Schallmo & Williams (2018) used this definition and considered digitalization as fundamental changes made to business operations and business models based on newly acquired knowledge gained via digitization initiatives that add value in new ways.

Digitalization has become the new and indispensable tool for companies of the entire value chain – from design to product development, manufacturing, transportation and logistics through to sales. It has the potential to drive economic and productivity growth and increase sales and economic competitiveness for companies all over the world (Abolhassan, 2017; OECD, 2016). Moreover, it creates potent digital affordances that likely have a transformative effect upon the organization of economic activity by supporting radical business model innovation (Nambisan et al., 2017).

This direct contribution of digitalization to economic growth is enhanced through the adoption, use and diffusion of ICT goods and services across the economy and in many businesses (OECD, 2016)

Digital Transformation

The big buzzword was business transformation. Now, no one cannot even talk about business transformation without considering digital transformation.

The digital transformation of today goes much further. Companies are now using information technologies to develop fundamentally new business models, products and services (smarter with new solutions). It is the process of changing and convert the way of doing business to handle the emerging disruption and challenges in one industry.

Because most of industries including media, retail, healthcare, pharma, finance, recruitment, communication among others are currently transforming at high speed due to digital technologies that have fundamentally changed the way of living, working and playing. No industry is immune.

There is now an imposing need to begin and master digital transformation in order not to run the risk of falling behind (Hanna, 2016; Caudron & Van Peteghem, 2014). Big tech players like Google, Amazon and Apple are dictating their own universal rules and new challengers like Uber, Nest (acquired by Google), Tesla, AirBnB and others are disrupting one industry after the other (Caudron & Van Peteghem, 2014).

Digital Transformation is everything and everywhere. Underpinning this transformation is the constantly evolution of technology and business models by competition.

Digitalization is no longer a choice but an imperative and the motto is digitalize or drown (Schreckling & Steiger, 2017) or even die (Mazzone, 2014).

Digital transformation is expected to change the whole process of value creation in industries (Fitzgerald et al., 2014; Liere-Netheler et al., 2018). In order to smash and conquer, businesses need a mindset that keeps their business model constantly digitally evolving and consequently the customer value creation and operation organization for delivery simultaneously refined (Schreckling & Steiger, 2017; Berman, 2012; Yoo et al., 2012).

The impacts of the digital transformation can begin to emerge on different levels: the profound and accelerating transformation of customer relations, business activities and operational processes, competencies and the revitalization of business models to fully leverage the changes and opportunities brought by digital technologies and their impact across society in a strategic and prioritized way (Westerman et al., 2011; Demirkan et al., 2016). It means the use of new digital technologies (social media, mobile, analytics or embedded devices) to enable major business improvements (such as enhancing customer experience, streamlining operations or creating new business models) (Fitzgerald et al., 2014; Henriette et al., 2015; 2016). It also means a new development in the use of digital artifacts, systems and symbols within and around organizations (Bounfour, 2016; Abolhassan 2017; Nambisan, 2017).

Rogers (2016) identified the 5 domains of strategy to address carefully digital transformation:

1. **Customers:** With the use of digital tools, they are dynamically connected and interacting in ways that are changing their relationships to business and to each other;
2. **Competition:** Businesses compete and cooperate (co-opte) with other firms via platform business models and then each seek to gain more leverage in serving the ultimate consumer;
3. **Data:** Generated from every conversation, interaction, or process inside or outside these businesses, utilized with new analytical tools and then turned into a truly strategic asset. Such digital tools allow firms to make new predictions, uncover unexpected patterns in business activity, and unlock new sources of value through rapid and iterative learning and experimentation;

4. **Innovation:** Enabled by digital technologies and the approach to innovation focused on careful and rapid experiments and on minimum viable prototypes that maximize learning while minimizing cost of failures and improving organizational learning;
5. **Value:** An adapted value proposition delivered to the customers.

Lastly, the digital transformation potentially brings value for firms and society at large if it is fundamentally mastered with a focus on strategic thinking more than on technology (Kane et al., 2015; Rogers, 2016). Digital technologies should be a means and not an end to drive the digital transformation.

Digital Disruption

The digital disruption refers to the use of new disruptive technologies to disrupt business models and industries. Rogers (2016, p. 195) defines disruption as "*disruption happens when an existing industry faces a challenger that offers greater value to the customer in a way that existing firms cannot compete with directly*". In introducing the most powerful theories of disruption, Rogers cited Joseph Schumpeter's creative destruction as an industry disruption and Clayton Christensen's Innovator's Dilemma as disruptive technology.

The power of digital disruption is that it can disrupt any aspect of any product or service, including processes deep within companies focused on physical things, processes that govern partnerships, data collection, pricing, and the management of labor or capital resources (Mc Quivey, 2013). In fact, its power multiplies precisely because it can apply to industries that are not even digital. In this way, digital disruption happens to and through digital things, which then accelerate the disruption of physical things. In sum, digital disruption is the transformative impact produced by digitalization on how business, economy, and the society operate.

According to Prinsloo & James (2015), scholars suggest that digital disruption has so far undergone three main waves of change: 1st wave with the dotcom era and the emergence of Internet-based companies such as Amazon, Google and Netflix in early 2000s; 2nd wave with the Web 2.0 era diminishing the importance of economies of scale and leaving room for networks, community and the enabling of open sourced developers; 3rd wave with the use of Internet of Things and big data technologies with the integration of multiple other applications and the hyper-scaling of global online businesses.

In this respect, Uber can be a great example of digital disruption in the Taxi industry. This industry has been disrupted by providing customers with substantially greater value, without even being in the same industry. The average taxi company will have less chance to replicate the same model.

The digitalization of every industry is leading to ever more battles between incumbents and digitally powered disrupters (Rogers, 2016).

Digital-to-disruptive technologies are rewriting the rules of business and creating new opportunities for the disruptors (e.g., iPhone in mobile phone manufacturing sector; iTune in music industry…).

Emerging Digital-to-Disruptive Technologies Creating the Wave of Innovation for Businesses

In order to help companies focus their emergent technology efforts, PricewaterhouseCoopers, a multinational professional services network, analyzed the business impact and commercial viability of more than 250 emerging technologies to zero in on what they called the "Essential Eight." These are the core technologies that matter most for business, across every industry, over the next three to five years. Figure 1 illustrates these essential emerging technologies: Internet of Things; Augmented reality; Virtual reality; Blockchain, Artificial intelligence; 3D printing; Drones; Robots.

These Essential Eight are considered as catalysts for digital transformation. They are shifting the capabilities and value proposition of global manufacturing and business. They will have a profound global impact on business, employees, and customers and across all the aspects of strategy, customer engagement, operations, talents and compliance. They will provide an unprecedented horizon of possibilities for who can master them (e.g., Sussan et al., 2016; Iansiti & Lakhani, 2014).

Digital Entrepreneurship and Innovation for a Global Prosperity

Entrepreneurship and innovation are considered as powerful engines and tools for a modern knowledge-based economy and for driving economic growth, job creation and prosperity (Al-Mubaraki et al., 2014; Curley & Salmelin, 2017; Denison, 2017; Acs et al., 2017c).

Technological innovations have become the new economic force that will fundamentally change the way we live and work. They bear enormous potential for economic prosperity and social progress. We are witnessing a reshaping of traditional business models, business strategies, processes and organizational structures caused by the emergence of digital technologies. Furthermore, digital entrepreneurship has been viewed as an engine for economic development, job creation, and innovations by several countries (Santana, 2017). For instance, entrepreneurship based on (disruptive) digital technologies has made significant influence on behaviors and actions of entrepreneurs, their entrepreneurial processes of development and on performance of their ventures.

Figure 1. Essential eight of emerging digital-to-disruptive technologies
Source: http://usblogs.pwc.com/emerging-technology/a-guide-to-the-essential-eight-emerging-technologies/

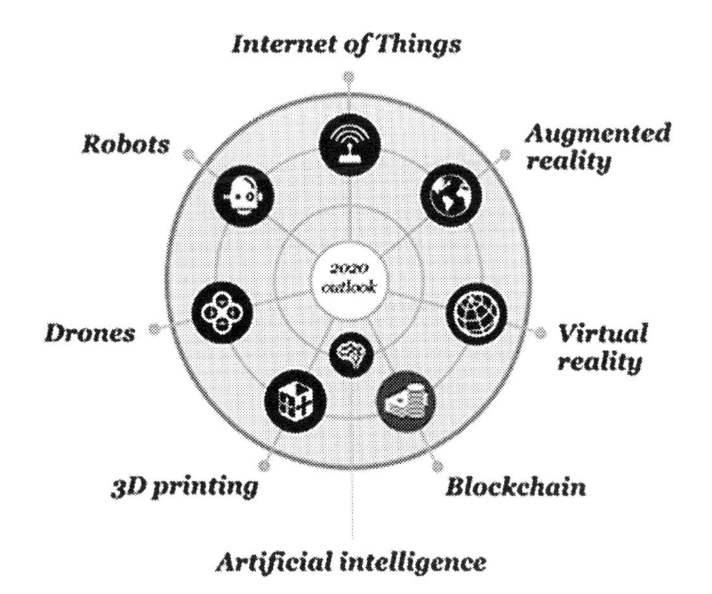

Moreover, several countries have intended to stimulate digital innovation for growth and inclusiveness (OECD, 2016) and prosperity and sustainability (Curley & Salmelin, 2017; Denison, 2017). They have relied on knowledge and information to transform their enterprises and industries (Andersson et al., 2009; European Commission, 2014; 2015).

In the next sections, the light will be shed on the increasingly important the phenomenon of digital entrepreneurship and innovation from a digital technology perspective (Nambisan, 2017; Davidson & Vaast, 2010).

DIGITAL ENTREPRENEURSHIP AND INNOVATION

Defining Entrepreneurship [1]

Research on entrepreneurship is still under a theoretical construction. Four dominant paradigms are emerging since the two last decades considering entrepreneurship as involving the recognition and the seizing of opportunities, transforming those opportunities into marketable goods or services, adding value through time and resources, assuming risk and realizing rewards (Shane & Venkataraman, 2000); or as a process of creating and emerging an organization (Gartner et al., 1992), or as an innovation by perceiving and creating new economic opportunities and by introducing new innovative ideas in the market (Schumpeter, 1934), or as a process of creation of new value by entrepreneur(s) (Bruyat & Julien, 2001).

Entrepreneurship may occur in a variety of settings, including new and old ventures, non-profit organizations and the public sector (Hull et al., 2007; Santana, 2017).

Several scholars argued that entrepreneurship has a complex and dynamic nature (Aloulou, 2016; Berger & Kuckertz, 2016; Gartner, 1992). With the presence of digital technologies and the flourishing digitalization era, entrepreneurship and innovation become more complex. The next paragraphs describe the phenomenon of digital entrepreneurship and innovation.

Digital Entrepreneurship: Definition and Types

While the phenomenon has been used for some researchers and practitioners and becomes a buzzword, its conceptualization remains obscure. The term was used broadly in recent studies, but dealing with a diversity of conceptions. Its conceptualization remains quite elusive (Zhao & Collier, 2016).

Mainly, digital entrepreneurship as a concept has two characteristics: digitalization and entrepreneurship. It is at the intersection of digital technologies and entrepreneurship and can be defined according to a digital technology perspective (Nambisan, 2017).

According to the most widely accepted definition, digital entrepreneurship is described as a specific type of technology entrepreneurship or entrepreneurial practice focused on the pursuit of new venture opportunities through new digital media, internet/information, digital-to-disruptive technologies on which new products and services are based (Davidson & Vaast, 2010; Giones & Brem, 2017; Nambisan, 2017). Their products or services can be entirely digital and have no physical substance (Berthon et al., 2014).

Digital entrepreneurship is a subcategory of entrepreneurship in which some or all of what would be physical in a traditional organization has been digitized. In fact, Hull et al. (2007) presented a framework that includes a typology of digital ventures and explores the potential of digitalization within the activities, processes, boundaries and relationships associated with the firm.

The degree of digitalization may be derived through: (1) the degree of digital marketing undertaken by a firm, (2) a firm's digital selling, (3) the digital nature of a firm's good or service, (4) the digital distribution potential of a good or service, (5) the potential digital interactions with key external stakeholders within the value chain, and (6) the digital potential of virtual internal activities associated with a firm's operation. Thus, digital entrepreneurship implies entrepreneurship involving digital goods or services, digital distribution, a digital workplace, a digital marketplace, or some combination of these. It relies on the use of ICT as the basic infrastructure.

When investigating the phenomenon, Davidson and Vaast, (2010) have argued that in digital economy entrepreneurship is a multi-faceted phenomenon based on three interrelated types that create opportunities: business, knowledge and institutional entrepreneurship. So, succeeding a digital entrepreneurial venture is a sociomaterial practice means combining business, knowledge and institutional opportunities sequentially, iteratively or synergistically. Such Sociomateriality is referred to the duality of social and material and interaction between human and material actors (Davidson and Vaast, 2010). It emerges from the mutual exploitation, adjustment and enactment of means-end relationships between human and non-human actors including the entrepreneur and the digital artifact combined in action. With such conception, the phenomenon is much closer to the information systems' concepts of artefacts, platforms, and information infrastructure (Nambisan, 2017).

Entrepreneurial actions entail close combination of technology affordances (internet, new media, mobile…) with social practices and the digital entrepreneurial phenomena are inextricably inscribed into a context that is not only socially and historically situated but also technically and materially defined (Davidson and Vaast, 2010, p.4).

Taking a broadly approach, the European Commission (2015) defined digital entrepreneurship as "creating new ventures and transforming existing businesses that drive economic and/or social value by developing and/or using novel digital technologies (particularly social, big data, mobile and cloud solutions) to improve business operations, invent new business models, sharpen business intelligence, and engage with customers and stakeholders through new (digital) channels".

European Commission (2014) identified five 'pillars' in its conceptual model of digital entrepreneurship, each of which is relevant in the analysis of digital entrepreneurship. The phenomenon rests on five pillars:

1. Digital knowledge base and ICT market: enhancing digital innovation, commercialization and the ICT sector;
2. Digital business environment: strengthening digital infrastructure, the regulatory framework and improving ease of doing business;
3. Access to finance: facilitating access to finance and enhancing digital investments;
4. Digital skills and e-leadership: fostering e-leadership skills through education and training;
5. Entrepreneurial Culture: creating supportive entrepreneurial culture.

The commission expressed its commitment to working towards the deployment and implementation of this five-pillar plan.

In their pioneering work, Hull et al., (2007) proposed a typology of digital entrepreneurship: mild; moderate; and extreme digital entrepreneurship. They advanced that the distinction between these different types can be seen in the context of ease of entry, ease of manufacturing and storing, ease of distribution in the digital marketplace, digital workplace, digital goods, digital service, and digital commitment.

In their work, Giones & Brem (2017) proposed the digital technology entrepreneurship as necessarily combining elements of technology and digital entrepreneurship. This type is focused on the identification and exploitation of opportunities based on scientific or technological knowledge through the creation of digital artefacts. Digital technology entrepreneurs build firms based on technologies on the one hand and on services on the other hand. They are embedded in an interconnected system when they aim to commercialize their solutions (digital platforms…) (Giones & Brem, 2017; Nambisan, 2017).

In a relatively recent literature (Baldwin & Von Hippel, 2011; von Briel et al., 2018; Chandra & Leenders, 2012; Kraus et al., 2018; McAdam et al., 2018; Ojala et al., 2018; Rippa & Secundo, 2018), other forms of digital entrepreneurship were identified: digital social entrepreneurship; Digital academic entrepreneurship; digital corporate entrepreneurship; digital User entrepreneurship; women's digital entrepreneurship, international digital entrepreneurship… This variety of forms proofs the most heterogeneous character of digital entrepreneurship that needs to be understood.

Digital Innovation [2]

Innovation is defined by the OECD (2005) as "*the implementation of a new or significantly improved product (good or service), or process, new marketing method, or new organisational method in business practices, workplace organisation or external relations*".

Digital innovation is becoming increasingly important in today's economy. They are developed not within organizations, but in innovation-driven entrepreneurial ecosystems (Li et al., 2017).

According to OECD (2016), digital innovation should be understood:

- In a narrow sense, as the implementation of a new or significantly improved ICT product, i.e. ICT product innovation;
- In a broader sense, as also including the use of ICTs for the implementation of a new or significantly improved product or process, a new marketing method, or a new organisational method in business practices, workplace organisation or external relations, or simply put as ICT-enabled innovation.

Digital innovation is meant "an innovation enabled by digital technologies that leads to the creation of new forms of digitalization" (Yoo at al., 2010), but occasionally also refers to any new and innovative ICT products (ICT-enabled innovation), internet-enabled service innovations or e-innovations that mainly occurs in the supply side (i.e. the ICT producing industries), or in the ICT demand side (across the economy).

Digital innovation can be disruptive, inducing the "creative destruction" of established businesses, markets and value networks, and challenging existing regulatory frameworks. It cannot be always disruptive or revolutionary, and may involve evolutionary improvements.

With respect to Brunswicker et al., (2015), Fichman et al., (2014) and OECD (2016), innovations as digital disruptions, that are based on established digital-to-disruptive technologies can be:

1. Data-driven innovation: digital innovations driven by data where the use of "big data" and analytics is driving knowledge and value creation across society (e.g., in sports sector when Addidas acquired and opened up Runtastic as an health and fitness online community and shut down its

online miCoach platform; or when Nike developed the Nike + ecosystem which connects these physical goods to the Internet where users track activities and share progress with their friends),
2. Digital innovation emerging under the label of the "sharing economy" that is enabled to a large extent by the diffusion of mobile smartphones and big data analytics (e.g., Zipcar, Uber, Airnbn…), and maybe even
3. The digitalisation of industrial production enabled in particular by the (Industrial) Internet of Things and also big data analytics (e.g., in manufacturing, with the use of interconnected production machine and products for the aim of monitoring them and analyzing their performance).

TOWARD AN ECOSYSTEM APPROACH OF DIGITAL ENTREPRENEURSHIP AND INNOVATION

In order to flourish in any country, entrepreneurship and innovation in the era of digitalization need an conducive environment where major stakeholders interact, collaborate and add value to customers and to the community as a whole.

Digital disruption can, and very likely will, shift consumer preferences and the very rules that govern these industries. Even the terminology of business is changing, with talk of 'ecosystems' and 'platforms' replacing the familiar 'value chain' (Prinsloo & James, 2015)

To achieve a digital and entrepreneurial revolution in the making (Isenberg, 2010; Parker et al., 2016; Schwab, 2017), there is a need to conceive and implement a digital entrepreneurship ecosystem as a new paradigm for the digital economic policy (Isenberg, 2011; European Commission, 2017). This digital entrepreneurship ecosystem approach lays in the focus on (productive) digital entrepreneurship as an output of the ecosystem (Isenberg, 2010; Acs et al., 2017b).

A well designed cluster with externalities and specific structural elements, digital entrepreneurship and innovation ecosystems are explicitly organized systems of digital entrepreneurial opportunity discovery and pursuit, of start-up and scale-up of new ventures (e.g., new venture accelerators, co-learning spaces, makerspaces, business angel networks, networking events, and so on) (Acs et al., 2014; Aldrich, 2014…). In such ecosystems, the dominant cluster-level benefit is business model innovation, and not process, product, or 'technology' push innovation, as is the case for flexible production systems and learning and innovation systems (Herrera et al., 2018; Stam & Spigel, 2018).

Digital Innovation Ecosystem

This "digital ecosystem" can be represented as a set of three layers from equipment providers (layer 1), telecom network operators (layer 2), content providers (layer 3), in the app industry for example. The pattern of innovation behaviour is different in each layer and all layers are interdependent and innovations in one layer impact innovations in another layer (Nepelski et al., 2017).

Sussan & Acs (2017) identified two foundation pillars of digital ecosystem: digital technologies and people. For them, digital ecosystem is an ecosystem in that digital technologies (e.g., mobile search engine) can be viewed as the non-living component, and the people who use these technologies (e.g., anyone who uses Google) are the living component, and the interactions of the living and the non-living and the dynamic and continuous changes resulting from the interactions of these two components form the behavior of an ecosystem.

Digital-to-disruptive technologies influence the process and speed of digital innovation and have led to the emergence of network effects beyond the large economies of scale or scope in the digital ecosystem.

Digital Entrepreneurship Ecosystem

Building their argument that entrepreneurship should be treated as a systemic phenomenon at the country level, Acs et al. (2014, p. 476) introduced the novel concept of National Systems of Entrepreneurship (NSE), which are fundamentally resource allocation systems that are driven by individual-level opportunity pursuit, through the creation of new ventures, with this activity and its outcomes regulated by country-specific institutional characteristics. NSE are driven by individuals, with institutions regulating who acts and the outcomes of individual action.

A National System of Entrepreneurship is the dynamic, institutionally embedded interaction between entrepreneurial attitudes, ability, and aspirations, by individuals, which drives the allocation of resources through the creation and operation of new ventures.

In exploring empirically the relationship between economic growth, factor inputs, institutions, and entrepreneurship, Acs et al., (2017a) argue that entrepreneurship operating in particular institutional contexts, perhaps via the National Entrepreneurial Ecosystem, impacts efficiency which measures how effectively given technology and factors of production are actually used.

Based on two dominant lineages of literature (regional development and strategy) rooted in ecological systems thinking, Acs et al., (2017b) outlined contributions to the entrepreneurial ecosystem approach and concluded with a promising new line of research to our understanding of the emergence, growth, and context of start-ups that have achieved great impact by developing new (physical/virtual) platforms centering on network externalities and employing a value chain approach.

In order to guide their understanding of entrepreneurship in the digital economy, Sussan & Acs (2017) integrated two existing ecosystem literatures: the literature surrounding digitization and digital ecosystems with its focus on digital *infrastructure* (socially embedded system) and *users* (technologists who directly interact with digital technologies as co-creators in order to develop new products and services for themselves and other users); and the entrepreneurial ecosystem with its focus on *agency* (see Figure 2).

Their framework is called Digital Entrepreneurial Ecosystem (DEE) and is composed of Schumpeterian entrepreneurs creating digital companies and innovative products and services for many users and agents in the global economy.

For Sussan & Acs (2017), DEE is defined as "*the matching of digital customers (users and agents) on platforms in digital space through the creative use of digital ecosystem governance and business ecosystem management to create matchmaker value and social utility by reducing transactions cost". Its components are related to the existence of a digital infrastructure, governance (policies process and implementations, digital infrastructure access to capital, non-digital infrastructure), digital user citizenship (entrepreneurial culture, mentoring networks), digital entrepreneurship (skills, entrepreneurial culture, life style, culture and creativity) and the digital marketplace (market, knowledge spillovers). The users are people using the Internet and the agents are entrepreneurs (Sussan & Acs, 2017, p. 63).*

The outcome of the DEE is a sustainable system by allowing the birth of new digital entrepreneurs to disrupt existing digital entrepreneurs, change the rule of the game and influence the process of the formation of new regulations regarding the digital economy. The sustainability of the DEE is also related to the increasing savviness of digital users worldwide who continuously become disruptive digital entrepreneurs (Sussan & Acs, 2017, p. 64). The continuous value co-creation between entrepreneur as ambitious, risk-taking and innovative agents and users as active participants in digital marketplace is another route to a sustainable DEE.

Digital affordances derive from the technical architecture of digital infrastructures, and they support an economy-wide redesign of value creation, delivery, and capture processes (Autio et al., 2017; Nambisan, 2017; Spigel, 2017; Stam & Spigel, 2018).

DEE is considered as a new form of organizing, collaborating and cooperating between various entrepreneurship related stakeholders within smart regions and cities to promote local economies and territories Carayannis et al., 2018; Chinta & Sussan, 2018; Carvalho, 2017; Li et al., 2018; dos Santos et al., 2017; Spigel, 2017).

When proposing an interdisciplinary framework to study digital entrepreneurship ecosystem, Sussan & Acs (2017) elucidated the impact of digital infrastructure on entrepreneurship, the role of users in digital ecosystems and the interaction of users and agents.

Digital affordances and infrastructure are made available from the digital ecosystem. They are possibilities to perform existing functions much more effectively than before, or perform entirely new functions with the use of digital-to-disruptive technologies (Autio et al., 2017). Moreover, dedicated support infrastructure is made also available from the entrepreneurship ecosystem (entrepreneurial spaces such as innovation, fab, and living labs and hubs, incubators and accelerators; collaborative spaces for coworkers, makerspaces / hackerspaces makers, and hackers; crowdsourcing and crowdfunding; venture capital; business angels; other special events such as meetups and Hackathon/days) (Aldrich, 2014; Costa & Turvani, 2013; Kelly & Firestone, 2016; Le Dinh et al., 2018).

Within the digital entrepreneurship ecosystem, players, resources and relationship are essential to the activation of such ecosystem under the principles of openness, collaboration, co-creation and reduction of transaction costs (Berthon et al., 2014; Fox & Stucker, 2009; Troxler & Wolf, 2017; van Welsum, 2016). The players in this ecosystem are diverse: digital entrepreneurs/digipreneurs, digital maker-entrepreneurs; digital users; firms and institutions as matchmakers and bridgemakers, buyers and sellers of goods, products and services. They are playing in different (digital) marketplaces, online platforms, or multisided platforms and building collaborative relationships and using matchmaking services (Evans & Schmalensee, 2016).

Figure 2. The integration of the two ecosystems - Digital Entrepreneurship Ecosystem DEE -
Source: Adapted from Sussan & Acs (2017)

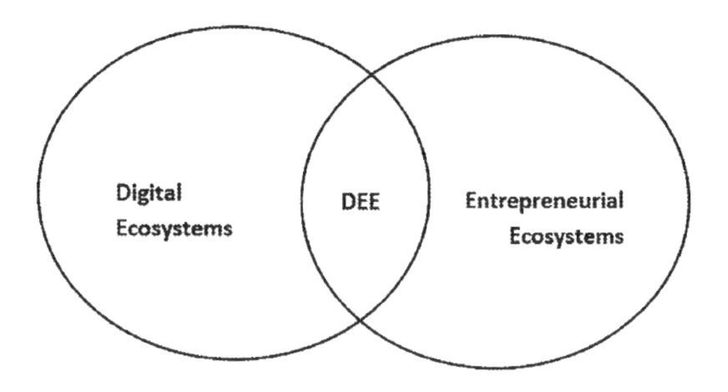

EMERGENT TRENDY PHENOMENA AND UNCHARTED TERRITORIES

There are emergent trendy phenomena that jointly transformed entrepreneurship and innovation in the digitalization era. Such phenomena rendered certain uncharted territories that need to be explored by scholars within digitalization.

Sharing Economy and Platformization

The sharing (shared, collaborative, shar-) economy refers to economic and social systems that enable shared access to physical goods/assets, services, data and talent. This sharing of goods, services, data and talents is commonly possible through online marketplaces, mobile apps/location services or other technology-enabled platforms (Schwab, 2017; Richter et al., 2017).

It's also called platform economy and its platformization is due to the increasing importance of digital platforms as a venue for value creation and delivery like Apple's App Store, Alibaba, eBay, PayPal, Facebook (Nambisan et al., 2018; Cusumano, 2017; Lin et al., 2017; de Reuver et al., 2018; Hsieh et al., 2018).

The platform revolution is transforming the economy and making the fundamental shift in how businesses relate to each other (from linear to more networked business models) and playing a strategic role of creating value in all types of businesses (asset-light, asset-heavy, and any variation in between as mixed systems) in a wide range of industries such as retail, media, advertising, finance, gaming, mobile computing, business software, hospitality and tourism, gaming, home appliance, transportation, education, recruiting and job search, freelance work, philanthropy and social initiaitves (Rogers, 2016; Schreckling & Steiger, 2017).

The sharing economy is considered as a precursor for entrepreneurship and innovation business models in the era of digitalization. It helps digital entrepreneurs with the evolving digital technologies as external enablers at an exponential pace to renovate long-standing industries by disrupting their business models and proposing new offers through web applications and digital platforms (von Briel et al., 2018; Curley & Salmelin, 2017; Evans & Schmalensee, 2016; Henriette et al., 2016; Kenney & Zysman, 2015; Parker et al., 2016; de Reuver et al., 2017; Richter et al., 2015; 2017; Stemler, 2017).

Research addressing the so-called 'sharing economy' is still in its infancy (Richter et al., 2017; 2015) and the connection between this up-and-coming field and entrepreneurship has not been scientifically proven until the work of Richter (2015) when this research gap was remedied.

Open Innovation for Collaboration, Co-Creation and Coopetition

The concept of open (digital) innovation involves a shift towards more open and distributed models of innovation beyond the own boundaries of organizations (Nambisan et al., 2018; West & Bogers, 2018). Like platform strategy, open innovation can facilitate entrepreneurial pursuits in different ways. It underlies openness when sharing of knowledge and assets via ecosystems, platforms and digital communities. It creates numerous entrepreneurial opportunities for new and established firms with demonstrated business models and well-established markets (Nambisan et al., 2018; Nambisan, 2017; Curley & Salmelin, 2017). It redefines the nature of partnerships and collaboration involved in entrepreneurial pursuits in order to co-create value (Cuc et al., 2018). It encourages the coopetition (mix of cooperation and competition) between entrepreneurial firms as fundamental in their collaborative dynamics.

Business Model Innovation

According to Westerman et al., (2014), we consider that digital-to-disruptive technologies are driving business model innovation in five different archetypes:

- Reinventing industries by responding to fundamentally new consumer behaviors or involving a substantial reshaping of an industry structure;
- Substitution is about substituting core products or services by a new digital format;
- New digital businesses involving the creation of new products and services that generate additional revenues;
- Reconfiguration of value delivery models by recombining products, services, and data to change the way a firm plays in the value chain;
- Rethinking of value proposition by using new digital capabilities to target unmet needs for existing or new customers.

These technologies are reshaping the competitive landscape as firms develop new means of value creation, delivery, and capture (Vendrell-Herrero et al., 2018).

Since digital entrepreneurs (especially new entrants) are facing business model innovation and disruption from disrupters, they are needing to gain strategic agility through pragmatic methods such as lean startup approach or Disruptive Business Model Map (Ghezzi & Cavallo, 2018; Fichman et al., 2014; Rogers, 2016).

In the manufacturing sectors, the impact of such technologies (e.g., industrial Internet of Things) is real. Manufacturers are changing their offerings through a (digital) servitization (smarter products and services, from product to solutions; or to platform). Entrepreneurial opportunities are mainly related to servitized businesses that need to enhance traditional nondigital products with the implementation of digital technologies (Vendrell-Herrero et al., 2018; Zhu & Furr, 2016).

Digital Startups and Ventures

Digital startups are acting as catalysts of a new digital entrepreneurial ecosystem. Different types of startups are emerging from this digitalization era: Startups of internet were booming since the dotcoms era early in 2000s; Upstarts as new trend in the types of digital business that investor prefer and they have higher valuations (Stone, 2017); Digital Scale-ups as having stabilized their digital business model and getting traction on users and customers, and typically becoming "Gazelle" companies; "Centaurus" as digital scale-ups reaching valuations of more than 100 million dollars; and ultimately "Unicorns" as new type of companies which is disrupting existing businesses while creating billions of dollars in wealth (Acs et al., 2017b; Grilo et al., 2017; Sussan & Acs, 2017).

The challenge of scholars is to understand these digital startups in the sharing economy (e.g., their process of founding, of pursuing digital opportunities, of growth…)

FUTURE RESEARCH DIRECTIONS

Several scholars claimed for exploring the emerging concept of digital entrepreneurship and innovation from multiple disciplinary perspectives (Davidson and Vaast, 2010; Nambisan, 2017; Nambisan et al., 2017; Zhao & Collier, 2016)

Under the technological complexity brought by the digitalization, digital entrepreneurship and innovation are certainly complex phenomena since that entrepreneurship and innovation are complex phenomena too (Aloulou, 2016; Berger & Kuckertz, 2016). This is challenging scholars to do research in digital entrepreneurship and innovation.

The intersection of digital technologies and entrepreneurship/or innovation provides ample research opportunities as digitization dissolves boundaries and shifts agency of traditional entrepreneurship (Nambisan, 2017) and innovation processes and outcomes (Nambisan et al., 2017). Scholars can benefit from the integration of a profound general entrepreneurship literature with information technology literature in order to do research on digital entrepreneurship (Kraus et al., 2018).

For understanding, exploring and explaining these new phenomena, there is a need for appropriate methods in the context of complexity in entrepreneurship and innovation (Berger & Kuckertz, 2016) and under the digital technology or sociomaterial perspectives (Nambisan, 2017; Nambisan et al., 2018; Davidson and Vaast, 2010). Nzembayie (2017) claimed for an insider action research study for understanding the nature of uncertainty surrounding the digital entrepreneurship and its process. Since entrepreneurship in the digital context is a highly dynamic and fluid process, such research study appears well-suited for use in researching the phenomenon in order to produce new knowledge.

Kraus et al., (2018) have identified and discussed six streams of research that deal with digital entrepreneurship: digital business models; digital entrepreneurship process; platform strategies; digital ecosystem; entrepreneurship education; and social digital entrepreneurship. They invited Entrepreneurship researchers to gain enhanced insights by undertaking cross-sectional and longitudinal studies, and varying the levels of analysis when focusing on digital organizations.

CONCLUSION

The chapter contributed to the literature of entrepreneurship and innovation in the digital era by defining the state-of-the art of the phenomenon of digital entrepreneurship and innovation and its ecosystem, precursors and enablers from the perspective of digital-to-disruptive technologies.

The digital artifacts, technologies and infrastructures have led to the democratization of entrepreneurship (Aldrich, 2014), innovation (Von Hippel, 2009) and digitalization (Curley & Salmelin, 2017). The democratization means the engagement of a greater number and diverse set of people in all stages of the entrepreneurial and innovation processes with designing, testing, funding and implementing new, disruptive and sustaining business models. Furthermore, one of the most effects of the digital revolution has been the democratization of the business world in general with an plenty of opportunities to access to capital and business connections.

The fluid characteristic of the knowledge and information society drives the democratization of entrepreneurship and innovation (in their different forms and archetypes) and induces the emergence of (open) entrepreneurship/innovation 2.0 (and even phenomena 3.0).

The trend toward democratization of entrepreneurship/innovation applies to physical and digitized information products and business processes. It demonstrates the universal empowerment of users, the huge power of such technologies and the emergence of new enterprising actors and new entrepreneurial practices from such democratization.

The reflections on this chapter suggest that several aspects of research in the phenomenon of digital entrepreneurship and innovation invite researchers to deepen conceptual and empirical investigations not only from a technological perspective (Nambisan. 2017; Manbisan et al., 2018), but also from a humane and sociomaterial/sociotechnical perspective (Davidson & Vaast, 2010).

REFERENCES

Abolhassan, F. (2017). Pursuing Digital Transformation Driven by the Cloud. In *The Drivers of Digital Transformation* (pp. 1–11). Cham: Springer. doi:10.1007/978-3-319-31824-0_1

Acs, Z. J., Autio, E., & Szerb, L. (2014). National Systems of Entrepreneurship: Measurement Issues and Policy Implications. *Research Policy*, *43*(1), 476–494. doi:10.1016/j.respol.2013.08.016

Acs, Z. J., Estrin, S., Mickiewicz, T., & Szerb, L. (2017a). *Entrepreneurship, Institutions and Productivity Growth: A Puzzle*. Available at SSRN: https://ssrn.com/abstract=3060982

Acs, Z. J., Stam, E., Audretsch, D. B., & O'Connor, A. (2017b). The lineages of the entrepreneurial ecosystem approach. *Small Business Economics*, *49*(1), 1–10. doi:10.100711187-017-9864-8

Acs, Z. J., Szerb, L., & Lloyd, A. (2017c). Entrepreneurship and the Future of Global Prosperity. In *Global Entrepreneurship and Development Index 2017* (pp. 11–27). Cham: Springer. doi:10.1007/978-3-319-65903-9_2

Al-Mubaraki, H. M., Muhammad, A. H., & Busler, M. (2014). *Innovation and entrepreneurship: Powerful tools for a modern knowledge-based economy*. Springer.

Aldrich, H. E. (2014). The democratization of entrepreneurship? Hackers, makerspaces, and crowdfunding. *Annual Meeting of the Academy of Management*.

Aloulou, W. (2016). Understanding entrepreneurship through the chaos and complexity perspective. In *Şefika Şule Erçetin and Hüseyin Bağcı, Handbook of Research on Chaos and Complexity Theory in the Social Sciences*. IGI Global.

Arthur, W. B. (2011). The second economy. *The McKinsey Quarterly*, *4*, 90–99.

Autio, E. (2017). *Digitalisation, Ecosystems, Entrepreneurship and Policy. Perspectives into Topical Issues Is Society and Ways to Support Political Decision Making. Government's Analysis, Research and Assessment Activities Policy Brief 20/2017*. Helsinki: Prime Minister's Office.

Autio, E., Nambisan, S., Thomas, L. D., & Wright, M. (2017). Digital affordances, spatial affordances, and the genesis of entrepreneurial ecosystems. *Strategic Entrepreneurship Journal*, *12*(1), 72–95. doi:10.1002ej.1266

Berger, E. S. C., & Kuckertz, A. (2016). The Challenge of Dealing with Complexity in Entrepreneurship, Innovation and Technology Research – An Introduction. In E. S. C. Berger & A. Kuckertz (Eds.), *Complexity in entrepreneurship, innovation and technology research – Applications of emergent and neglected methods* (pp. 1–9). Cham, Switzerland: Springer International Publishing.

Berman, S. J. (2012). Digital transformation: Opportunities to create new business models. *Strategy and Leadership*, *40*(2), 16–24. doi:10.1108/10878571211209314

Berthon, B., Hintermann, F., & Berjoan, S. (2014). *The promise of digital entrepreneurs*. Accenture.

Bounfour, A. (2016). *Digital Futures, Digital Transformation*. Cham: Springer. doi:10.1007/978-3-319-23279-9

Brunswicker, S., Bertino, E., & Matei, S. (2015). Big data for open digital innovation–a research roadmap. *Big Data Research*, *2*(2), 53–58. doi:10.1016/j.bdr.2015.01.008

Bruyat, C., & Julien, P. A. (2001). Defining the field of research in entrepreneurship. *Journal of Business Venturing*, *16*(2), 165–180.

Brynjolfsson, E., & McAfee, A. (2014). *The Second Machine Age: Work, Progress, and Prosperity in a Time of Brilliant Technologies*. New York: WW Norton & Company.

Carayannis, E. G., Grigoroudis, E., Campbell, D. F., Meissner, D., & Stamati, D. (2018). The ecosystem as helix: An exploratory theory-building study of regional co-opetitive entrepreneurial ecosystems as Quadruple/Quintuple Helix Innovation Models. *R & D Management*, *48*(1), 148–162. doi:10.1111/radm.12300

Carvalho, L. C. (2017). Entrepreneurial Ecosystems: Lisbon as a smart start-up city. In L. C. Carvalho (Ed.), *Handbook of Research on Entrepreneurial Development and Innovation Within Smart Cities*. IGI Global. doi:10.4018/978-1-5225-1978-2.ch001

Caudron, J., & Van Peteghem, D. (2014). *Digital transformation. A model to master digital disruption*. Duval Union Consulting.

Chandra, Y., & Leenders, M. (2012). User Innovation and Entrepreneurship in the Virtual World: A Study of Second Life Residents. *Technovation*, *32*(7-8), 464–476.

Chinta, R., & Sussan, F. (2018). A triple-helix ecosystem for entrepreneurship: a case review. In *Entrepreneurial Ecosystems* (pp. 67–80). Cham: Springer. doi:10.1007/978-3-319-63531-6_4

Costa, C., & Turvani, M. (2013). *New industrial spaces as sustainable communities: the case of digital incubators. ERSA conference papers*. European Regional Science Association.

Cuc, J., Paredes, M., & Ventura, R. (2018). Online platform business models for value co-creation within a digital entrepreneurial ecosystem in B2B settings. In *2018 CBIM International Conference* (p. 150). Academic Press.

Curley, M., & Salmelin, B. (2017). Digital Disruption. In *Open Innovation 2.0: The New Mode of Digital Innovation for Prosperity and Sustainability* (pp. 15–25). Cham: Springer.

Cusumano, M. A. (2017). The sharing economy meets reality. *Communications of the ACM*, *61*(1), 26–28. doi:10.1145/3163905

Davidson, E., & Vaast, E. (2010). Digital entrepreneurship and its sociomaterial enactment. In *Proceedings of the 43rd Hawaii International Conference on System Sciences*. IEEE Computing Society.

de Reuver, M., Sørensen, C., & Basole, R. C. (2018). The digital platform: A research agenda. *Journal of Information Technology*, *33*(2), 124–135. doi:10.105741265-016-0033-3

Demirkan, H., Spohrer, J. C., & Welser, J. J. (2016). Digital innovation and strategic transformation. *IT Professional*, *18*(6), 14–18. doi:10.1109/MITP.2016.115

Denison, E. (2017). *Driving prosperity in the digital era. The role of digital talent, innovation and entrepreneurship*. Retrieved from https://www2.deloitte.com/de/de/pages/about-deloitte/articles/prosperity-digital-era.html

dos Santos, D. A. G., Zen, A. C., & Schmidt, V. K. (2017). Entrepreneurship ecosystems and the stimulus to the creation of innovative business: A case in the App industry in Brazil. *Journal of Research in Business, Economics and Management*, *8*(5), 1537–1543.

European Commission. (2014). *Fuelling digital entrepreneurship in Europe, Background paper*. Retrieved from http://ec.europa.eu/geninfo/query/resultaction.jsp?QueryText=EU+vision%2C+strategy+and+actions&query_source=GROWTH&swlang=en&x=18&y=8

European Commission. (2015). *Digital Transformation of European Industry and Enterprises; A report of the Strategic Policy Forum on Digital Entrepreneurship*. Available from: http://ec.europa.eu/DocsRoom/documents/9462/attachments/1/translations/en/renditions/native

European Commission. (2017). *Enterprise and industry directorate-general, "strategic policy forum on digital entrepreneurship."* Available from: https://ec.europa.eu/growth/industry/policy/digital-transformation/strategic-policy-forum-digital-entrepreneurship_en

Evans, D. S., & Schmalensee, R. (2016). *Matchmakers: the new economics of multisided platforms*. Harvard Business Review Press.

Fichman, R. G., Dos Santos, B. L., & Zheng, Z. E. (2014). Digital innovation as a fundamental and powerful concept in the information systems curriculum. *Management Information Systems Quarterly*, *38*(2), 329–343. doi:10.25300/MISQ/2014/38.2.01

Fitzgerald, M., & Kruschwitz, D. B., & Welch, M. (2014). Embracing digital technology: A new strategic imperative. *MIT Sloan Management Review*, *55*(2), 1–12.

Fox, S., & Stucker, B. (2009). *Digipreneurship: New types of physical products and sustainable employment from digital product entrepreneurship*. VTT Working Papers. Finland, VTT Technical Research Centre of Finland: 36.

Gartner, W. B., Bird, B. J., & Starr, J. A. (1992). Acting as if: Differentiating entrepreneurial from organizational behavior. *Entrepreneurship Theory and Practice*, *16*(3), 13–32. doi:10.1177/104225879201600302

Ghezzi, A. (2018). Digital startups and the adoption and implementation of Lean Startup Approaches: Effectuation, Bricolage and Opportunity Creation in practice. *Technological Forecasting and Social Change*. doi:10.1016/j.techfore.2018.09.017

Ghezzi, A., & Cavallo, A. (2018). Agile business model innovation in digital entrepreneurship: Lean Startup approaches. *Journal of Business Research*. doi:10.1016/j.jbusres.2018.06.013

Giones, F., & Brem, A. (2017). Digital Technology Entrepreneurship: A Definition and Research Agenda. *Technology Innovation Management Review*, *7*(5).

Gobble, M. M. (2018). Digitalization, Digitization, and Innovation. *Research Technology Management*, *61*(4), 56–59. doi:10.1080/08956308.2018.1471280

Grilo, A., Romero, D., & Cunningham, S. (2016), Digital Entrepreneurship: Creating and Doing Business in the Digital Era, Call for papers, Technological Forecasting and Social Change, Retrieved December 21, 2017, from https://www.journals.elsevier.com/technological-forecasting-and-social-change/call-for-papers/digital-entrepreneurship-creating-and-doing-business-in-the

Hanna, N. K. (Ed.). (2016). Mastering digital transformation: Towards a smarter society, economy, city and nation. In Hanna, N. K. (Ed.), Mastering Digital Transformation: Towards a Smarter Society, Economy, City and Nation (i-xxvi). Emerald Group Publishing Limited. doi:10.1108/978-1-78560-465-220151009

Hartmann, P. M., Hartmann, P. M., Zaki, M., Zaki, M., Feldmann, N., Feldmann, N., ... Neely, A. (2016). Capturing value from big data–a taxonomy of data-driven business models used by start-up firms. *International Journal of Operations & Production Management*, *36*(10), 1382–1406. doi:10.1108/IJOPM-02-2014-0098

Henriette, E., Feki, M., & Boughzala, I. (2015). The shape of digital transformation: a systematic literature review. *MCIS 2015 Proceedings*, 431-443.

Henriette, E., Feki, M., & Boughzala, I. (2016). Digital transformation challenges. *MCIS 2016 Proceedings, 33*. Retrieved from http://aisel.aisnet.org/mcis2016/33

Herrera, F., Guerrero, M., & Urbano, D. (2018). Entrepreneurship and Innovation Ecosystem's Drivers: The Role of Higher Education Organizations. In *Entrepreneurial, Innovative and Sustainable Ecosystems* (pp. 109–128). Cham: Springer. doi:10.1007/978-3-319-71014-3_6

Hsieh, Y.-J., & Wu, Y. (2018). Entrepreneurship through the platform strategy in the digital era: Insights and research opportunities. *Computers in Human Behavior*, 1–9.

Hull, C. E., Hung, Y.-T. C., Hair, N., Perotti, V., & DeMartino, R. (2007). Taking advantage of digital opportunities: A typology of digital entrepreneurship. *International Journal of Networking and Virtual Organisations*, *4*(3), 290–303. doi:10.1504/IJNVO.2007.015166

i-SCOOP. (2016). *Digitization, digitalization and digital transformation: the differences*. Retrieved from https://www.iscoop.eu/digitization-digitalization-digital-transformation-disruption/

Iansiti, M., & Lakhani, K. R. (2014). Digital Ubiquity: How Connections, Sensors, and Data Are Revolutionizing Business. *Harvard Business Review*, *92*(11), 91–99.

Isenberg, D. (2011). The entrepreneurship ecosystem strategy as a new paradigm for economic policy: Principles for cultivating entrepreneurship. *Presentation at the Institute of International and European Affairs*. Retrieved from https://doc.uments.com/download/s-the-entrepreneurship-ecosystem-strategy-as-a-new-paradigm-for.pdf

Isenberg, D. J. (2010). How to start an entrepreneurial revolution. *Harvard Business Review*, *88*(6), 40–50.

Kane, G. C., Palmer, D., Phillips, A. N., Kiron, D., & Buckley, N. (2015). Strategy, not technology, drives digital transformation. MIT Sloan Management Review and Deloitte University Press.

Kelly, K. (2016). *The Inevitable: Understanding the 12 Technological Forces That Will Shape Our Future*. Viking.

Kenney, M., & Zysman, J. (2015, June). Choosing a future in the platform economy: the implications and consequences of digital platforms. *Kauffman Foundation New Entrepreneurial Growth Conference*.

Knickrehm, M., Berthon, B., & Daugherty, P. (2016). *Digital disruption: The growth multiplier.* Accenture Strategy, Tech. Rep.

Kraus, S., Palmer, C., Kailer, N., Kallinger, F. L., & Spitzer, J. (2018). Digital entrepreneurship: A research agenda on new business models for the twenty-first century. *International Journal of Entrepreneurial Behavior & Research*. doi:10.1108/IJEBR-06-2018-0425

Lanzolla, G., & Anderson, J. (2008). Digital transformation. *London Business School Review*, *19*(2), 72–76.

Le Dinh, T., Vu, M. C., & Ayayi, A. (2018). Towards a living lab for promoting the digital entrepreneurship process. *International Journal of Entrepreneurship*, *22*(1), 1–17.

Li, F. (2018). The digital transformation of business models in the creative industries: A holistic framework and emerging trends. *Technovation*. doi:10.1016/j.technovation.2017.12.004

Li, W., Du, W., & Yin, J. (2017). Digital entrepreneurship ecosystem as a new form of organizing: The case of Zhongguancun. *Frontiers of Business Research in China*, *11*(1), 5. doi:10.118611782-017-0004-8

Liere-Netheler, K., Packmohr, S., & Vogelsang, K. (2018, January). Drivers of Digital Transformation in Manufacturing. *Proceedings of the 51st Hawaii International Conference on System Sciences*. 10.24251/HICSS.2018.493

Mazzone, D. M. (2014). *Digital or death: digital transformation: the only choice for business to survive smash and conquer*. Smashbox Consulting Inc.

McAdam, M., Crowley, C., & Harrison, R. T. (2018). “To boldly go where no [man] has gone before”-Institutional voids and the development of women’s digital entrepreneurship. *Technological Forecasting and Social Change*.

McQuivey, J. (2013). *Digital disruption: Unleashing the next wave of innovation*. Academic Press.

Meltzer, J. P., & Pérez, C. (2016). *Digital Colombia: Maximizing the global internet and data for sustainable and inclusive growth*. Washington, DC: Global Economy and Development at Brookings. Working Paper No. 96.

Nambisan, S. (2017). Digital entrepreneurship: Toward a digital technology perspective of entrepreneurship. *Entrepreneurship Theory and Practice*, *41*(6), 1029–1055. doi:10.1111/etap.12254

Nambisan, S., Lyytinen, K., Majchrzak, A., & Song, M. (2017). Digital innovation management: Reinventing innovation management research in a digital world. *Management Information Systems Quarterly*, *41*(1), 223–238. doi:10.25300/MISQ/2017/41:1.03

Nambisan, S., Siegel, D., & Kenney, M. (2018). On Open Innovation, Platforms, and Entrepreneurship. *Strategic Entrepreneurship Journal*, *12*(3), 354–368. doi:10.1002ej.1300

Nepelski, D., Bogdanowicz, M., Biagi, F., Desruelle, P., De Prato, G., Gabison, G., . . . Van Roy, V. (2017). *7 ways to boost digital innovation and entrepreneurship in Europe*. Available at: http://publications.jrc.ec.europa.eu/repository/bitstream/JRC104899/jrc104899_formatted_final_20170426.pdf

Ngoasong, M. Z. (2017). Digital entrepreneurship in a resource-scarce context: A focus on entrepreneurial digital competencies. *Journal of Small Business and Enterprise Development*. doi:10.1108/JSBED-01-2017-0014

Nzembayie, K. F. (2017). Using insider action research the study of digital entrepreneurial processes: a pragmatic design choice? In *European Conference on Research Methodology for Business and Management Studies.* Academic Conferences International Limited.

OECD. (2016). Stimulating Digital Innovation for Growth and Inclusiveness: The Role of Policies for the Successful Diffusion of ICT. In *OECD Digital Economy Papers, No. 256*. Paris: OECD Publishing.

Ojala, A., Evers, N., & Rialp, A. (2018). Extending the international new venture phenomenon to digital platform providers: A longitudinal case study. *Journal of World Business*, *53*(5), 725–739. doi:10.1016/j.jwb.2018.05.001

Parker, G., Alstyne, M., & Choudary, S. (2016). *Platform Revolution: How Networked Markets are Transforming the Economy, and How to Make Them Work for You*. W. W. Norton & Company.

Piccinini, E., Hanelt, A., Gregory, R., & Kolbe, L. (2015). Transforming industrial business: the impact of digital transformation on automotive organizations. *ICIS 2015 Conference Proceedings*.

Prinsloo, C., & James, I. (2015). *Digital Disruption: Changing the rules of business for a hyper-connected world*. Gordon Institute of Business Science, University of Pretoria.

Richter, C., Kraus, S., Brem, A., Durst, S., & Giselbrecht, C. (2017). Digital entrepreneurship: Innovative business models for the sharing economy. *Creativity and Innovation Management*, *26*(3), 300–310. doi:10.1111/caim.12227

Richter, C., Kraus, S., & Syrjä, P. (2015). The shareconomy as a precursor for digital entrepreneurship business models. *International Journal of Entrepreneurship and Small Business*, *25*(1), 18–35. doi:10.1504/IJESB.2015.068773

Rippa, P., & Secundo, G. (2018). Digital academic entrepreneurship: The potential of digital technologies on academic entrepreneurship. *Technological Forecasting and Social Change*. doi:10.1016/j.techfore.2018.07.013

Rogers, D. L. (2016). *The digital transformation playbook: rethink your business for the digital age*. Columbia University Press. doi:10.7312/roge17544

Rosenbaum, H., & Cronin, B. (1993). Digital entrepreneurship: Doing business on the information superhighway. *International Journal of Information Management*, *13*(461-463).

Santana, M. (2017). *Digital entrepreneurship: expanding the economic frontier in the mediterranean. Iemed*. European Institute of the Mediterranean.

Schallmo, D. R., & Williams, C. A. (2018). *Digital Transformation Now! Guiding the Successful Digitalization of Your Business Model*. Springer International Publishing.

Schreckling, E., & Steiger, C. (2017). Digitalize or Drown. In G. Oswald & M. Kleinemeier (Eds.), *Shaping the Digital Enterprise. Trends and use cases in digital innovation and transformation* (pp. 3–27). Cham: Springer.

Schumpeter, J. A. (1934). *The theory of economic development*. Cambridge, MA: Harvard University Press.

Schwab, K. (2017). *The fourth industrial revolution*. Crown Business.

Shane, S., & Venkataraman, S. (2000). The promise of entrepreneurship as a field of research. *Academy of Management Review*, *25*(1), 217–226.

Spigel, B. (2017). The relational organization of entrepreneurial ecosystems. *Entrepreneurship Theory and Practice*, *41*(1), 4. doi:10.1111/etap.12167

Stam, E., & Spigel, B. (2018). Entrepreneurial Ecosystems. In R. Blackburn (Ed.), *The SAGE Handbook of Small Business and Entrepreneurship*. London: SAGE Publications Ltd. doi:10.4135/9781473984080.n21

Stemler, A. (2017). The myth of the sharing economy and its implications for regulating innovation. *Emory Law Journal*, *67*, 197.

Stone, B. (2017). *The Upstarts: How Uber, Airbnb and the Killer Companies of the New Silicon Valley are Changing the World*. Random House.

Sussan, F., & Acs, Z. J. (2017). The digital entrepreneurial ecosystem. *Small Business Economics*, *49*(1), 55–73. doi:10.100711187-017-9867-5

Sussan, F., Autio, E., & Kosturik, J. (2016). *Leveraging ICTs for Better Lives: The Introduction of an Index on Digital Life*. Academic Press.

Troxler, P., & Wolf, P. (2017). Digital maker-entrepreneurs open design: What activities make up their business model? *Business Horizons*, *60*(6), 807–817. doi:10.1016/j.bushor.2017.07.006

van Welsum, D. (2016). *Enabling Digital Entrepreneurs. WDR 2016 Background Paper*. Washington, DC: World Bank. Retrieved at www.openknowledge.worldbank.org/handle/10986/23646

Vendrell-Herrero, F., Parry, G., Bustinza, O. F., & Gomes, E. (2018). Digital business models: Taxonomy and future research avenues. *Strategic Change*, *27*(2), 87–90. doi:10.1002/jsc.2183 von Briel, F., Davidsson, P., & Recker, J. (2018). Digital technologies as external enablers of new venture creation in the IT hardware sector. *Entrepreneurship Theory and Practice*, *42*(1), 47–69. doi:10.1177/1042258717732779

Von Hippel, E. (2009). Democratizing innovation: The evolving phenomenon of user innovation. *International Journal of Innovation Science*, *1*(1), 29–40. doi:10.1260/175722209787951224

West, J., & Bogers, M. (2017). Open innovation: Current status and research opportunities. *Innovation*, *19*(1), 43–50. doi:10.1080/14479338.2016.1258995

Westerman, G., Bonnet, D., & McAfee, A. (2014). *Leading digital: Turning technology into business transformation*. Harvard Business Press.

Westerman, G., Calméjane, C., Bonnet, D., Ferraris, P., & McAfee, A. (2011). *Digital Transformation: A Roadmap for Billion-Dollar Organizations*. MIT Center for Digital Business and Capgemini Consulting.

Yoo, Y., Boland, R. J., Jr., Lyytinen, K., & Majchrzak, A. (2012). Organizing for Innovation in the Digitized World. *Organization Science*, *23*(5), 1398-1408.

Zhao, F., & Collier, A. (2016). Digital entrepreneurship: Research and practice. *9th Annual Conference of the EuroMed Academy of Business*.

Zhu, F., & Furr, N. (2016). Products to Platforms: Making the Leap. *Harvard Business Review*, 1.

KEY TERMS AND DEFINITIONS

Business Model: Describes a holistic view of how a business creates/designs value, delivers it to the market, and captures value in return.

Digital Entrepreneurship Ecosystem: Refers to the matching of digital customers on platforms in digital space through the creative use of digital ecosystem governance and entrepreneurship/business ecosystem management to create matchmaker value and social utility by reducing transactions cost.

Digital-to-Disruptive Technology: New digital technologies hat displaces an established technology and shakes up the industry or a ground-breaking product/service that creates a completely new industry, change dramatically the costumer expectation and reshape the marketplace.

Entrepreneurship: A dynamic process of vision, change, and creation of wealth by either an individual or a team who identifies a business opportunity and acquires and deploys the necessary resources required for its exploitation along with risks to be assumed.

Entrepreneurship Ecosystem: Refers to the factors—individuals, groups, firms, organizations, and institutions (micro ecosystem); and cultural, social, and material attributes (macro ecosystem)—outside the individual entrepreneur—that are conducive to, or inhibitive of, the choice and decision of a person or group of persons to become entrepreneur or enterprising.

Innovation: Any new idea, product, service, device, process, method, or solution to be designed and applied to meet new requirements, un-met, unsatisfied, unarticulated needs, or existing market need.

Platform Economy: Is economic and social activity facilitated by platforms online matchmakers or technology frameworks.

ENDNOTES

[1] As of August 2018, Google delivers more than 89 million hits for the term "entrepreneurship" but just a little more than 131 000 for "entrepreneurship ecosystem", 162 000 for "digital entrepreneurship", and 4270 for "digital entrepreneurship ecosystem". Google Scholar returns 1.4 million, 2600, 1230 and 36 results, respectively.

[2] As of September 2018, Google delivers more than 551 million hits for the term "innovation" but just a little more than 739 000 for "innovation ecosystem", 3 millions 570 thousands for "digital innovation", and 19 thousands for "digital innovation ecosystem". Google Scholar returns around 4 million, 13100, 13300 and 52 results, respectively.

Section 4
Digitalization Experiences and Applications

Chapter 12
Display Ads Effectiveness:
An Eye Tracking Investigation

Dionysia Filiopoulou
University of Patras, Greece

Maria Rigou
University of Patras, Greece

Evanthia Faliagka
Western Greece University of Applied Sciences, Greece

ABSTRACT

The average web user receives numerous advertising messages while browsing online and the formats of such digital marketing stimuli are constantly increasing in number and degree of intrusiveness. This chapter investigates the effectiveness of different types of display advertising by means of an eye-tracking study combined with a pre- and a post-test questionnaire with the purpose of collecting quantitative and qualitative data concerning ad visibility and interaction. Eye gaze data are particularly revealing when examining visual stimuli and they become more valuable when associated with asking users to recall seeing an advertising message we know they fixated on. Moreover, the study aimed to look into whether banner blindness still applies regardless of the type of display ad used, whether the visual pattern remains F-shaped, the effect of placing ads below the fold, how effective trick banners are, and which types of ads are annoying to users.

INTRODUCTION

Marketing is an integral part of any business and is defined as "... *the management process responsible for identifying, anticipating and satisfying customer requirements profitably" by the* UK Chartered Institute of Marketing's (CIM) while the American Marketing Association (AMA) provided a more extended definition in 2013: "Marketing is the activity, set of institutions, and processes for creating, communicating, delivering, and exchanging offerings that have value for customers, clients, partners, and society at large" (Johnson, 2015). The related literature indicates that marketing definitions cover a wide spectrum

DOI: 10.4018/978-1-5225-7262-6.ch012

and span a large period, with the first definitions documented in 1918 (Brunswick, 2014). Advertising is part of the marketing strategy (typically along with market research, media planning, pricing, public relations, community relations, customer support, and sales strategy) and is a means of communication with the users of a product or service aimed at informing people or influencing their buying behavior by promoting a product, service or company. According to Nicosia (1974) simply put, advertising means to give notice, to inform, to notify, or to make known. Dunn and Barban (1986) defined advertising as "*a paid, nonpersonal message from an identifiable source delivered through a mass-mediated channel that is designed to persuade*" (Sheehan, 2014, p. 2). Throughout the years marketing has been affected by technological developments and in many cases, it has acted as an early adapter of technological innovations. This does not imply that technology is the kernel of marketing; it is the people that marketing targets and revolves around, but marketing has always depended on technological developments that allow for new, faster or more efficient ways of coming to contact with people and establishing a communication channel with them (Ryan, 2016). Typical advertising channels included television and radio, print ads, direct mail. The advent of the web, its widespread adoption and availability almost everywhere, anytime provided marketing with unpreceded potential, giving it easily accessible channels large audiences and new ways of formatting attractive promotional messages. Online advertising (also called digital, online or web advertising), refers to the type of marketing strategy that involves the use of the internet for delivering promotional messages to users. It includes placing or dispatching advertisements on websites, through e-mail and ad supported software, in the form of text/multimedia messaging, in social platforms, as well as on mobile web-enabled devices (tablets, smartphones). This newly formed scenery is very dynamic and has also empowered the recipients of marketing messages (users) with wide access to peers when it comes to word-of-mouth marketing, online ratings and blogging. Current online marketing has evolved drastically and uses advanced computational techniques (Busch, 2016) able to manage large amounts of data integrated from various sources. The mere purpose of this ad-serving technology is sending the correct messages to the correct users, at the correct time to maximize sales, build brand identity, and serve customer needs as efficiently as possible.

Display advertising is one of the most important types of online advertising that visually conveys its advertising message using text, logos, animations, videos, photos, as well as hyperlinks. The main purpose of a display ad is to provide generic ads and brand messages to site visitors.

Advertisements presented as display ads, appear on third-party sites or on search engine results pages leading to websites or social media. Display ads aim to support brand awareness and help to increase consumer purchasing intent (Smith, 2013). The types of display advertising that we have examined in our experiment are: text ads, banner ads, pop-ups/pop-unders, floating/overlay ads, expandable ads, trick banners, news feed ads and interstitial ads.

To evaluate the types of display ads and to investigate the behavior of users towards these advertisements, an eye tracking experiment was conducted. The purpose of this experiment was to study how participants reacted to various types of display ads available in the set of selected real-life websites and with the help of eye tracking (visualizations and metrics) to investigate whether the banner blindness phenomenon still holds, the F-pattern is still followed, to what extend the stimuli placed above the fold override, what is the effect of trick banners and which types of ads are annoying to users. The overall aim of the work was to examine how modern advertising promotional messages work as visual stimuli for their recipients in the web environment and to address the factors that affect their effectiveness. This approach differs from typical experimental and empirical studies found in the related literature, as eye tracking offers valuable insights through the subconscious and unbiased visual attention data it captures.

This allows knowing whether an advertising messages has drawn the visual attention of web users, for how long and if this led to a click-through, as well as which was the visual behavior that preceded the click.

In this chapter the authors describe the main concepts of eye-tracking technology explaining what kind of data (in terms of values to specific metrics) can be provided and what kind of interpretations can be given about those data referring to the visual stimuli at hand. Section 3 describes the methodology of the study (purpose, participants and scenarios) and the equipment used. Section 4 presents and discusses the main findings of the study along with supporting eye-tracking data and visualizations and the chapter concludes with guidelines for designing effective display ads and current trends in the field.

BACKGROUND

The newly introduced potential of the web environment and its rapidly growing technologies have proven very tempting for marketers and business owners even to the point of adopting intrusive and aggressive ad delivery. This in effect contributed to accumulated user discomfort or annoyance and degraded online experiences. Over the years of web adaption, research in the domain of marketing and the domain of HCI (Human Computer Interaction) and UX (User Experience) has indicated that there are significant negative effects of advertising reported by web users (Rohrer & Boyd, 2004; Calder, Malthouse & Schaedel, 2009; Agarwal, Shrivastava, Jaiswal & Panjwani, 2013; Lin & Kim, 2016; Hernandez, Wang, Sheng, Kalliny & Minor, 2017) including difficulties in reading, scanning, and browsing information online. Users may become confused and tired and face an increased cognitive load by imposed advertising messages, they may experience frustration, irritation, and may develop negative affect, emotions or moods towards the advertising messages, the brand promoted by them, as well as the websites hosting them (Brajnik & Gabrielli, 2010). A recent study (Fessenden, 2017) aimed at identifying the advertising techniques that are most disruptive and detrimental for web users among 15 ad types, concluded that most disliked (and annoying) ads include modal ads, autoplay video ads, intracontent ads (i.e. ads that shuffle page content as they load), and deceptive links (which look like content but are in fact ads). It is worth noticing that these ad types are annoying for both desktop and mobile users. Among the least annoying ads were right rail ads (ads placed on the right side of the page) for desktop and related links for both desktop and mobile. With the availability of numerous ways to impose advertising messages on the web, marketers and digital businesses need to balance between increasing advertising revenue without compromising a good user experience. These two goals may seem contradicting at first, but they do not have to be; marketing is about telling a fascinating story and in storytelling the audience must be willing to listen and not feel obliged to do so. To this end, HCI and UX may provide marketers with valuable guidance for assuring effective advertising.

The assessment of online advertising effect on web users can be significantly supported by eye tracking, a tool used often by HCI and UX experts. Eye tracking is a technology by which we can accurately know where the user is looking at ('fixating') in real time and how she moves her gaze between successive fixations. Eye tracking is therefore a method of recording eye focus. This process is accomplished through a device known as the eye-tracker. By adequate interpretation of eye tracking data, scientists can measure attention, interest, and arousal, and reach interesting conclusions for human behavior research applied in a variety of fields such as Cognitive Studies, Psychology, Medicine, Neurology, UX and HCI, Marketing, Engineering and so on. In the fields of marketing and consumer behavior research (which is the domain of the current chapter), eye-tracking allows capturing subconscious and unbiased data implicitly thus offering the benefit of objective measurements of consumers' attention and spontaneous responses to marketing messages. Collected data when analyzed and adequately interpreted can help optimize the design and placement of ads and the research conducted by domain scientists is rich (Beelders & Bergh, 2014; Wästlund, Shams & Otterbring, 2018; Mou & Shin, 2018; Pham et al., 2017). Understanding patterns of visual attention can provide guidance towards more effective advertising results particularly in the online environment, without compromising a positive user experience.

Human vision may be giving the impression that it is stable and smooth but as a process it is divided into sequences of fixations and saccades. The basic measure is the gaze point, which equals one raw sample captured by the eye tracker. Fixations aggregate a series of gaze points and represent a period in which the eyes are locked toward a specific point and become relatively stationary. Between fixations the eyes make quick movements between fixations, called saccades. Saccades are the rapid moves of the eye from one fixation to another to help the eye compose and create a complete scene of what a person sees. By analyzing user scanpaths, we can understand which visual elements were mostly fixated and in which order. The ordered set of fixations points (depicted by circles) connected by saccades (depicted by lines) is a scanpath (or gazetrail). Another prominent visualization of eye tracking recordings is the heatmap representing data from several participants. On a heatmap colors or opacity vary with the density of the number of fixations or their duration.

Eye tracking data can be extracted on the basis of a sub region of the displayed stimuli (e.g., the time spent looking at a particular object), defined as an AOI (area of interest) and are generally defined either before the experiment or after it, during the analysis process performed by the researcher who determines the most relevant areas of the stimuli (such as specific images, blocks of text, calls to action, etc.). AOIs can be grouped to allow for accumulation and analysis of data concerning more than one stimulus area or across stimuli. The basic visual effort metrics recorded by Tobii Studio are based on fixations and clicks and include (Tobii Studio 2.X User Manual, 2010):

- **Time to First Fixation (TFF):** Time (in secs) from displaying the stimulus until the beginning of the first fixation within an Area Of Interest (AOI). A fast TFF on an object or an area means that it attracted the visual attention of the participant fast.
- **Fixation Duration** or **Fixation Length:** The length (duration) of fixations within an AOI (or a group of AOIs) in seconds. The longer the fixations suggests difficulty in locating relative information.
- **First Fixation Duration:** Duration of first fixation in seconds.
- **Fixation Count:** Number of fixations in an AOI or group of AOIs. More fixations in an area denote that it attracts the user attention and is considered more important.
- **Fixations Before:** The number of fixations before the participant fixated within an AOI for the first time.
- **Visit Duration or Observation Length:** The total time in seconds for every time a person has looked within an AOI, starting with a fixation within the AOI and ending with a fixation outside the AOI (or groups of AOIs).
- **Observation Count:** The number of visits and revisits to an AOI.
- **Visit Count:** Number of visits in an AOI or group of AOIs.
- **Percentage Fixated-% (or Participant %):** Percentage of participants that have fixated at least in an AOI or a group of AOIs.

- **Mouse Click Count:** Number of times a participant clicks in an AOI or a group of AOIs.
- **Time to First Mouse Click:** Time elapsed (in seconds) until the first mouse click.

Since that data is provided for each AOI or group of AOIs defined on each "media element" (generally pages, when talking in context of web sites), it is challenging to select the metrics that can support scientific findings. It is not meaningful to force insights out of every eye tracking metric; to the contrary, it is necessary to select metrics that will lead to actionable and informed insights taking into account the context of a specific task (for instance, is it expected/desired that the user clicks in a specific AOI, or that she fixates in an AOI, or that she follows a certain scan path, or that she reads a block of text, etc.). The interpretation of eye tracking metrics remains a crucial issue. It was shown that fixation related metrics might be used to define observer's engagement (Bylinskii & Borkin, 2015). Long fixation duration is correlated with high cognitive workload and higher cognitive effort (Shojaeizadeh, Djamasbi & Trapp, 2016). This metric was also found as related to the difficulty of the visual content of the performed task (Fitts, Jones & Milton, 2005). Scan path related metrics are usually interpreted based on the assumption that an ideal scanpath contains a short in time and straight-line saccade to the specific target (Conversy, Hurter & Chatty, 2010). Usually a deviance from this ideal straight line is interpreted as poor search (Goldberg & Kotval, 1999). Popular scanpath-based metrics are scanpath duration, scanpath length, spatial density, transition matrix, scanpath regularity (repeatability), scanpath direction (search strategies), saccade/ fixation ratio (Blascheck et al., 2014). AOI related metrics are also useful as they are related to high importance parts of stimulus, defined usually based on the semantic information of the stimulus. Typical metrics related to AOIs are based on a transition, which is defined as a movement between two AOIs. Typical AOI-based tasks cover exploration of images or multimedia which should be noticed by an observer in a specific context (Borys & Plechawska-Wójcik, 2017).

Eye-tracking is a technology that reveals large amounts of detailed data concerning user reactions to visual stimuli and can be interpreted in numerous ways by experts in related domains but there are some crucial limitations to bear in mind. More specifically, recording may be prevented in case of participants with contact lenses, glasses and long eyelashes and is sensitive to head position in relation to the recording area. In addition, the calibration process is time consuming, but most importantly, eye-tracking records foveal vision, which only accounts for 8% of the overall picture we perceive and does not record parafoveal and peripheral vision.

Related research has indicated that the viewing pattern while looking at advertisements is influenced by the goal of the viewer (Radach et al., 2003) and the nature of the advertisement (Rayner, Miller & Rotello, 2008). Viewers tend to look at pictures first (Pieters & Wedel, 2004) and then the text (Rayner, Miller & Rotello, 2008), but they look at color advertisements first and for more time than at advertisements without color (Lohse, 1997). In terms of digital advertising media, eye tracking was used to evaluate viewing behavior on digital advertising screens mounted in public transport vehicles. Results indicate that many passengers paid attention to the screen, but that there was no correlation between fixation time and content (Höller et al., 2010). In terms of online advertising present in content pages, eye-tracking has been used to determine that text-based advertisements are often not looked at by users of the page, particularly if the text-based advertisement is situated on the right-hand side of the page (Owens, Chaparro & Palmer, 2011).

Advertisements on the web are typically distinguished as:

- **Text ads:**They contain text and links (rather than an image) often in a box area (Cambridge Dictionary, n.d.). Text ads download almost instantly and are not affected by ad blocking software. They are widely spread in keyword search advertising.
- **Frame ad or (Traditional) banner ads:** Banner ads are graphical web advertising units placed by website publishers in a rectangular space on the page. The dimensions and placing options of the frame displaying banner ads have been standardized (Interactive Advertising Bureau, n.d.). They typically contain a single .html, .gif, .png, or .jpg file and allow tracking a single click-through link. Banner ads were the first form of web banners but nowadays the term web banners refers to more visually engaging forms (namely, rich media).
- **Rich Media ads:** They are in a sense enriched forms of banner ads and include more forms of data (like video, audio, animations, polls, send-to-a-friend, gallery, or other interactive elements, alone or combined, to increase the degree of user engagement). Rich media ads also offer many options in the way they are 'revealed' to web users as they may be displayed alongside with page loading in the browser window but may also float, expand, etc. and are constructed using Java applets, HTML5, Adobe Flash, and relevant technologies. Rich media ads can offer detailed tracking metrics on user interactivity and track multiple click-through links (DoubleClick Creative Solutions, n.d.). Rich media ads can generally have the form of in-page ads (placed in various areas of a web page in the form of a rectangle or preset banner), out-of page ads (such as floating ads, pop-up ads and expandable ads) and in-stream ads (appearing as pre or post roll videos). Their engaging power results in being frequently considered as interruptive because they distract user attention from the actual contents of the webpage and impose their message to web users.

In this study the experiment concerned all the above formats and more specifically in the case of rich media ads the following types:

- **Pop-Ups/Pop-Under ads:** A pop-up ad appears in a new browser window that opens above the browser of a site visitor. A pop-under ad is displayed in a new browser window that opens beneath a site visitor's original browser window (Smith, 2013). Pop-under ads, which are less invasive than pop-ups, appeared on the internet in the 1990s and are still used by some ad networks to increase traffic to websites. In many cases users do not see pop-unders before they close all overlying browser windows but the 'silent' way of their display often annoys them and this creates a negative impression about the advertised brand.
- **Floating or Overlay Ads:** They either float over the page contents or appear as a full-screen ad during natural transition points such as loading and disappear or become less imposing after a preset period (typically 5-30 seconds).
- **Expandable (or Expanding) ads:** They expand over the top of the page containing them on the event of a user interaction (for instance, click or mouse-over) or automatically on page load or after a pre-set period (auto-expand). Expandable ads can expand in any direction, shape, or with fading effects and can collapse on user interaction or automatically (after a certain period). This type of ads allow fitting additional information in an area that is initially more restrained.
- **Interstitial ads:** They are a type of floating ads where ads do not float but are placed within the contents of the webpage. They appear and disappear the same way floating ads do and almost always they delay page loading.

- **Trick Banners:** They are banner ads that mimics (resembles) certain widely used screen elements such as message windows from the operating system or a button labelled in a way that correlates with the typical user task on the webpage and misleads users to clicking them and generate an ad impression. These ads are also known as bait-and-switch as they do not look like advertisements, they get clicked more often that typical banner ads but they are not effective as users respond negative to the stimuli due to realizing that they were deceived (Smith, 2013).
- **News Feed ads:** They are also named "Sponsored Stories" or "Boosted Posts" and are typically placed in social network platforms that offer a steady stream of information updates in homogeneous display formatting. These ads are intertwined with non-promoted news feeds and unlike banner ads (that are distinct), their form merges seamlessly into actual news updates or featured content. News feed ads gain high click through rates but they are in fact similar to trick banners (i.e. their success is based on user deception).

Apart from the formatting of the stimuli that carries the advertising message on the web, web advertising and its effectiveness is also closely related with the typical behavior of web users when they access information online. Research in the field of HCI has been significant and has indicated a series of patterns recorded in typical web user behavior. Among them feature the banner blindness phenomenon, the F-shaped pattern and the role of page fold when placing content on a page (the effect of placing content above the fold). All three phenomena are part of the research questions used in this study.

Eye-tracking studies have indicated that web users often ignore webpage regions either because these regions are most likely to contain ads, or users view the banner ad but because it they quickly forget its contents as it is irrelevant to their current task; a phenomenon referred to as "Banner Blindness" (Wedel & Pieters, 2008; Resnick & Albert, 2013; McCoy et al., 2007). On the other hand, studies show that even if these ads are "ignored" by users, they may influence the user subconsciously (Lee & Ahn, 2014). The term "Banner Blindness" has been coined by Benway & Lane (1998).

A long scale eye tracking study conducted by Nielsen (2006) revealed that web users share a common basic behavior when reading text online regardless of the specific website or their intended task (browsing or searching). This dominant behavior pattern is shaped like an F and demonstrates the importance of content that is placed in the various areas of the page. It is important though to stress that the F pattern applies when users encounter a content area and want to assess its content. Pernice (2017) stresses that web users scan in an F-shape when all 3 conditions hold: a page or a section of a page includes text that has little or no formatting for the web, the user is trying to be efficient and is not very committed or interested to read every word.

Even though the F-shaped pattern might suggest positioning advertisements at the topmost part of a webpage, a combination of the two phenomena (banner blindness and F-shaped pattern) has led to web users applying the F pattern slightly lower on the page leaving information above that area unnoticed (Wojdynski & Evans, 2015) as they expect advertising to be toward the right side or top of a web page (Shaikh & Lenz, 2006).

The term "above the fold" (ATF) was borrowed from the printed media domain and refers to the upper half of the front page of a newspaper or tabloid where important news stories where placed and were visible without unfolding the newspaper. The term when used is the web domain (also known as "above the scroll") refers to the portion of the webpage that is visible without scrolling. Studies have showed that web users focus more on information above the fold (Nielsen, 2005; Nielsen, 2010) and this is an observation that was initially explained by the fact that early web users were reluctant to scrolling.

Nowadays, web users are quite accustomed to scrolling but still spend significantly more time observing the area above the fold (Djamasbi et al, 2011; Fessenden, 2018).

The study aims to investigate the effectiveness of online display ads and poses a series of research questions regarding whether (a) users reacted to various types of display ads available in the set of selected real-life websites, (b) the banner blindness phenomenon still holds, (c) the F-pattern is still followed, (d) the stimuli placed above the fold override and to what extent, and (e) some types of display ads are annoying to users.

PARTICIPANTS AND EXPERIMENT SETUP

For addressing the questions above mentioned a series of user testing sessions were executed based on a specific scenario comprising 5 existing websites that include at least one display advertising type. The test was conducted using the Tobii T120 Series Eye Tracker with the Tobii Studio 2.2 support software.

The experiment took place in the Usability Engineering Lab of the Computer Engineering and Informatics Department (University of Patras). Users were seated in the testing room and used the computer that ran the Tobii Studio Logger, while the test moderator used the Live Viewer software on a PC in a short distance from the user that allowed her to talk to the user and respond to questions and remarks (inspection room).

In the first step, each user had to complete an entry form in order to gather the necessary demographic data for later analysis. After that, the script's instructions prompted users to a specific target on the site that would appear each time. Between each scenario there were some questions asked to users before moving on and forgetting the details of their recent interaction. At the end of the process there was a printed questionnaire where, from the answer that everyone gave us, we got information that helped us complete our research.

The selection and number of participants in the experimental process play a catalytic role in the outcome and findings. After examination, we decided that 20 participants were enough for our case, based on analogous studies. For sample uniformity, we chose ten (10) men and ten (10) women. Also, the participants' age ranged from eighteen (18) to thirty-five (35).

Through the participation form we collected important information about each user's experience with computers and the internet in general and we also examined the level of their education. Most participants were university or technological institution students or graduates. More specifically, 60% of the users were undergraduate students, 20% were graduate students, 10% were postgraduate students and 5% held an MSc degree. Finally, 5% of the participants held secondary education diploma. Concerning the daily usage of the internet, all participants stated they use the internet on a daily basis, with 55% of users stating that they use the internet for 6 or more hours per day, 25% that they use the internet for 3 up to 6 hours per day, and 20% that they usually use the internet for up to 3 hours daily. Through these percentages we conclude that most users were experienced web users.

Users were asked to think aloud and freely express their questions or opinions during the recording. After signing a hardcopy with the terms of participation of the experiment, the moderator explained the eye-tracking process and users were asked to provide demographic information on a printed questionnaire. The moderator helped users through the calibration phase and upon successful completion, the users were assigned specific tasks of browsing and searching in each given website (Table 1). Scenarios did not literary make any references or directly concerned the advertising messages that appeared during task execution and this was intentional in order to resemble the real-life situation of users browsing a website. Along with the answers that were used as a measure of successful completion of the task, users upon task completion were asked specific questions regarding the display ads they had just interacted with.

SENARIOS AND ANALYSIS

Scenario 1

At the first site (http://download.cnet.com/windows/), users were asked to download the PDF Reader for Windows 10 to see whether they watched advertisements when given a specific task, if they were affected by trick banners and how they responded to expandable ads.

In Figure 1, the image on the left (a) shows the placement of the trick banners and the download buttons as they were marked in the analysis phase in the form of AOIs and (b) shows the participants' gazeplots where the circles represent fixations.

In addition to the visualizations, recorder eye tracking metrics indicated that banner_1 has the highest number of fixations compared to other banners but has the shortest fixation duration (0.19 sec), which means that the test participants understood that this is a trick banner and did not give any further attention. Moreover, clicks count clearly shows that none of the participants clicked on any of the trick banners to download the requested file. Some may have been confused by the presence of this form of display ads and fixated on it, but they all eventually decided not to click on them and managed to successfully accomplish their goal.

Then using the metrics of Tobii Studio 2.2, we examined the fixations for the three different download options available on the webpage and the results are shown below (Figure 2).

Based on the Fixation Count metric for the three different download buttons (Figure1.a) the following observations were made: 85% fixed on the download_1 within 17.32 sec from the moment the site was displayed, 35% of users were really disturbed by trick banners, 50% of the participants were hardly affected by the presence of trick banners and all participants successfully accomplished their goal.

After triggering the download (clicking on one of the download buttons) and until the loading of the setup file was completed, a mandatory video with audio ad (rich media) began to play and after that a second video. When users scrolled, this ad moved to the bottom and right of the screen (expandable ad) and was still visible and playing. The large percentage of participants (95%) noticed this ad and fixed their gaze on it almost immediately after it was displayed (0.76sec). Some users (35%) were bothered so

Table 1. Websites and ad types included

Scenarios Websites	Display Ads
1. Software downloading and technology related website	Trick banners, expandable ads
2. News portal with information on citizen transactions with public services	Text ads, Riche media banner ads, news feed ads, expandable ads
3. News portal (using content syndication)	Interstitial, rich media banners, news feed, text ads
4. Sports news portal	Banners ads, rich media banners
5. Sports news vortal (tennis)	Banner ads, rich media banners

Figure 1. Scenario 1 AOIs(a) and gazeplots (b)

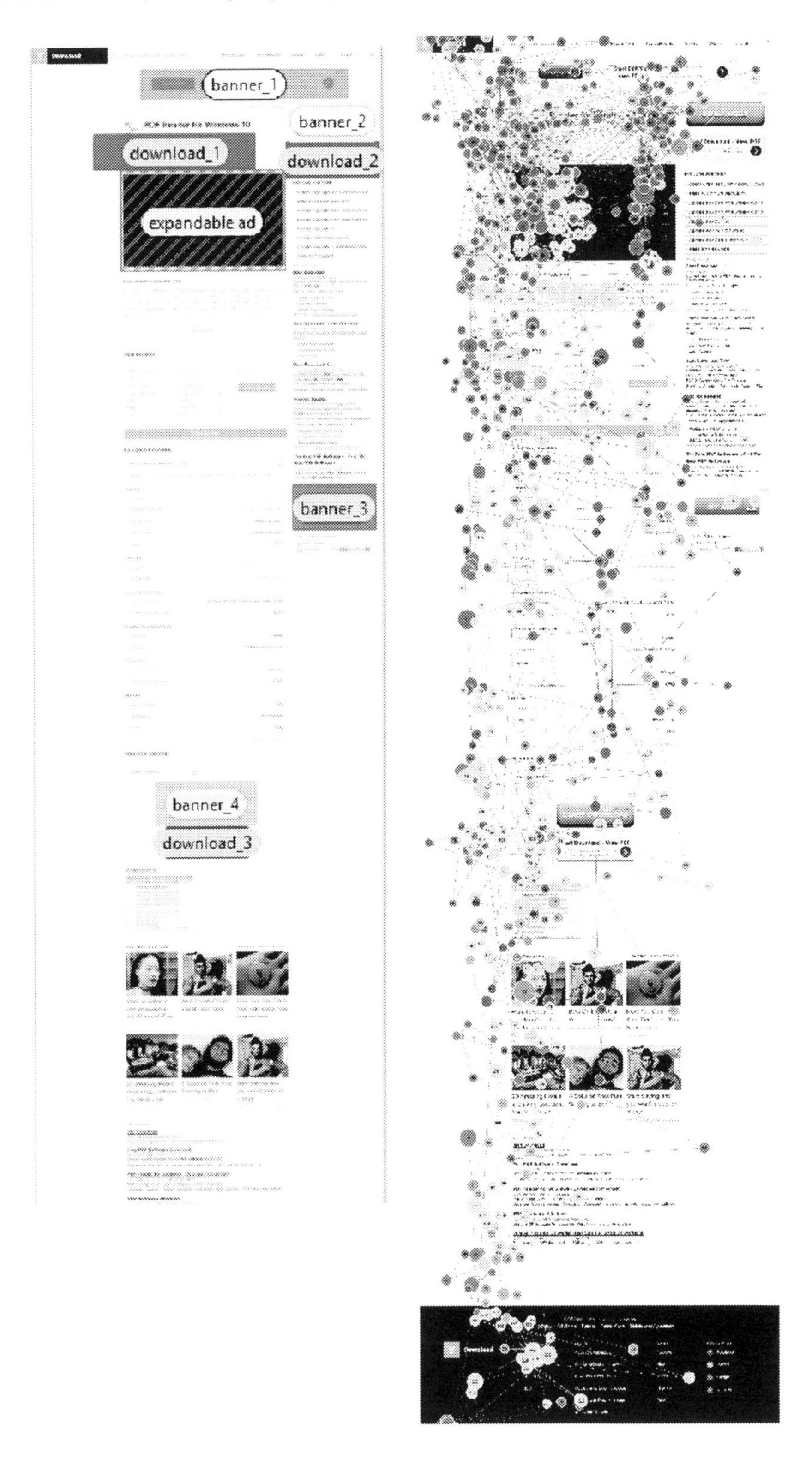

Table 2. Scenario 1 Time to First Fixation (in seconds)

	download_1			download_2			download_3		
Recording	*N (count)*	*Mean*	*Sum*	*N (count)*	*Mean*	*Sum*	*N (count)*	*Mean*	*Sum*
REC 01	1	0.00	0.00	1	5.85	5.85	-	-	-
REC 02	1	0.29	0.29	-	-	-	-	-	-
REC 03	1	6.25	6.25	1	7.36	7.36	-	-	-
REC 04	1	0.62	0.62	1	21.85	21.85	1	11.02	11.02
REC 05	1	0.84	0.84	-	-	-	1	12.98	12.98
REC 06	1	2.40	2.40	-	-	-	-	-	-
REC 07	1	0.30	0.30	-	-	-	-	-	-
REC 08	1	2.83	2.83	-	-	-	-	-	-
REC 09	-	-	-	-	-	-	-	-	-
REC 10	-	-	-	-	-	-	-	-	-
REC 11	1	28.29	28.29	-	-	-	-	-	-
REC 12	1	39.76	39.76	-	-	-	-	-	-
REC 13	1	49.74	49.74	-	-	-	-	-	-
REC 14	1	18.92	18.92	-	-	-	-	-	-
REC 15	1	56.05	56.05	-	-	-	-	-	-
REC 16	-	-	-	-	-	-	-	-	-
REC 17	1	8.49	8.49	-	-	-	-	-	-
REC 18	1	34.90	34.90	-	-	-	-	-	-
REC 19	1	24.74	24.74	-	-	-	-	-	-
REC 20	1	20.10	20.10	-	-	-	-	-	-
All Recordings	**17**	**17.32**	**294.51**	**3**	**11.69**	**35.06**	**2**	**12.00**	**24.00**

much that they paused the video because they were annoyed by the sound of the advertisement. At the end of the process, users were asked about the content of the ad, but only two users recalled its content (which was the brand and the product advertised and which was the message).

Scenario 2

Users were led to the webpage depicted in Figure 2 (https://www.dikaiologitika.gr/), which is a news portal that also facilitates citizens' transactions with public services. The portal homepage offers rich content along with display ads placed in non-typical locations. Specifically, at the time of the experiment ads had the form of a rich media banner, a text ad, an expandable ad and news feed ads under "READ ALSO". Users were asked to make a free browse to the site to see how they react to available types of advertising and whether non-traditional placement is effective or not.

It is worth noticing that, as shown in Table 3, 60% of the participants saw the text ad and fixed their gaze on it 62 times in total with average fixation duration of 0.26sec. The rich media ad received a total of 39 fixations from 50% of the participants and average time for each individual fixation of 0.23 sec.

Finally, the expandable ad was seen a total of 17 times by 30% of participants with an average fixation duration of 0.27sec.

Users were asked at the end of the process if they could recall any of the ads on the previous page. 35% of participants said they did not remember any ads. 5 out of 10 people who viewed the rich media banner ad on the left side of the page (25% of all users), were able to recall it. Also, only 1 person out of 6 (5% of the total) who saw the expandable video advertisement remembered it at the end of the process and 7 out of 14 (35% of all participants) could recall the text ad that appeared at the upper left corner of the content block of the page. Users placed limited fixations (as shown in the heat map) at the bottom of the page. To see whether they noticed the news feed ads at the bottom of the page and whether they confused them with suggested content articles within the same website, they were explicitly asked at the end of the session if articles in the "READ ALSO" section were content articles, promotional items, or both. 55% of participants responded correctly that in that section there were both content articles and ads, 25% thought all articles were content articles and 20% had the impressions that all articles were promotional. It is though interesting that all users had noticed this particular section and had made an assumption about its contents. This scenario indicated that users are significantly annoyed by ads comprising sound and also that the placement of ads plays a very crucial role regarding their visibility.

Table 3. Scenario 2 Fixation Durations (in seconds)

	Expandable Ad			**Rich Media**			**Text Ad**		
Recording	*N (count)*	*Mean*	*Sum*	*N (count)*	*Mean*	*Sum*	*N (count)*	*Mean*	*Sum*
REC 01	-	-	-	-	-		2	0.32	0.64
REC 02	4	0.14	0.57	1	0.23	0.23	3	0.32	0.97
REC 04	1	0.17	0.17	8	0.21	1.71	14	0.27	3.78
REC 05	-	-	-	-	-	-	-	-	-
REC 06	4	0.29	1.14	10	0.29	2.90	2	0.34	0.67
REC 07	4	0.40	1.58	-	-	-	3	0.21	0.64
REC 08	2	0.25	0.49	5	0.25	1.26	3	0.28	0.85
REC 10	-			-	-	-	4	0.26	1.04
REC 11	-			-	-	-	-	-	-
REC 12	-			-	-	-	3	0.21	0,62
REC 13	-			3	0.21	0.64	3	0.26	0,77
REC 14	-			-	-	-	-	-	-
REC 15	-			3	0.14	0.42	5	0.23	1,13
REC 16	-			1	0.17	0.17	-	-	-
REC 17	2	0.32	0.65	-	-	-	1	0.07	0.07
REC 18	-			2	0.15	0.30	3	0.23	0.69
REC 19	-			2	0.22	0.43	15	0.26	3.93
REC 20	-			4	0.25	0.98	1	0.33	0.33
All Recordings	**17**	**0.27**	**4.60**	**39**	0.23	**9.05**	**62**	**0.26**	**16.13**

Figure 2. Scenario 2 AOIs (a) and heatmap for all participants (b)

Scenario 3

The home page depicted in Figure 3 (http://huggy.gr/) is a news portal (based on a content syndication model) and at the time of the experiment included interstitial text ads, a rich media banner on the right side, and news feed ads at the bottom. To examine whether participants noticed the existing ads, but also to see how they absorbed the content of the site and which items tend to attract their attention, user were asked to do a free browsing across the entire webpage and then to locate and read a specific article.

Figure 3 presents the gazeplot (a) and the heatmap (b) of a single user where it is evident that the user did not read the article word for word, but generally read horizontally at the top of the web page, then a second horizontal movement slightly lower on the web page and eventually scanned vertically down the left side of the screen, following the F shaped pattern. Interestingly, users when given a specific task (to locate an article with a specific heading) did not fixate much on the prevailing images of the page but skipped to headlines and text blocks.

After the end of the scenario and the completion of their goal, participants responded to two questions related to the ads. More specifically, 60% of the respondents replied that they did not notice the advertisements cited, as opposed to the 40% who saw them but could not remember their content. In addition, when asked whether they were annoyed by the presence of interstitial ads (on a scale from 1-not at all, to 5-extremely) as depicted in Figure 4, 14 out of 20 (70%) stated that they were not bothered or were slightly bothered (5 out of 20), while none declared being annoyed (5 or 4 on the scale).

Among the three ads that were served in between the actual content of the page more fixations were observed on the ad at the end of the textual content users were asked to locate. The two first interstitial ads were noticed by 35% of the participants with average fixation duration 1.11sec for the first and 0.86sec for the second. One can conclude that users with a specific task in mind do not notice interstitial ads even if they scan them. The ad placed immediately after the actual content requested, gathered a much longer fixation duration (30.38 sec in total by all users), clearly due to its placement. The news feed ads at the bottom of the page was seen by 25% of users. When users were asked at the end of the session if they can recall the message of the rich media ad on the top right corner of the page 12 out of 20 responded positively indicating that probably this is due to the high relevance of the advertising message with the content of the article they were asked to read.

Scenario 4

The stimulus in Figure 5 depicts a sports news site (http://www.sport24.gr/) which was displayed to users with the purpose of locating and opening one of the articles on the home page. Based on the TFF metric, it turned out that all participants in the test (100%) fixed their gaze on the main banner of the page almost immediately, after 0.36 sec (on average). For 3 users the topmost banner was the item they fixated on immediately (the eye tracker recorded 0.00 for TTFF). The rich media banner ad on the right of the site was noticed afterwards by 13 of the 20 participants in the experiment (65%), 4.52 (on average) after the stimulus was displayed. There were two text ads at the bottom right of the page, which gathered the two lower fixation percentages, since 15% and 10% of the participants fixated on them after 5.11sec and 10.74 sec respectively.

Fixation duration metrics showed that the main banner collected a total of 358 fixations with an average fixation duration of 0.27sec. The rich media banner ad on the right attracted 43 fixations with 0.24 sec average fixation duration. Finally, the two last text ads at the bottom of the page accrued 5 and 13 fixations respectively with 0.23sec and 0.30sec (average duration). All this information is also shown by the heat map shown below.

Concluding, all participants viewed the top site content unlike the bottom site content, indicating that the area above the fold is significantly more visible than the area below it. Although organic content will always be (apparently) more visible than ads, ads placed in the correct locations can be much more visible than marginal content placements and traditional ad placements.

Scenario 5

In the final scenario of the experiment, participants were asked to view the home page of a tennis news website (http://tennisnews.gr/) which was displayed to them as an image (no interactivity) and users were asked to click on all the ads they could discover. The objective was to study in which areas of the page the gaze moved faster to look for the ads.

From the heat map in Figure 6 it is clear that when users were asked to find the advertisements on the page, they looked up, right and down. This fact suggests that users are familiar with the traditional placement of advertisements and the resulting banner blindness. It is also very important that some users also clicked on frames with actual content, as they thought they were advertising messages due to their location.

At the end of the session, participants were asked if there is any place on webpages they usually avoid looking because they know there are advertisements. 80% of participants replied that they almost never look at the rightmost vertical region. 35% stated that they do not look at the bottom of webpages, 10% do not look at the top, and 35% do not look at the left of the main content of the website, because they expect to find advertisements. None of them mentioned the center of the page, which means that they do not expect to see ads mixed in between content blocks.

Post-Test Questionnaire

As a final step of the experiment, participants responded to a short questionnaire (see Appendix) to see if users' opinion on ads agree with their behavior during the experiment. The main observations made are summarized as follows:

- Rich media ads attract much more attention than simple text ads and banners. This does not agree with the results of scenario 2 where 70% looked at the text ads while rich media ads and expandable ads received less fixations.
- 65% responded that they are only paying attention to advertisements that may interest them or are related to their current task. This is in line with the results of the experiments (scenario 1), where users did not pay much attention to trick banners as they were not interested to them.
- 15% replied that all ads are inconvenient, 35% said they were disturbed by sound ads, 40% by ads appearing in front of the screen (pop ups and float ads) and 35% by advertisements that have mandatory display duration. This was also shown by the experiments and especially in scenario 1 where 35% were bothered so much that they paused the video.
- 90% stated that when they saw floating or pop up ads they were looking for a way to close them.
- 30% never clicks on ads even though they notice many of them, while 70% rarely clicked on them, which was also shown in the experiments (scenario 1)

Figure 3. Scenario 3 AOIs and heatmap (single participant)

Figure 4. Responses on whether users were disturbed by interstitial ads

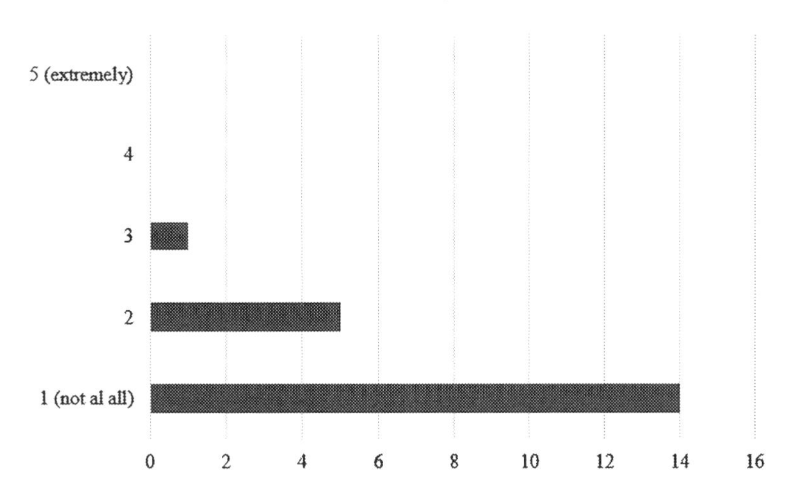

- 60% of participants use software or have configured the browser to block pop-ups and the remaining 40% do not use software because they simply do not know how to.

SOLUTIONS AND RECOMMENDATIONS

After completing the experimental process and studying the collected qualitative and quantitative data, we have concluded that Internet users have become very experienced concerning current advertising and have developed skills to avoid ads on web pages, as in general they consider them annoying. Specifically, when Internet users first visit a website to look for specific information, they are not concerned with ads, especially if they are completely irrelevant to the actual content of the page and their current task (or information needs). This means that current web users confirm the banner blindness phenomenon. In fact, Internet users have developed a 'mechanism' for ignoring advertising stimuli, either consciously or unconsciously, which means that not only do they refuse to click on display ads but also they are not exposed to them at all (they do not receive the advertising message by ignoring it).

The display advertising industry needs to adapt to break the banner blindness circle and regain the attention of its target audience on the web. Many advertisers chose to impose promotional messages that users cannot (or cannot easily) avoid but this approach is in the wrong direction, as the result is user annoyance, as well as rejection of the message and negative image of the promoted brand.

The eye tracking study conducted indicated that display ad effectiveness may be accomplished by using non-traditional placements, by placing ads on sites with relevant content, and placing them above the fold. In addition, captivating interactive rich media ads draw faster and maintain longer the visual attention of users when compared to text ads and static web banners. These recommendations are in line with what Resnick & Albert (2013) recommend for tackling the banner blindness problem. If the reason for it is that the banner is not viewed at all, then the solution to marketers could be to use design techniques such as salience, animation, or other attention-grabbing techniques and they can use non-traditional page placements (Bayles, 2002). If the banner area grabs the attention of the user but the user cannot recall its contents it is most likely that the content of the ad was irrelevant for the user or the

Figure 5. Scenario 4 heatmap (all participants, display ads in red frames) (a) and clusters (b)

Figure 6. Heatmap upper part (a) and bottom part (b) of tennisnews.gr home page (all participants, with mouse clicks)

tasks at hand. In this case, a matching between the ad and the needs or preferences of the user (using user profiling or search query matching) might help serving more relevant ads.

Copying with the F-shaped pattern according to Pernice (2017) can be accomplished by formatting webpage contents in a way that directs users to the desired destinations. According to the theory of visual hierarchy, the design of the page is likely to have an impact on users' viewing pattern. In particular, the visual hierarchy of a page is likely to influence users' tendency to view it in an F-shaped pattern (Faraday, 2000; Djamasbi et al., 2011). Some practical directions include putting prioritized content in the first two paragraphs on the page, use headings and subheadings, place information-rich words upfront, use visual clues (color, size and even frames or shapes) to communicate importance and highlight content, name links with meaningful descriptions, and decrease the length text blocks as reading online is very different that reading a printed text.

The role of page fold is another issue marketers should consider in combination with banner blindness and web reading patterns. Current web pages (especially in information-rich websites such as portals and blogs) tend to be long and include negative space, according to the flat design approach. Web users may be more inclined to scroll than in the past, but they still spend most of their page viewing time above the fold (Fessenden, 2018). This, in marketing terms means that ads should preferably be placed above the fold but in the case that this is not possible, then strong visual indicators should let users know that there is more content to discover bellow the fold.

CONCLUSION AND FUTURE RESEARCH DIRECTIONS

Digital advertising is one of the application domains of web technology that adapts and incorporates latest advances at a remarkable pace. Marketing has been transformed by web technology basically in terms of three dimensions; speed, relevance and reach of campaigns. The main challenge for brands though remains the same and it is to efficiently connect with customers through computer screens, tablets and smartphones in real time and create campaigns that work across social media, display advertising and e-commerce. The aim of this chapter was to examine how modern display advertising promotional messages work in the web environment as visual stimuli for their recipients and to address the factors that determine their effectiveness by means of an eye-tracking study combined with a pre and a post test questionnaire. Eye gaze data are particularly revealing when examining visual stimuli and they become more valuable when associated with asking users to recall seeing an advertising message we know they fixated on. Moreover, the chapter aimed to study whether banner blindness still applies regardless of the type of display ad used, whether the visual pattern remains F-shaped, the effect of placing ads below the fold, how effective trick banners are and which types of ads are annoying to users.

Conclusions reached have confirmed that web users still demonstrate banner blindness regardless of the increased captivating power of current rich media ads technology and are negative towards promotional messages that do not relate to their needs. They still follow the F pattern when reading textual content online and still appreciate informational content formatted adequately to support easy absorption (understanding) and are more willing than before to scroll below the fold if they have visual clues indicating that there will be valuable content to access. In this newly forming scenery, marketers need to combine the creative side of the discipline (using powerful narratives to tap into people's wishes and aspirations) with software engineering tools and analytics and deliver compelling stories people want to engage with and will not consider as a barrage of unnecessary, unwanted and irrelevant messages (Accenture, n.d.).

REFERENCES

Accenture. (n.d.). *The Future of Digital Advertising*. Retrieved from https://www.accenture.com/us-en/~/media/Accenture/next-gen/pulse-of-media/pdf/Accenture-Future-Of-Advertising-POV.pdf

Agarwal, L., Shrivastava, N., Jaiswal, S., & Panjwani, S. (2013). Do not embarrass: re-examining user concerns for online tracking and advertising. In *Proceedings of the Ninth Symposium on Usable Privacy and Security (SOUPS '13)*. ACM. 10.1145/2501604.2501612

Bayles, M. J. (2002). Designing online banner advertisements: Should we animate? *Proceedings of CHI*, 363–366.

Beelders, T., & Bergh, L. (2014). Age as differentiator in online advertising gaze patterns. In *Proceedings of the Southern African Institute for Computer Scientist and Information Technologists Annual Conference 2014 on SAICSIT 2014 Empowered by Technology (SAICSIT '14)*. ACM. doi:10.1145/2664591.2664595

Benway, J., & Lane, D. (1998). Banner blindness: Web searchers often miss 'obvious' links. *Internetworking*, *1*, 3.

Blascheck, T., Kurzhals, K., & Raschke, M. (2014). State-of-the-art of visualization for eye tracking data. *Proc EuroVis*.

Borys, M., & Plechawska-Wójcik, M. (2017). Eye-tracking metrics in perception and visual attention research. *European Journal of Medical Technologies*, *3*(16), 11–23. Retrieved from http://www.medical-technologies.eu/upload/2_eye-tracking_metrics_in_perception_-_borys.pdf

Brajnik, G., & Gabrielli, S. (2010). A review of online advertising effects on the user experience. *International Journal of Human-Computer Interaction*, *26*(10), 971–997. doi:10.1080/10447318.2010.502100

Brunswick, G. (2014). A Chronology Of The Definition Of Marketing. *Journal of Business & Economics Research*, *12*(2), 105–114. doi:10.19030/jber.v12i2.8523

Busch, O. (2016). *Programmatic Advertising: The Successful Transformation to Automated, Data-Driven Marketing in Real-Time*. Cham, Switzerland: Springer International. doi:10.1007/978-3-319-25023-6

Buscher, G., Cutrell, E., & Morris, M. R. (2009). What Do You See When You're Surfing? Using EyeTracking to Predict Salient Regions of Web Pages. *Proceedings of the 27th International Conference on Human Factors in Computing Systems*, 21–30.

Bylinskii, Z., & Borkin, M. A. (2015). *Eye fixation metrics for large scale analysis of information visualizations*. ETVIS Work. Eye Track. Vis.

Calder, B., Malthouse, E., & Schaedel, U. (2009). An Experimental Study of the Relationship between Online Engagement and Advertising Effectiveness. *Journal of Interactive Marketing*, *23*(4), 321–331. doi:10.1016/j.intmar.2009.07.002

Cambridge English Dictionary. (n.d.). *Text ad meaning*. Retrieved from https://dictionary.cambridge.org/dictionary/english/text-ad

Conversy, S., Hurter, C., & Chatty, S. (2010). A descriptive model of visual scanning. Proc. 3rd BELIV'10 Work. BEyond time errors Nov. Eval. methods. *Information Visualization, 2010*, 35–42.

Djamasbi, S., Siegel, M., & Tullis, T. (2011). Visual Hierarchy and Viewing Behavior: An Eye Tracking Study. In J. A. Jacko (Ed.), Lecture Notes in Computer Science: Vol. 6761. *Human-Computer Interaction. Design and Development Approaches. HCI 2011*. Berlin: Springer.

DoubleClick Creative Solutions. (n.d.). *How rich media works: Rich media creative types*. Retrieved from https://support.google.com/richmedia/answer/2417545?hl=en

Faraday, P. (2000). Visually Critiquing Web Pages. *Proceedings of the 6th Conference on Human Factors and the Web*, 1–13.

Fessenden, T. (2017). *The Most Hated Online Advertising Techniques*. Retrieved from https://www.nngroup.com/articles/most-hated-advertising-techniques/

Fessenden, T. (2018). *Scrolling and Attention*. Retrieved from https://www.nngroup.com/articles/scrolling-and-attention/

Fitts, P.M., Jones, R.E., & Milton, J.L. (2005). Eye movements of aircraft pilots during instrument-landing approaches. *Ergon Psychol Mech Model*, 3-56.

Goldberg, J. H., & Kotval, X. P. (1999). Computer interface evaluation using eye movements: Methods and constructs. *International Journal of Industrial Ergonomics*, *24*(6), 631–645. doi:10.1016/S0169-8141(98)00068-7

Hernandez, M., Wang, Y., Sheng, H., Kalliny, M., & Minor, M. (2017). Escaping the corner of death? An eye-tracking study of reading direction influence on attention and memory. *Journal of Consumer Marketing*, *34*(1), 1–10. doi:10.1108/JCM-02-2016-1710

Höller, N., Schrammel, J., Tscheligi, M., & Paletta, L. (2010). The perception of information and advertisement screens mounted in public transportation vehicles – Results from a mobile eye-tracking study. *Lecture Notes in Informatics*, 4007-4021. Retrieved from https://www.tobiipro.com/fields-of-use/marketing-consumer-research/

Interactive Advertising Bureau. (n.d.). *Standards, Guidelines, and Best Practice Documents in Public Comment*. Retrieved from https://www.iab.com/guidelines/iab-standards-guidelines-best-practice-documents-in-public-comment/

Johnston, M. (2015). *What is (and isn't) marketing?* Retrieved from http://regions.cim.co.uk/south-east/home/news/2015/january/what-is-marketing

Lee, J., & Ahn, J.-H. (2014, December 8). Attention to Banner Ads and Their Effectiveness: An Eye-Tracking Approach. *International Journal of Electronic Commerce*, 119–137.

Lin, C., & Kim, T. (2016). Predicting user response to sponsored advertising on social media via the technology acceptance model. *Computers in Human Behavior, 64*, 710-718. doi:10.1016/j.chb.2016.07.027

Lohse, G. L. (1997). Consumer eye movement patterns on yellow pages advertising. *Journal of Advertising*, *26*(1), 61–73. doi:10.1080/00913367.1997.10673518

McCoy, S., Everard, A., Polak, P., & Galletta, D. F. (2007). The effects of online advertising. *Communications of the ACM*, *50*(3), 84–88. doi:10.1145/1226736.1226740

Mou, J., & Shin, D. (2018). Effects of social popularity and time scarcity on online consumer behaviour regarding smart healthcare products: An eye-tracking approach. *Computers in Human Behavior*, *78*, 74–89. doi:10.1016/j.chb.2017.08.049

Nicosia, F. (1974). *Advertising, Management, and Society: A Business Point of View*. New York: McGraw-Hill.

Nielsen, J. (2005). *Scrolling and Scrollbars*. Retrieved from https://www.nngroup.com/articles/scrolling-and-scrollbars

Nielsen, J. (2006). *F-Shaped Pattern For Reading Web Content*. Retrieved from https://www.nngroup.com/articles/f-shaped-pattern-reading-web-content-discovered

Nielsen, J. (2010). *Scrolling and Attention (Original Research Study)*. Retrieved from https://www.nngroup.com/articles/scrolling-and-attention-original-research/

Owens, J. W., Chaparro, B. S., & Palmer, E. M. (2011). Text advertising blindness: The new banner blindness? *Journal of Usability Studies*, *6*(3), 172–197

Pernice, K. (2017). *F-Shaped Pattern of Reading on the Web: Misunderstood, But Still Relevant (Even on Mobile)*. Retrieved from https://www.nngroup.com/articles/f-shaped-pattern-reading-web-content

Pham, C., Rundle-Thiele, S., Parkinson, J., & Li, S. (2017). Alcohol Warning Label Awareness and Attention: A Multi-method Study. *Alcohol and Alcoholism (Oxford, Oxfordshire)*, 1–7. doi:10.1093/alcalc/agx087 PMID:29136096

Pieters, R., & Wedel, M. (2004). Attention capture and transfer in advertising: Brand, pictorial and text-size effects. *Journal of Marketing*, *68*(2), 36–50. doi:10.1509/jmkg.68.2.36.27794

Radach, R., Lemmer, S., Vorstius, C., Heller, D., & Radach, K. (2003). Eye movements in the processing of print advertisements. In J. Hyönä, R. Radach, & H. Deubel (Eds.), *The Mind's Eyes: Cognitive and Applied Aspects of Eye Movement Research* (pp. 609–623). Oxford, UK: Elsevier Science. doi:10.1016/B978-044451020-4/50032-3

Rayner, K., Miller, B., & Rotello, C. M. (2008). Eye movements when looking at print advertisements: The goal of the viewer matters. *Applied Cognitive Psychology*, *22*(5), 697–707. doi:10.1002/acp.1389 PMID:19424446

Resnick, M., & Albert, W. (2013). The Impact of Advertising Location and User Task on the Emergence of Banner Ad Blindness: An Eye-Tracking Study. *International Journal of Human-Computer Interaction*, *30*(3), 206–219. doi:10.1080/10447318.2013.847762

Rohrer, C., & Boyd, J. (2004). The rise of intrusive online advertising and the response of user experience research at Yahoo! In CHI'04 Extended Abstracts on Human Factors in Computing Systems (pp. 1085-1086). ACM.

Ryan, D. (2016). *Understanding Digital Marketing: Marketing Strategies for Engaging the Digital Generation*. New York: Kogan Page Publishers.

Shaikh, D., & Lenz, K. (2006). Where's the Search? Re-Examining User Expectations of Web Objects. *Usability News*, *8*(1), 14. Retrieved from http://usabilitynews.org/wheres-the-search-re-examining-user-expectations-of-web-objects/

Sheehan, K. (2014). *Controversies in contemporary advertising (2nd ed.)*. Thousand Oaks, CA: SAGE Publications, Inc.

Shojaeizadeh, M., Djamasbi, S., & Trapp, A. C. (2016). Density of Gaze Points Within a Fixation and Information Processing Behavior. *Int. Conf. Univers. Access Human-Computer Interact.*, 465–71.

Smith, P. (2013). *Business Development, Marketing and Sales*. Retrieved from https://bizzmaxx2012.wordpress.com/2013/12/24/types-of-advertising-in-e-commerce

Tobii Studio 2.X User Manual. Manual release 1.0, September 2010.

Wästlund, E., Shams, P., & Otterbring, T. (2018). Unsold is unseen … or is it? Examining the role of peripheral vision in the consumer choice process using eye-tracking methodology. *Appetite*, *120*, 49–56. doi:10.1016/j.appet.2017.08.024 PMID:28851559

Wedel, M., & Pieters, R. (2008). A Review of Eye-Tracking Research in Marketing. Review of Marketing Research.

APPENDIX

Pre-Test Questionnaire

Gender:

- Male
- Female

Age:

- 18-23
- 24-29
- 30-35

Educational level:

- Highschool graduate
- University undergraduate student
- University graduate
- Postgraduate student
- MSc holder

Daily internet use:

- None
- 0-3 hours
- 3-6 hours
- More than 6 hours

Post-Test Questionnaire

1. What kind of ads you believe that draw more your attention, those that contain text or those that contain animation, sound and are interactive;
2. Which kind of web ads you consider as the most irritating?
3. Is there some part of a webpage you almost ignore and never look at because you know that it contains ads?
4. When do you notice an ad on a webpage? Do you generally pay no attention to ads on the web and you avoid them?
5. When you see an ad covering the contents of the page you requested which is your first reaction to it?

6. How often do you consciously click on an ad?
7. Is clicking on an ad something you do often by accident?
8. Has it ever happened to you to notice advertising windows open after you close your current browser window? If yes which is your reaction?
9. Do you use ad blocking software or you have made certain settings in your browser to block pop-ups? If not, why?
10. Are you familiar with cookies and how they work? Do you allow websites to store cookies in your PC?

Chapter 13
Does the Customers' Use of Mobile Technologies Influence the Number of Both Recruitments and the Branches in the Banking Sector in Tunisia?

Amira Sghari
University of Sfax, Tunisia

ABSTRACT

Digital determines new practices of companies and customers while touching all sectors of activity. Adaptation to the digital is imperative for banks. In this framework, this chapter explores the question of the influence of the use of mobile technologies by the customers on recruitment in the banking sector and on the number of the branch banking networks. Particularly, the authors seek to answer the following question: What are the effects of the use of mobile technologies by customers on recruitments and the number of branches in the banking sector in Tunisia? In order to answer this question, they analyze the annual reports of the last seven years published by Tunisia's Professional Association of Banks and Financial Institutions. Contrary to the findings observed in foreign countries concerning the reduction of the number of branches and recruitments following the digital transformation in the banking sector, in Tunisia the number of branches and recruitments has not stopped increasing.

DOI: 10.4018/978-1-5225-7262-6.ch013

INTRODUCTION

Traditional business strategies are fundamentally reshaped by digital technologies (Subra maniam and Venkatraman, 2001; Straub and Watson, 2001; Wheeler, 2002; Samba murthy et al., 2003; Tan riverdi and Venkatraman, 2005; Banker et al., 2006; Ettlie and Pavlou, 2006; Kohliand Grover, 2008; Rai et al., 2012; Bharadwaj et al., 2013) and the structure of social relationships between consumer and enterprise is transformed (Susarla and Tan, 2012).

The banking sector is characterized by competitive volatility, market uncertainty, and technology uncertainty (Pousttchi and Schurig, 2004). Mobile commerce, a new paradigm of an emerging information technology (IT) artefact has been made possible thanks to the convergence of Internet, wireless technologies, and mobile devices (Luo et al., 2010). In this context, "mobile banking is an innovative method meant to access banking services via a channel whereby the customer interacts with a bank via a mobile device (e.g., mobile phone or personal digital assistant). A great deal of promise in the ability to provide anywhere anytime banking is offered by Mobile banking (Luo et al., 2010, p. 222). According to Shaikh et al. (2015), the use of smartphones in the accomplishment of banking transactions has become popular.

In the Tunisian context, on the occasion of the 2nd edition of the International Forum on Innovative Digital Financial Instruments in April 2018[1], the President of Tunisia's Professional Association of Banks and Financial Institutions announced that; since affected by the new financial technologies 40% of the banking functions will be metamorphosed in the coming years.

Adaptation to the digital is an urgent imperative for banks. Retail banking is affected by the implications of digital transformation (Schuchmann and Seufert ; 2015) Indeed, banking services which used to be evaluated according to their intangible nature by clients in various aspects such as the relationship-based interaction between the contactor frontline employee and the customer (Berry and Parasuraman, 1993; Nguyen and Leblanc, 2002; Ivens, 2004; Durif et al., 2013) will no longer be applied since the operations are carried out remotely. In addition, the execution of remote banking operations by customers could lead to a reduction in the number of employees, as a result, the reduction of recruitments, and bank branches.

There is a dearth of research focusing on the possible impact of the use of mobile banking on the number of recruitments and that of the branches.This paper considers the overarching question "*Does the use of mobile technologies by customers influence the number of recruitments and that of the branches in the banking sector in Tunisia?*"

Accordingly, the paper first analyses prior annual reports of the last seven years published by Tunisia's Professional Association of Banks and Financial Institutions[2]; a descriptive study aimed at studying the evolution of the use of mobile banking in relation to the evolution of recruitment and the number of branches.

In what follows, the paper presents the evolution of the customers' use of mobile technologies in the execution of the banking operations and their impacts on the banks in the world. It then describes the same study on the Tunisian market.

MOBILE BANKING ADOPTION

The financial services sector has been particularly adept at connecting mobile users to their mobile technologies (Bansal and Bagadia, 2018, p. 51). Mobile banking refers to the provision of banking services with the help of mobile devices (Birch, 1999; Gefen et al., 2000; Hoehleand et al., 2012; Zhou, 2012; Illia et al., 2015). Offered Banking services include monitoring account balance, transferring funds between accounts, paying bills, remoting check deposit and performing various other transactions (Birch, 1999; Gefen et al., 2000; Illia et al., 2015). Mobile banking is defined as "activities that result in an entity's access to the range of banking products (related to savings or credit) by using mobile devices such as cell phones" (Porteous, 2007). It allows managing finance transactions, connecting perfectly anytime and anywhere (Riivari, 2005). Nevertheless, mobile banking contains uncertainty and risk because of the virtuality and lack of control (Yang, 2009; Zhou, 2012) which may prevent its use. Among the various inhibitors of mobile banking adoption, note the lack of ease of use (Bamoriya and Singh, 2012; Safeena et al., 2012), the lack of usefulness of services (Luo et al., 2012; Safeena et al., 2012) and the lack of knowledge (Luo et al., 2012; Safeena et al., 2012).

Studies of adoption and use of mobile banking has been explored by many researchers e.g.(Yang, 2009 ; Pantano and Di Pietro, 2012 ; Mathew et al., 2014 ; Lotfizadeh and Ghorbani, 2015 ; Illia et al., 2015, Gupta et al., 2017 ; Shareef et al., 2018).The consumers' acceptance of new technologies and their intention to use it is explained by several behavioural theories such as the Technology Adoption Model (TAM) (Davis, 1989), Theory of Planned Behavior (TPB) (Ajzen, 1991), Unified Theory of Acceptance and Use of Technology (UTAUT) (Venkatesh et al., 2003), Task Technology Fit model (TTF) (Goodhue and Thompson, 1995), and Diffusion of Innovation theory (DOI) (Rogers, 1995).

The works dealing with the factors affecting adoption of mobile banking were summarized (Ecer, 2018). These factors are presented in the table 1.

Traditionally, banking activities are performed by clients in face to face branch. The use of mobile banking allows customers to run multiple online transactions, which will change the traditional way of distributing banking products and services. Mobile banking allows to bypass the traditional branch-based retail banking (Bansal and Bagadia, 2018, p. 50)

Indeed, the use of mobile banking reduces the customers' visits to the branches.

Obviously, mobile banking influences the traditional branch channel. However, there is a dearth of research focused on the possible impact of the use of mobile banking on the number of recruitments and that of the branches. Accordingly, in order to understand the effect of consumers' use of mobile banking on banks, we have used professional documentation such as prior annual reports.

During the last few years, mobile banking has produced a surge in banking transactions in many countries through the increased use of smart phones (Gupta et al., 2017).

THE EVOLUTION OF THE USE OF MOBILE TECHNOLOGIES BY CUSTOMERS OF BANKS AROUND THE WORLD

Mobile technologies allow advanced access to Internet-based services through multiple functions (Lien et al., 2015). Its rapidly growing use, even in developing countries, has improved the customers' interaction with their banks (Lien et al., 2015).The rapid proliferation of mobile devices has resulted in the extensive diffusion of mobile banking (Gupta and al.,2017, p.127).

Table 1. Recent studies on mobile banking adoption (Ecer, 2018, p. 673)

Authors	Adoption Factors
Luarn and Lin (2005)	usefulness, convenience, credibility, self-efficacy, cost
Laukkanen (2007)	efficiency, convenience, safety
Lee *and al.* (2007)	Perceived risk, perceived usefulness, trust
Laukkanen and Cruz (2008)	usage, value, risk, tradition and image
Kim *and al.* (2009)	relative benefits, trust, structural assurances
Gu*and al.* (2009)	usefulness, convenience, trust
Crabbe *and al.* (2009)	Perceived credibility, facilitating conditions
Puschel*and al.* (2010)	compatibility, convenience, relative benefit, visibility, demonstrability, image, triability, perceived behavioral control, facilitation condition, subjective norm, testability, intention
Koenig-Lewis *and al.* (2010)	compatibility, perceived usefulness, risk, trust, cost
Cruz *and al.* (2010)	cost, risk, perceived advantage, complexity
Wessels and Drennan (2010)	usefulness, risk, convenience, financial cost, compatibility
Zhou *and al.* (2010)	Task characteristics, technology characteristics, convenience conditions, task technology fit, performance expectancy, effort expectancy, social influence
Riquelme and Rios (2010)	intention, perceived relative advantage, perceived risk, social norms, convenience, usefulness
Singh *and al.* (2010)	usefulness, ease of use, subjective norms, self-efficacy, cost, safety, trust
Lin (2011)	Perceived advantage, ease of use, compatibility, competence, benevolence, integrity
Negash (2011)	usefulness, convenience, enjoyment, network quality, security, privacy, trust, awareness, regulation, compliance
Akturan and Tezcan (2012)	risk, ease of use, usefulness, benefit
Zhou (2012)	structural assurance, ubiquity, ease of use, personal innovativeness
Chen (2013)	advantage, concurrency, trialability, complexity, different risk types, attitude, intention to use, brand image, brand awareness
Hanafi zadeh *and al.* (2014)	usefulness, ease of use, the need for interaction, risk, cost, compatibility with life style, credibility, trust
Bidar *and al.* (2014)	usefulness, ease of use, security, privacy, compatibility, social influence, facilitating conditions, cost

The connection with the digital media (computer, tablet, and Smartphone) as well as the connection to the banks' social networks will substitute traditional channels (go to a bank branch, call customer service ...) (survey[3] TNS Sofres). These uses appear to crumble because of hyper connectivity.

Another survey[4] conducted in 2014 on Americans using mobile phones to access to banking services, showed that 71% of American adults have smartphones (with internet connection) and 52% use them to carry out banking operations (checking account balances, transferring money, receiving an alert (text, message, notification, e-mail)from their bank, issuing checks, paying bills via an online system or application).

Similarly in Europe, Smartphones are used as a means of payment, for example, the Cityzi project in Nice and other cities in France (Ebben .,2013),. More than 2 million downloads of banking applications have been made in the Netherlands. These applications cover several features including funds transfer and invoice payments (Ebben, 2013).

In its press release on May 27th, 2015, Societe Generale[5] published the results of a survey[6] on the analysis of the French new behaviours concerning the management of their budgets in a bank digitization context. The main lessons of this study are:

- More and more French people are connected to their bank accounts and their management. As a result, online consultation of bank accounts is beginning to replace traditional means.
- People aged 25 to 34 are currently the main users of these services and act as technophile prescribers of mobile uses. Therefore, a progression can potentially be considered in the use of budget management applications, with the continued growth of smartphone and tablet equipment rate.
- The use of a Budget application is an activity that seduces its users and that is very beneficial for the banking universe image. It gives users a sense of simplicity, good time management and trust that reflects on the relationship with their banks.

In France[7] estimates show around 25 million online banking users in 2013. More than 86% of the incoming contacts of Societe Generale, for example, were executed via digital channels. Surveys also show that customers are increasingly interested in managing their accounts, 100% remotely. 42% of customers of an online bank consider it as their main establishment.

A recent study by the Commonwealth Bank in Australia predicts that smartphones are going to replace physical portfolios by 2021 (Bott and Milkau, 2014). It is essential to understand the consumer digital behaviour, preferences and choices that have significant consequences on banks that need continue existing services while developing strategies to manage the shift in the mix (Schuchmann and Seufert, 2015).

Digitalization is also disrupting African banks. HEBDO[8] of July 13th, 2018, explains with statistics, presented below, the effect of digitization on African banks.

At the end of 2017, sub-Saharan Africa had 338.4 million registered mobile money accounts.

In October 2016 a mobile application (Ecobank Mobile App) was launched by The pan-African banking group Ecobank Transnational Incorporated (ETI) It allows customers not only to open an account from their smart phones and instantly transfer funds in the the continent's 33 countries but also to take out loans, save money, access insurance products, block credit cards and access their accounts remotely.

1.95 million in 2017 and over 2.1 million customers in the first quarter of 2018 that was the customers' number who downloaded the Ecobank mobile app. In 2017, the value of transactions made via Ecobank Mobile App reached $ 604 million. Until March 31st, 2018, a billion dollar of transactions have already been made via this mobile platform.

A good part of the operations of Equity Bank and the majority of big Kenyan banks such as Kenya Commercial Bank, Cooperative Bank, Barclays Bank of Kenya and Commercial Bank of Africa have also been digitized. On March 16th, 2018, Timiza app, a mobile application allowing customers to take out micro-credits and save money was launched by Barclays Bank of Kenya. In less than two weeks, the app has been downloaded by over 10,000 people.

In South Africa, Standard Bank transactions via mobile platform have grown at an average of 100% per year since 2015. In West Africa and North of the continent, a mobile platform called "UBA Connect" was launched in August 2018 by the Nigerian banking group United Bank for Africa (UBA), which has subsidiaries in 19 countries in sub-Saharan Africa, and that to enable merchants and enterprises to carry out cross-border transactions via mobile phones. An entirely digital bank in Ivory Coast was launched in March 2018 by the Standard Chartered Bank.

The Moroccan group Attijari wafa Bank has already reshaped the dividends of the success of "Banka Lik", its 100% mobile platform.

The Effects of the Customers' Use of Mobile Technologies on the Banks

According to Laukkanen (2005), the banking sector is among the leading sectors in adopting the Internet and mobile technologies for consumer markets. Indeed, the emergence of the Internet had a significant impact on the diffusion of electronic banking which has changed retail banks business (Laukkanen and Lauronen, 2005; Laukkanen, 2005). For example, multiple electronic channels have made it possible to create new kinds of added value for customers (Coelho and Easingwood, 2003; Suoranta et al., 2005).

Luo et al. (2010) claim that the sociotechnical interaction between banking service providers and customers has been reshaped by mobile technologies. Consequently, the customers have become less willing to visit traditional branches and more receptive to new electronic channels (Suoranta et al., 2005).

Luo et al. (2010) presume that banks must expand the capabilities of their services in order to maintain and potentially create a competitive advantage. To this end, the banking industry is leading the way into mobile access (Mallat et al., 2004). The financial services via mobile devices have allowed the emergence of mobile banking, which can be implemented through techniques such as downloadable applications (Luo et al., 2010). Thus, mobile banking is considered as an innovative method which offers a great deal of promise in its ability to provide anywhere anytime banking (Luo et al., 2010).

In their study, Dubois et al. (2011) observed that the Internet makes the customer increasingly informed and makes him/her collaborate with the banker as to find a solution. Thus, the face-to-face service encounter is becoming increasingly interactive and customized requiring both highly cognitive hard skills and effective soft skills (Värlander and Julien, 2010).

The Impact of the Customers' Use of Mobile Technologies on the Job of the Banker

The customers' use of the Internet requires new employee skills (Farquhar and Rowley, 1998; Hunter et al., 2001; Hogg et al., 2003; Laing et al., 2005; Ployhart, 2006; Amadi-Echendu, 2007). Goldkind and Wolf (2015), affirmed that the development of individual competencies is necessary due to the digitalization changes in the industries. In the banking sector, many authors found that any person would need new skills in order to cope with the technological issues (Durkin and Howcroft, 2003; Värlander and Julien, 2010). The impact of technological change on individual competencies is challenging (McKelvie and Picard, 2008; Bartosova, 2011).

- The impact of the customers' use of mobile technologies on the number of staff in banks and on the branch banking networks[9]

Online banks have reduced the number of branches as well as the number of employees. In Lagos, Johannesburg, Nairobi, and Accra, a sharp downsizing has accompanied the digitization of bank offers. For instance, Kenyan banks have removed 2517 jobs in 2016 and 2036 in 2015. They also closed 39 branches during 2017.

Due to the digitization of the banking industry, 1620 employees were laid off and 39 branches were closed in 2017 by Banks listed on the Nairobi Securities Exchange.

Banks that have closed branches are:

- Bank of Africa: 12 branches
- Ecobank: 9 branches
- Equity Bank: 7 branches
- Barclays Bank of Kenya: 7 branches
- Standard Chartered Bank Kenya: branches

In South Africa, as part of its digital transformation, the Nedbank Group announced in March 2018 the forthcoming elimination of 3,000 jobs. 600 jobs has already been cut in The South African group FirstRand Bank while Ecobank has reduced its workforce by 16,000 since 2015. During the same period, 159 branches were closed: 74 in Nigeria, 75 in Eastern and Southern Africa (Cesa), and 10 in Ghana.

As part of France's network transforming project, Societe Generale has confirmed the closure of around 400 of its 2221 branches by 2020. It justifies this project by the eruption of the digital in the customers' behaviour and their relationship with their bank (Societe Generale, Press release May 27th, 2015).

EFFECTS OF THE CUSTOMERS' USE OF MOBILE TECHNOLOGIES ON TUNISIAN BANKS

In Tunisia[10], mobile access (smartphones and tablets) increased from 16% in 2013 to 39% in 2015 and traffic from the mobile increased from 21% in 2014 to 30% in 2015 (24% Smartphone and 6% Tablet).

Tunisian banks offer their customers several services that can be performed remotely. In this section, the paper first describes the Tunisian banking sector. It then presents the evolution of the consumers' use of mobile technologies. Ultimately the paper concludes with the effect of the customers' use of mobile technologies on the number of staff in Tunisian banks and on the branch banking networks.

Presentation of the Banking Sector

The banking system and financial institutions in Tunisia are composed of the Central Bank of Tunisia, 21 credit institutions having the quality of bank, 2 investment banks, 8 off-shore banks, 8 representative offices of foreign banks, 3 factoring companies and 10 leasing companies.

The banking system has suffered in recent years a certain disengagement from the state, the entry of foreign investors into the shareholding of banks as well as from the access of foreign banks to the local market. Hence, it has become a more balanced shareholder structure, classifying the bulk of banks into three categories[11]:

- Banks with high state ownership (mainly BNA, STB, and BH).
- Banks with private Tunisian capital (BIAT, BT and Amen Bank).
- Banks with the foreign majority (UIB-SG, UBCI-BNP Paribas, Attijari Bank and ATB).

Banks and financial institutions are required to publish periodic financial communications and annual reports. Among other sources of information on the development of the banking sector is Tunisia's Professional Association of Banks and Financial Institutions.

Created in 1972, Tunisia's Professional Association of Banks and Financial Institutions is a professional organization comprised of banks and financial institutions. Currently, it has 25 universal banks, 2 merchant banks, 8 leasing companies and 3 factoring companies. The association defends its members interests, informs them of the regulatory decisions concerning the exercise of their activities.

Tunisia's Professional Association of Banks and Financial Institutions is a key player in the implementation of an active information policy aiming at raising the banking profession. It publishes press releases, annual reports, statistics and mini-guides, and brochures.

Evolution of the Consumers' Use of Mobile Technologies

Tunisian banks offer several online services. For example, the BIATNET MOBILE was launched in 2014by the first private Tunisian Bank BIAT. Accessible on tablets and smartphones, BIATNET MOBILE offers several operations: Account statement, Search operation, Prepaid card loading, Card opposition, Transfers (transfers to BIAT accounts or other banks in Tunisia), Download transactions and documents, check book order, messaging for the processing of complaints, RIB / IBAN printing, consultation of the prices of major currencies and request for information or making appointments with an advisor in branch.

Mobile banking applications are available on download platforms of leading mobile app distributors: App store, Google Play store, Windows Phone store. Mobile banking applications of different Tunisian banks have been the subject of several downloads, as shown in table 2.

We notice the existence of a very large number of customers who use mobile banking in Tunisia. Sghari and al. (2017) have conducted an exploratory research on the study of the influence of the customers' use of mobile technologies on the activities and skills required from the account managers. The research has been carried out by a qualitative study based on semi-structured interviews with 9 account

Table 2. Number of downloads of mobile banking applications of some Tunisian Banks in 2018[12]

Examples of Tunisian Banks	Number of Downloads of Mobile Banking Applications (More Than)
Attijari Bank	100 000
BIAT	50 000
UIB	10 000
AMEN BANK	10 000
ZITOUNA	10 000
BH	10 000
STB	10 000
UBCI	1000

managers and 2 heads of branches of a leading bank in the Tunisian market. The results show that the customers 'use of mobile technologies has little or no effect on account managers. Indeed, it is important for account managers to have face-to-face contact with customers.

As part of this research, the paper investigates whether there is an effect of the clients' use of mobile technologies on recruitment and the number of branch networks.

Does the Customers' Use of Mobile Technologies Influence the Number of Staff in Tunisian Banks and That of the Branch Banking Networks?

As part of this descriptive study, the annual reports published by Tunisia's professional association of banks and financial institutions during the last seven years were analysed in order to understand the functioning of the banking sector, the different activities and to determine the evolution of recruitments and the number of new branches in recent years.

Evolution of the Number of Branches Over the Period 2010-2016

At the end of 2016, the network of bank branches has 1827 branches against 1739 branches in 2015, an evolution of 88 branches. In table 3 below, the evolution of the number of branches from 2010 to 2016 is presented.

From these data, it is noticed that the strategy of the banks foresees a strategy of development based on the extension of their networks. Indeed, the number of branches has increased by 201 branches between 2014 and 2016.

This can be explained by the fact that certain banking operations can be performed only in the branch. These activities are essential: "customer prospecting", "opening accounts", "sales of products", "credits" and "advising".

Customer prospecting can only be exercised in face-to-face. As for the "sale of products" activity only a simple presentation of the different products on the bank's website allows the customer to have a general idea about the products. In order to choose the best product, the customer has to contact his/her branch to advise and help him/her make the best choice.

Table 3. Evolution of the number of branches from 2010 to 2016 (Annual reports of Tunisia's professional association of banks and financial institutions (2010-2016))

Years	Number of Branches	Evolution of the Number of Branches
2010	1331	63
2011	1 394	61
2012	1455	57
2013	1512	114
2014	1626	113
2015	1739	88
2016	1827	-

Concerning the account opening and credit activities, customers find all the information about these activities, such as the requested paperwork and the procedure to follow, which gives them better visibility and understanding.

Thus, customers become better informed without however having the ability to purchase products, obtain credits or open an account completely online. Indeed, it is imperative to go and see their account managers at the branch. This is mainly explained by the Tunisian administrative and regulatory complications that prevent the realization of these activities completely online. Indeed, to grant credit, purchase products or opening an account, the law requires originals of papers and original signatures.

The effect of customers' use of mobile technologies will soon begin to impact the number of bank branches. Indeed, in the electronic press[13], UBCI Bank declares that it will close some 20 agencies by 2020. This decision is explained by the importance of the progressive adaptation to the new technologies of the banking industry.

Evolution of the Workforce in the Banking Sector During the Period 2010-2016

The increase in new branch openings each year is leading to an increase in recruitment. Below table 4 presents the evolution of the workforce in the banking sector during the period 2010-2016.

The overall workforce in the banking sector increased from 21478 employees in 2015 to 22073 in 2016, an increase of 595 people. This evolution is explained by recruitments as well as by departures for retirement for example. Analysis of the annual reports from 2012 to 2016 of Tunisia's professional association of banks and financial institutions identified the number of hirings and retirements in the banking sector, which is presented in Table 5 below.

The change in the workforce of the banking population is mainly due to the opening of new branches each year.

Table 4. Evolution of the workforce in the banking sector from 2010 to 2016 (Annual reports of Tunisia's professional association of banks and financial institutions (2010-2016))

Years	The Overall Workforce in the Banking Sector	Evolution of the Workforce
2010	18933	968
2011	19901	366
2012	20.267	512
2013	20779	398
2014	21177	301
2015	21478	595
2016	22073	-

Table 5. Number of hirings and retirements in the banking sector from 2012 to 2016 (Annual reports of Tunisia's professional association of banks and financial institutions (2012-2016))

Years	Recruitment	Departures
2012	700	374
2013	768	409
2014	1151	532
2015	880	588
2016	1348	577

CONCLUSION

The descriptive study builds on the question of the possible impact of the costumers' use of mobile banking on the number of recruitments and the branch banking networks. For this purpose, the annual reports of the last seven years published by Tunisia's Professional Association of Banks and Financial Institutions were analysed.

Unlike the observed findings in foreign countries concerning the reduction of the number of branches and recruitments following the digital transformation in the banking sector, it has been observed that in Tunisia the number of branches and recruitments has not stopped increasing.

Until now, the results show that the customers 'use of mobile technologies has no effect on recruitments and on the number of the branch banking networks. This result can be explained by the fact that only low value-added operations can be performed online. Several essential activities can only be performed in face-to-face, for example, opening accounts, sales of products" and credits" activities since Tunisian regulations require originals of papers and original signatures. The results show the existence of a gap between the discourse of practitioners and the media, which conveys the upheaval of trades following the digital transformation and the reality that shows the absence of a remarkable change.

In several countries, a change in the banking profession, a decrease in the staff and the closure of a number of bank branches has been led to by digital transformation. It is recommended to plan new training programs on the skills required by the transformation of the banker's profession and begin to prepare their employees for the possible reduction in workforce and closure of a number of bank branches.

This study has certain limitations. Firstly, the paper did not take into consideration the Tunisian cultural aspect. Secondly, the descriptive aspect of the study makes it neglect certain elements that characterize the Tunisian context demanding the importance of the opening of branches and the increase of recruitment in the context of regional development. It will be interesting in future researches to conduct an explanatory study in order to understand the impact of the use of mobile banking on the number of recruitments and that of the branch banking networks.

REFERENCES

Ahluwalia, P., & Varshney, U. (2009). Composite quality of service and decision making perspectives in wireless networks. *Decision Support Systems*, *46*(2), 542–551. doi:10.1016/j.dss.2008.10.003

Ajzen, I. (1991). The theory of planned behavior. *Organizational Behavior and Human Decision Processes*, *50*(2), 179–211. doi:10.1016/0749-5978(91)90020-T

Akturan, U., & Tezcan, N. (2012). Mobile banking adoption of the youth market: Perceptions and intentions. *Marketing Intelligence & Planning*, *30*(4), 444–459. doi:10.1108/02634501211231928

Amadi-Echendu, J. E. (2007). Thinking styles of technical knowledge workers in the systems of innovation paradigm. *Technological Forecasting and Social Change*, *74*(8), 1204–1214. doi:10.1016/j.techfore.2006.09.002

Bamoriya, D., & Singh, P. (2012). Mobile banking in India: Barriers in adoption and service preferences. *Journal of Management*, *5*(1), 1–7.

Banker, R. D., Bardhan, I. R., Chang, H., & Lin, S. (2006). Plant Information Systems, Manufacturing Capabilities, and Plant Performance. *Management Information Systems Quarterly*, *30*(2), 315–337. doi:10.2307/25148733

Bansal, A., & Bagadia, P. (2018). The Effect of Financial Risk Tolerance on Adoption of Mobile Banking in India: A Study of Mobile Banking Users. *IUP Journal of Bank Management*, *17*(1), 50–76.

Bartosova, D. (2011). The future of the media professions: Current issues in media management practice. *International Journal on Media Management*, *13*(3), 195–203. doi:10.1080/14241277.2011.576963

Berry, L. L., & Parasuraman, A. (1993). Building a new academic field-The case of services marketing. *Journal of Retailing*, *69*(1), 13–60. doi:10.1016/S0022-4359(05)80003-X

Bharadwaj, A., El Sawy, O. A., Pavlou, P. A., & Venkatraman, N. (2013). Digital Business Strategy: Toward a Next Generation of Insights. *Management Information Systems Quarterly*, *37*(2), 471–482. doi:10.25300/MISQ/2013/37:2.3

Bidar, R., Fard, M. B., Salman, Y. B., Tunga, M. A., & Cheng, H. I. (2014). Factors affecting the adoption of mobile banking: sample of Turkey. *16th International Conference on Advanced Communication Technology (ICACT)*. 10.1109/ICACT.2014.6779165

Birch, D. G. W. (1999). Mobile financial services: The Internet isn't the only digital channel to consumers. *Journal of Internet Banking and Commerce*, *4*(1), 20–29.

Bott, J., & Milkau, U. (2014). Mobile wallets and current accounts: Friends or foes? *Journal of Payments Strategy and Systems*, *8*(3), 6–19.

Chen, C. (2013). Perceived risk, usage frequency of mobile banking services. *Managing Service Quality*, *23*(5), 410–436. doi:10.1108/MSQ-10-2012-0137

Coelho, F., & Easingwood, C. (2003). Multiple channel structures in financial services: A framework. *Journal of Financial Services Marketing*, *8*(1), 22–34. doi:10.1057/palgrave.fsm.4770104

Crabbe, M., Standing, C., Standing, S., & Karjaluoto, H. (2009). An adoption model for mobile banking in Ghana. *International Journal of Mobile Communications*, *7*(5), 515–543. doi:10.1504/IJMC.2009.024391

Cruz, P., Neto, L. N. F., Munoz-Gallego, P., & Laukkanen, T. (2010). Mobile banking rollout in emerging markets: Evidence from Brazil. *International Journal of Bank Marketing*, *28*(5), 342–371. doi:10.1108/02652321011064881

Davis, F. D. (1989). Perceived usefulness, perceived ease of use, and user acceptance of information technology. *Management Information Systems Quarterly*, *13*(3), 319–340. doi:10.2307/249008

Dubois, M., Bobillier Chaumn, M. E., & Retour, D. (2011). The impact of development of customer on-line banking skills on customer adviser skills. *New Technology, Work and Employment*, *26*(2), 156–173. doi:10.1111/j.1468-005X.2011.00266.x

Durif, F., Geay, B., & Graf, R. (2013). Do key account managers focus too much on commercial performance? A cognitive mapping application. *Journal of Business Research*, *66*(9), 1559–1567. doi:10.1016/j.jbusres.2012.09.019

Durkin, M. G., & Howcroft, J. B. (2003). Relationship marketing in the banking sector: The impact of new technologies. *Marketing Intelligence & Planning*, *21*(1), 61–71. doi:10.1108/02634500310458162

Ebben, W. (2013). Will smartphones contribute to making payments easier and more efficient? *Journal of Payments Strategy and Systems*, *7*(1), 11–17.

Ecer, F. (2018). An integrated fuzzy ahp and aras model to evaluate mobile banking services. *Technological and Economic Development of Economy*, *24*(2), 670–695. doi:10.3846/20294913.2016.1255275

Ettlie, J., & Pavlou, P. A. (2006). Technology-Based New Product Development Partnerships. *Decision Sciences*, *37*(2), 117–148. doi:10.1111/j.1540-5915.2006.00119.x

Farquhar, J. D., & Rowley, J. (1998). Enhancing the customer experience: Contribution from information technology. *Management Decision*, *36*(5), 350–357. doi:10.1108/00251749810220568

Gefen, D., Straub, D., & Boudreau, M. (2000). Structural equation modeling techniques and regression: Guidelines for research practice. *Communications of AIS*, *4*(7), 2–76.

Goldkind, L., & Wolf, L. (2015). A digital environment approach: Four technologies that will disrupt social work practice. *Social Work*, *60*(1), 85–87. doi:10.1093wwu045 PMID:25643579

Goodhue, D. L., & Thompson, R. L. (1995). Task-technology fit and individual performance. *Management Information Systems Quarterly*, *19*(2), 213–236. doi:10.2307/249689

Gu, J. C., Lee, S. C., & Suh, Y. H. (2009). Determinants of behavioral intention to mobile banking. *Expert Systems with Applications*, *36*(9), 11605–11616. doi:10.1016/j.eswa.2009.03.024

Gupta, S., Haejung, Y., Xu, H., & Kim, H. W. (2017). An exploratory study on mobile banking adoption in Indian metropolitan and urban areas: A scenario-based experiment. *Information Technology for Development*, *23*(1), 127–152. doi:10.1080/02681102.2016.1233855

Hanafizadeh, P., Behboudi, M., Abedini Koshksaray, A., & Jalilvand Shirkhani Tabar, M. (2014). Mobilebanking adoption by Iranian bank clients. *Telematics and Informatics*, *31*(1), 62–78. doi:10.1016/j.tele.2012.11.001

Hoehle, H., Scornavacca, E., & Huff, S. (2012). Three decades of research on consumer adoption and utilization of electronic banking channels: A literature analysis. *Decision Support Systems*, *54*(1), 122–132. doi:10.1016/j.dss.2012.04.010

Hogg, G., Laing, A., & Winkelman, D. (2003). The professional service encounter in the age of the Internet: An exploratory study. *Journal of Services Marketing*, *17*(5), 476–494. doi:10.1108/08876040310486276

Hunter, L. W., Bernhardt, A., Hughes, K. L., & Skuratowicz, E. (2001). It's not just the ATMs: Technology, firm strategies, jobs, and earnings in retail banking. *Industrial & Labor Relations Review*, *54*(2), 402–424. doi:10.1177/001979390105400222

Illia, A., Ngniatedema, T., & Huang, Z. (2015). A conceptual model for mobile banking adoption. *Journal of Management Information and Decision Sciences*, *18*(1), 111–122.

Ivens, B. S. (2004). How relevant are different forms of relational behavior? An empirical test based on Macneil's exchange framework. *Journal of Business and Industrial Marketing*, *19*(5), 300–309. doi:10.1108/08858620410549929

Kim, G., Shin, B., & Lee, H. G. (2009). Understanding dynamics between initial trust and usage intentions of mobile banking. *Information Systems Journal*, *19*(3), 283–311. doi:10.1111/j.1365-2575.2007.00269.x

Koenig-Lewis, N., Palmer, A., & Moll, A. (2010). Predicting young consumers' take up of mobile banking services. *International Journal of Bank Marketing*, *28*(5), 410–432. doi:10.1108/02652321011064917

Kohli, R., & Grover, V. (2008). Business Value of IT: An Essay on Expanding Research Directions to Keep up with the Times. *Journal of the Association for Information Systems*, *9*(1), 23–39. doi:10.17705/1jais.00147

Laing, A., Hogg, G., & Winkelman, D. (2005). The impact on professional relationships: The case of health care. *Service Industries Journal*, *25*(5), 675–687. doi:10.1080/02642060500101021

Laukkanen, T. (2005). Comparing Consumer Value Creation in Internet and Mobile Banking. In *International Conference on Mobile Business*. Sydney, NSW, Australia: IEEE. 10.1109/ICMB.2005.28

Laukkanen, T. (2007). Internet vs mobile banking comparing customer value perceptions. *Business Process Management Journal*, *13*(6), 788–797. doi:10.1108/14637150710834550

Laukkanen, T., & Cruz, P. (2008). Barriers to mobile banking adoption: a cross-national study. *Proceedings of the International Conference on E-Business*, 1, 26-29.

Laukkanen, T., & Lauronen, J. (2005). Consumer value creation in mobile banking services. *International Journal of Mobile Communications*, *3*(4), 325–338. doi:10.1504/IJMC.2005.007021

Lee, K. S., Lee, H. S., & Kim, S. Y. (2007). Factors influencing the adoption behavior of mobile banking: A South Korean perspective. *Journal of Internet Banking and Commerce*, *12*(2), 1–9.

Lien, J., Hughes, L., Kina, J., & Villasenor, J. (2015). Mobile money solutions for a smartphone-dominated world. *Journal of Payments Strategy and Systems*, *9*(3), 341–350.

Lin, H. F. (2011). An empirical investigation of mobile banking adoption: The effect of innovation attributes and knowledge-based trust. *International Journal of Information Management*, *31*(3), 252–260. doi:10.1016/j.ijinfomgt.2010.07.006

Lotfizadeh, F., & Ghorbani, A. (2015). A Multi-dimensional Model of Acceptance of Mobile Banking, *International Journal of Management. Accounting and Economics*, *2*(5), 414–427.

Luarn, P., & Lin, H. (2005). Toward an understanding of the behavioral intention to use mobile banking. *Computers in Human Behavior*, *21*(6), 873–891. doi:10.1016/j.chb.2004.03.003

Luo, X., Lee, C. P., Mattila, M., & Liu, L. (2012). An exploratory study of mobile banking services resistance. *International Journal of Mobile Communications*, *10*(4), 366–385. doi:10.1504/IJMC.2012.048136

Luo, X., Li, H., Zhang, J., & Shim, J. P. (2010). Examining multi-dimensional trust and multi-faceted risk in initial acceptance of emerging technologies: An empirical study of mobile banking services. *Decision Support Systems*, *49*(2), 222–234. doi:10.1016/j.dss.2010.02.008

Mathew, M., Sulphey, M. M., & Prabhakaran, J. (2014). Perceptions and Intentions of Customers towards Mobile Banking Adoption, *Journal. Contemporary Management Research*, *8*(1), 83–101.

McKelvie, A., & Picard, R. G. (2008). The growth and development of new and young media firms. *Journal of Media Business Studies*, *5*(1), 1–8. doi:10.1080/16522354.2008.11073458

Negash, S. (2011). Mobile banking adoption by under-banked communities in the United States: adapting mobile banking features from low-income countries. *11th International Conference on Mobile Business (ICMB)*.

Nguyen, N., & Gaston, L. (2002). Contact personnel, physical environment and the perceived corporate image of intangible services by new clients. *International Journal of Service Industry Management*, *13*(3), 242–262. doi:10.1108/09564230210431965

Pantano, E., & Di Pietro, L. (2012). Understanding Consumer's Acceptance of Technology-Based Innovations in Retailing. *Journal of Technology Management & Innovation*, *7*(4), 1–19. doi:10.4067/S0718-27242012000400001

Ployhart, R. E. (2006). Staffing in the 21st century: New challenges and strategic opportunities. *Journal of Management*, *32*(6), 868–897. doi:10.1177/0149206306293625

Porteous, D. (2007). *Just how transformational is m-banking?* Commissioned by Finmark. Retrieved September 1, 2018, from https://www.microfinancegateway.org/sites/default/files/mfg-en-paper-just-how transformational-is-m-banking-feb-2007.pdf

Pousttchi, K., & Schurig, M. (2004). Assessment of Today's Mobile Banking Applications from The View of Consumer Requirements. *37th Hawaii International Conference on System Sciences.*

Puschel, J., Mazzon, J. A., & Hernandez, J. M. C. (2010). Mobile banking: Proposition of an integrated adoption intention framework. *International Journal of Bank Marketing, Vol.*, *28*(5), 389–409. doi:10.1108/02652321011064908

Rai, A., Pavlou, P. A., Im, G., & Du, S. (2012). Inter firm IT Capability Profiles and Communications for Cocreating Relational Value: Evidence from the Logistics Industry. *Management Information Systems Quarterly*, *36*(1), 233–262.

Riivari, J. (2005). Mobile banking: A powerful new marketing and CRM tool for financial services companies all over Europe? *Journal of Financial Services Marketing*, *10*(1), 11–20. doi:10.1057/palgrave.fsm.4770170

Riquelme, H. E., & Rios, R. E. (2010). The moderating effect of gender in the adoption of mobile banking. *International Journal of Bank Marketing*, *28*(5), 328–341. doi:10.1108/02652321011064872

Rogers, E. M. (1995). *Diffusion of Innovations*. New York: The Free Press.

Safeena, R., Date, H., Kammani, A., & Hundewale, N. (2012). Technology adoption and Indian consumers: Study on mobile banking. *International Journal of Computer Theory and Engineering*, *4*(6), 1020–1024. doi:10.7763/IJCTE.2012.V4.630

Sambamurthy, V., Bharadwaj, A., & Grover, V. (2003). Shaping Agility Through Digital Options: Reconceptualizing the Role of Information Technology in Contemporary Firms. *Management Information Systems Quarterly*, *27*(2), 237–263. doi:10.2307/30036530

Schuchmann, D., & Seufert, S. (2015). Corporate Learning in Times of Digital Transformation: A Conceptual Framework and Service Portfolio for the Learning Function in Banking Organizations. *International Journal of Advanced Corporate Learning*, *8*(1), 31–39. doi:10.3991/ijac.v8i1.4440

Sghari, A., Chaabouni, J., & Schiopoiu Burlea, A. (2017). Effets de l'usage des technologies mobiles par les clients sur le métier des chargés de clientèle dans les banques: Etude de cas. In *Colloque international La transformation numérique des entreprises & les modèles prédictifs sur Big Data*. Université de Med BOUDIAF.

Shaikh, A., Karjaluoto, H., & Chinje, N. B. (2015). Continuous mobile banking usage and relationship commitment-A multi-country assessment. *Journal of Financial Services Marketing*, *20*(3), 208–219. doi:10.1057/fsm.2015.14

Shareef, M. A., Baabdullah, A., Dutta, S., Kumar, V., & Dwivedi, Y. K. (2018). Consumer adoption of mobile banking services: An empirical examination of factors according to adoption stages. *Journal of Retailing and Consumer Services*, *43*, 54–67. doi:10.1016/j.jretconser.2018.03.003

Singh, S., Srivastava, V., & Srivastava, R. K. (2010). Customer acceptance of mobile banking: A conceptual framework. *SIES Journal of Management*, *7*(1), 55–64.

Straub, D. W., & Watson, R. T. (2001). Transformational Issues in Researching IS and Net-Enabled Organizations. *Information Systems Research*, *12*(4), 337–345. doi:10.1287/isre.12.4.337.9706

Subramaniam, M., & Venkatraman, N. V. (2001). Determinants of Transnational New Product Development Capability: Testing the Influence of Transferring and Deploying Tacit Overseas Knowledge. *Strategic Management Journal*, *22*(4), 359–378. doi:10.1002mj.163

Suoranta, M., Mattil, M., & Munnukka, J. (2005). Technology-based services: A study on the drivers and inhibitors of mobile banking. *International Journal of Management and Decision Making*, *6*(1), 33–46. doi:10.1504/IJMDM.2005.005964

Susarla, A., Oh, J. H., & Tan, Y. (2012). Social Networks and the Diffusion of User-Generated Content: Evidence from YouTube. *Information Systems Research*, *23*(1), 123–141. doi:10.1287/isre.1100.0339

Tanriverdi, H., & Venkatraman, N. V. (2005). Knowledge Relatedness and the Performance of Multibusiness Firms. *Strategic Management Journal*, *26*(2), 97–11. doi:10.1002mj.435

Värlander, S., & Julien, A. (2010). The Effect of the Internet on Front-line Employee Skills: Exploring Banking in Sweden and France. *Service Industries Journal*, *30*(8), 1245–1261. doi:10.1080/02642060802350979

Venkatesh, V., Morris, M. G., Davis, G. B., & Davis, F. D. (2003). User Acceptance of Information Technology: Toward a Unified View. *Management Information Systems Quarterly*, *27*(2), 425–478. doi:10.2307/30036540

Wessels, L., & Drennan, J. (2010). An investigation of consumer acceptance of M-banking. *International Journal of Bank Marketing*, *28*(7), 547–568. doi:10.1108/02652321011085194

Wheeler, B. C. (2002). NEBIC: A Dynamic Capabilities Theory for Assessing Net-Enablement. *Information Systems Research*, *13*(2), 125–146. doi:10.1287/isre.13.2.125.89

Yang, A. S. (2009). Exploring Adoption Difficulties in Mobile Banking Services. *Canadian Journal of Administrative Sciences*, *26*(13), 136–149. doi:10.1002/cjas.102

Zhou, T. (2012). Examining mobile banking user adoption from the perspectives of trust and flow experience. *Information Technology Management*, *13*(1), 27–37. doi:10.100710799-011-0111-8

Zhou, T., Lu, Y., & Wang, B. (2010). Integrating TTF and UTAUT to explain mobile banking user adoption. *Computers in Human Behavior*, *26*(4), 760–767. doi:10.1016/j.chb.2010.01.013

ENDNOTES

[1] Newsletter - Tunisia's professional association of banks and financial institutions N° 1- April 2018.

[2] Professional organization comprised of Tunisian banks and financial institutions.

[3] A Survey carried out online on behalf of the online banking ING Direct, between 19 and 24 December 2012, with a representative panel of 1630 individuals aged 18 years and over.

[4] A Survey conducted The Federal Reserve Board's Division of Consumer and Community Affairs.

[5] One of the leading European financial service groups.

[6] The study is conducted with a sample of 1051 people, representative of the French population aged 18 and over, constituted according to the quota method, with regard to the criteria of gender, age, socio-occupational category, category agglomeration and region of residence. The sample is queried online on Cawi (Computer Assisted Web Interview) system. The interviews are conducted on April 8 and 9, 2015.

[7] Statistics provided by the Observatory of banking professions, http://www.observatoire-metiers-banque.fr/index.do.

[8] Swiss magazine published in Lausanne belonging to the Ringier press group, https://www.agenceecofin.com/la-une-de-lhebdo/2704-56487-la-digitalisation-de-banques-africaines-sacrifie-des-milliers-de-salaries-et-dessine-le-futur-de-l-industrie-bancaire.

[9] The statistics presented in this section are taken from HEBDO of July 13, 2018.

[10] WMC Portail (webmanagercenter (FR) – Directinfo (FR) – Almasdar (AR) -Online Magazines.

[11] Amen Invest (intermediate in stock Exchange) report (March 2011), http://www.businessnews.com.tn/pdf/Secteur-bancaire0311.pdf

[12] Google Play store.

[13] Kapitalis of December 21, 2017.

Chapter 14
Results-Oriented Influencer Marketing Manual for the Tourism Industry

Carlos de-Laguno-Alarcón
University of Malaga, Spain

Plácido Sierra-Herrezuelo
University of Malaga, Spain

María-Mercedes Rojas-de-Gracia
University of Malaga, Spain

ABSTRACT

This chapter aims to provide a better tool for implementing the marketing technique known as influencer marketing in the tourism industry. To do so, a results-oriented influencer marketing manual for the tourism industry has been created. Despite the success of influencer marketing, the few previous studies in this field do not include verified measures to ensure its effectiveness. For this reason, the approach that is presented here could be crucial to support these marketing activities. As this topic is new and often little-understood, the data compiled was based on the case study methodology. This chapter proposes the following phases: (1) campaign planning, (2) search for influencers, (3) evaluating the best profiles, (4) contacting influencers, (5) proposing a project, (6) execution, and (7) analyzing the results. This work could help companies considering influencers as a new communication channel to successfully run their campaigns.

INTRODUCTION

Although the era of digitalization may be perceived as relatively new, research has long stressed its importance in studies about the need for companies to use new technologies to improve their business (Lindh and Rovira, 2018). Autonomous and collaborative robots, the Internet of things, additive manufacturing, cloud computing, augmented reality, and Big Data are some of the advances that companies are using to increase their productivity (Özüdoğru, Ergün, Ammari & Görener, 2018).

DOI: 10.4018/978-1-5225-7262-6.ch014

Furthermore, the way in which companies relate to customers has changed dramatically, with new figures such as influencers appearing on the scene. The term influencer is not new, but its use on social media is, where this concept is proving to be an important marketing tactic. Thanks to the Internet, this tool has become powerful and easy to use. Influencers allow brands to reach customers in a more social, direct, and effective way. In fact, influencer marketing perfectly meets companies' need to face increasing skepticism of advertisement among their customers. Moreover, purchase decisions are now based on comments from users who customers see as similar (Castelló, del Pino & Tur, 2016; Kanellopoulos & Panagopoulos, 2008; Pan, MacLaurin & Crotts, 2007). Thus, according to Nielsen (2015), 66% of the global audience trust consumers posting opinions online and 83% trust recommendations from people they know. Therefore, encouraging different neutral players to create positive content about a company's products or services has become a new and extremely effective communication tool.

Specifically, social media have changed the rules of the tourism industry. This phenomenon, known as Travel 2.0, has made it possible for people other than the companies themselves to share information about a specific service on a website; users can now share and rely on this information (Casaló, Flavián, Guinalíu & Ekinci, 2015; Wöber, 2006; Xiang, Wöber & Facemailed, 2008). In this context, influencers have become new tour guides. The fact they share their experience in the first person provides a new point of view that is characterized by naturalness and trust. This is why campaigns with influential people can be extremely beneficial for tourist brands, since this allows them to reach a much more specific user profile at a lower cost than other conventional advertising campaigns (Hernández, Fuentes & Marrero, 2012).

In fact, there are already several tourist destinations that are using influencers to raise their profile in certain areas. This is the case of the Spain China Project, from 2018, through which the Spain Brand has organized an important campaign to promote Spain in China through influencers. In total, 10 Chinese influencers with more than a combined 30 million followers on social networks are visiting several cities in Spain, which allows them to share their experience and promote Spain's tourism, culture and gastronomy in the Chinese market. Reciprocally, and almost at the same time, the China Influencer Project was also implemented, through which 150 Spanish influencers, with more than 50 million followers combined, visited China and also shared their experiences on social networks.

However, conducting a successful marketing campaign with influencers is not easy if a company has no prior experience in the subject. On the other hand, there are many companies linked to tourism, such as hotels, airlines, etc., that could benefit from having influencers in their marketing campaigns. In this sense, although there are currently some guidelines on how to implement this type of strategy, there are no solid proposals that offer a results-oriented approach or indicators that help to better evaluate the campaign. This is precisely the objective of this paper: to create a results-oriented manual for influencer marketing campaigns in the tourism industry and, potentially, in any other industry. To achieve this objective, this manual is based on a practical case, that is, it is based on field experience, which gives it greater practical value.

THEORETICAL FRAMEWORK

Before delving into the project, it is necessary to clearly and precisely define the terms that will be used in this manual. This is vital, considering that the field of influencers is still new and unknown to a large part of the population. Specifically, there are three key elements in this study: influencers, influencer marketing, and influence inside tourism.

It is also necessary to indicate that when we talk about brands, although many of the concepts presented here can be used by all brands in a general sense, this manual is focused on tourism companies. Therefore, it should be understood that the concept of brand encompasses not only tourism company brands but also tourism destination brands. Although often managed by public bodies, these brands can still be used for the content that will be developed in this manual.

What Is an Influencer?

The word influencer has not always been used correctly. According to Chae (2017), social media influencers are online celebrities who exhibit their personal lives to many followers via social media. When this term is used in social media, there is a tendency to think of someone with a high number of followers and likes on their social networks. However, the reality is that not everyone with a big account can be considered to be an influencer. It should be remembered that popularity and influence are not synonymous. Therefore, more qualities must be included to realistically outline the meaning of influence over an audience. In other words, an influencer should be someone who is considered a powerful communication tool.

In this sense, according to the Cambridge dictionary, an influencer must have the power to change the way people behave. Therefore, a high number of followers alone does not make someone an influencer; rather, it is their real ability to affect users' decision-making processes. This concept of influencer is also shared by Castelló-Martínez and del Pino-Romero (2015). These authors consider that an influencer is a person who, thanks to their personality, the fact that they belong to a specific organization, or to their knowledge, generates influence via their opinions, ratings, or purchase decisions.

On the other hand, Armano (2011) argues that influence is based on six dimensions or pillars:

- **Reach:** Traditionally limited to conventional media, today social media allows influencers to expand their horizon and grow their reach exponentially.
- **Proximity:** The closer people are to the influencer, the more likely they are to agree and act according to their recommendation. Small networks facilitate proximity.
- **Expertise:** Experts giving opinions on their subject of expertise are key for achieving influence. On social media, there are also experts. However, these experts do not become such thanks to degrees or courses, but thanks to approval and positive assessment from their social system.
- **Relevance:** The influence is as effective as the relevance or relationship the subject has with the topic to which the influence is related.
- **Credibility:** This is the determining aspect of the influence. The activities carried out and the transparency shown help build reputation.
- **Trust:** The reason why people trust their friends, even if there is a lack of expertise or credibility, is because they believe them, share the same interests or, simply, because they know them. The social media environment creates a new dynamic, like a private environment, where trust is developed to a certain degree, even if not all the members of the network are known.

It is important to note that influencers on social media tend to focus on a particular category and they seek interaction with other users by incentivizing them to share their opinions, thoughts or ideas, too (Booth & Matic, 2011). Taking this into account, to summarize, social media influencers can be defined as those who generate information about products, services or about any contemporary topic, mainly through a social media platform, achieving great acceptance and respect from a portion of the users of the social network on which they operate.

Therefore, for the purpose of this project, influencers are born from their social networks and the trust that the audience gives them makes them ambassadors. However, other possible meanings should be excluded, such as specific recommenders in small circles or celebrities outside of social media. To avoid confusion, in this paper, the term influencers will be used for people whose origins are purely in social media. Conversely, all those with an influence on social media, but whose main sector and origin of this reach and ability to influence is outside of social media, in the offline world, will be called celebrities.

Influencers as Prosumers

It is important to mention the prosumer aspects of every influencer. Toffler (1980) defines prosumers as people who produce goods and services for their own consumption. Thus, this term is not new either. However, it acquires a whole new dimension thanks to social media. Richter (2014) establishes a chronological timeline of the term prosumer. According to this author, the most traditional version of prosumer is related to the first wave, which relates the term to ideas such as people making their own clothes or hunting their own food. The second wave, the factory boom, separates consumption and production. This brings about dramatic growth in consumption and a shift to a consumer's perspective, where some people consume and others produce. This unbalanced situation of prosumers versus consumers leaves society with few recorded cases of prosumers and an undervalued image of the prosumer. It is the third wave that brings back the prosumer concept. People leave mass consumption in favor of individualization. These users have the need to produce what they consume and this has great implications in marketing.

The concept of prosumer is implicitly linked to the concept of service experience (Bitner, Faranda, Hubbert & Zeithaml, 1997; Gebauer, Johnson & Enquist, 2010) and co-creation (Vargo & Lusch, 2004; Xie, Bagozzi & Troye, 2008), given that today, consumers not only consume, they also need to share their opinions and offer information about products or services. As a result of this evolution, the term prosumer applies to creative people providing content for others, particularly in the context of social media (Kosnik, 2018). They create content around their experience and social media offers them the means to share this experience. In this sense, an influencer can be a prosumer. If a company makes a user, and, specially, an influencer, happy or unhappy, their experience can be shared among millions of users almost instantaneously (Wang, Yu & Fesenmaier, 2002).

Influencer Marketing

Today, traditional word of mouth (WOM) has become eWOM, allowing for the emergence of influencer marketing. This is a form of marketing in which advertising focuses on specific individuals with the ability to influence their followers (Chatzigeorgiou, 2017). Thus, a first approach to influencer marketing is the art and science of involving influential people online to share brand messages with their audiences in the form of sponsored content (Sammis, Lincoln & Pomponi, 2016).

The use of influencers in marketing is essentially the evolution of earlier celebrity or sport star sponsorships, thanks to social media. This evolution is aligned with the boom in online comments and reviews and the role they play in users' decision-making processes. The Internet, or more precisely social media, has therefore become one of the most relevant channels to influence consumer decisions (Marine-Roig, 2017). Thus, brands seek to generate the credibility or the trust they obtained before from sponsorship and celebrities via users. These users generate feelings among their potential or current customers because of their experience or knowledge on that specific matter. This type of marketing, among other things, increases sales (Brown & Hayes, 2008). Therefore, a new technique used by brands emerges. It is called influencer marketing.

Brown & Hayes (2008) state that influencer marketing encompasses the process of identifying, researching, engaging and supporting the people who create high-impact conversations with customers about your brand, products or services. According to these authors, this marketing activity is a relationship-based strategy that allows companies to align and combine their public relations, sales, product, and digital marketing activities. This will drive results that must be analyzed in accordance with the objectives set by the company. Therefore, when working with influencers, it is extremely important for organizations to know the different types of influencer, allowing them to select the profile that best fits their objectives. Knowing the different types of influencers will help organizations narrow down their selection of influencers and make them more valuable, long-lasting, and effective (Barone, 2013).

There are many ways of classifying the different types of influencers, such as by the type of content published, prestige, or number of followers. Following the classification system created by the Word of Mouth Marketing Association-WOMMA (2013), there are five types of influencers:

- **Advocate or Defender:** Can be defined as a person with positive feelings towards a brand. This user tends to defend the organization and would not hesitate to share content or experiences related to it. Sometimes, this love of the brand may lead to a loss of trust. This type of influencer is frequently used to share their user experience and reach or maintain brand equity.
- **Ambassador or Brand Ambassador:** These are people who share the values and philosophy of the company, such as employees or partners, and, therefore, have the role of representing it. By signing a contract and receiving the corresponding monetary compensation, the ambassador's main tasks are to represent the brand publicly; therefore, companies should see this type of influencer as an extension of their brand. They are characterized by their ability to influence others and they are mainly used with the aim of growing the brand or making it stronger.
- **Citizen:** People who share their opinion or ideas, either positive or negative, with other users. In this case, there is no type of contract or relationship with the brand, which increases trust among users, as they are seen as more independent than other types of influencers. Although their intention is not originally to influence others, some audiences, such as millennials, take the recommendations of other users into serious consideration, and, therefore, they can have significant impact on a brand.
- **Professionals or Experts on the Topic:** These are people who have influence because of their job, experience, or knowledge about a certain topic, such as bloggers or journalists. Their ability to influence catches the attention of brands and this generates part of their income. Through their social position, media channels, events, etc., they are able to have an impact and influence on the audience. The objective when working with this type of influencers is to increase value and knowledge of the brand, sales, or subscriptions.
- **Celebrity:** Celebrities are those who, thanks to the fame they have achieved outside the online world, can influence a large population. Therefore, they are the type of influencers with the biggest reach, but also imply a higher cost. The relationship between the two parties requires a formal contract. On top of that, their actions are easy to measure. The main objective pursued with this

type of influencer is to use their fame to have an impact on the organization's results. They are frequently used to increase sales, spread a message, support causes, increase brand awareness, etc.

Each category of influencer requires a specific program, while the type of influence achieved is different according to the audience. It must be taken into account that whether an individual belongs to one influencer type or another depends, to a large extent, on the situation and the context. Therefore, an influencer can fall into more than one category. For example, someone who derives their income from one specific activity may be considered a professional. However, if this person recommends the product of a company with which they have no affiliation, they can be an advocate.

Tourism and Influencers

Social media has come to change the rules of the tourism industry. Search engines have conventionally been where travelers were presented with information in a coherent way by tourism suppliers (Wöber, 2006; Xiang, Wöber & Fesenmaier, 2008). However, the traditional approach to search engines is threatened by the impact of social media, because of how frequently social media is updated and the high amount of content and links that it generates (Gretzel, 2006). Social media is not only a new player in search engine optimization, but it is also an alternative itself for finding relevant content. For these reasons, today, it is not enough to promoting a service just with words. The tourism industry relies heavily on visuals, where lots of images or videos can be used to show their products and services (Fatanti & Suyadnya, 2015). Within this visual content, pictures are key when looking for references about a particular tourism service. Consumers are gradually becoming more demanding and they want to see more abundant, exact, and complete information. The video format is the most complete, since it includes all the necessary information about a product or service and puts it within the reach of potential users (Stylidis & Cherifi, 2018).

In this context, influencers have become the new tour guides. The fact that they share their experience in the first person provides a new point of view, characterized by a naturalness and trust that is lacking in traditional formats. This is why campaigns with influencers can be extremely beneficial for tourism brands: they allows them to reach a much more targeted user profile at a lower cost than other conventional campaigns (Suciati, Maulidiyanti & Lusia, 2018). In this sense, destinations should work closely with the influencers, not only in order to create innovative content, but also to better understand their potential customers by attracting tourists that feel identified with the values and experiences offered by the destination (Iyiola & Akintunde, 2011). These influence leaders can greatly assist tourism companies in achieving their business objectives at different levels, such as positioning their brands or destination and working on their brand equity, image, and trust (Hernández, Fuentes & Marrero, 2012).

According to Axon Marketing & Communication (2018), the benefits that can be obtained from using influencers for marketing campaigns in the tourism sector are:

- **Broader Visibility of the Brand or Destination:** Working with popular people increases destination virality and exposure.
- **Greater Impact on Tourists of Interest:** Instagram, Facebook, Twitter, Snapchat, and YouTube are some of the social media channels most followed by millennials, and most millennials are travelers; therefore, tourism companies should pay attention to these followers of celebrities on social networks.

- **A message in Tune With Advertising-Skeptic Tourists:** When a celebrity recommends a product, people will likely think that it is a commercial. However, this changes when considering the promotion of a destination, since the influencer also travels voluntarily and nobody can prove that their photos on Instagram or Facebook are part of a campaign.
- **Organic Impact on Media:** Very often, influencers' activities are the focus of different types of press. This means free and organic exposure in traditional media such as newspapers, magazines, TV shows, and radio.

However, conducting a successful marketing campaign with influencers is not easy if the organization has no prior experience in the subject. Lack of experience with influencer marketing can leave some growing companies behind (Cave & Jenkin, 2012). On the other hand, there are many companies linked to tourism, such as hotels, airlines, etc., that could benefit from using influencers in their marketing campaigns. In this sense, although there are currently some guidelines on how to implement this type of strategy, there are no solid proposals that offer a results-oriented approach or indicators that help to better evaluate the campaign. This is precisely the objective of this paper: to create a results-oriented manual on influencer marketing campaigns in the tourism industry and, potentially, any other industry. Specifically, this manual seeks to offer a procedure that allows organizations to measure the efficiency of campaigns and learn from them, identifying where and why they have been inefficient. This proposal should help to establish a long-term relationship with influencers that allows companies to co-create more valuable content and get closer to their target audience.

THE CASE STUDY

To achieve the objectives proposed in this work, the case study methodology was chosen. This technique is appropriate when a comprehensive, in-depth investigation is required and for cases with a limited amount of prior research or data compiled in studies. Specifically, as Yin (1994) states, the case study is an empirical investigation that studies a contemporary phenomenon in its own context in real life, especially when the boundaries between the context and the phenomenon are not clear enough. An investigation based on a case study successfully addresses a situation in which there are more variables to analyze or define than observed data. As a result, it is based on multiple sources that should coincide;. in other words, using more than one method to collect data on the same topic, the results should be similar. This allows the researcher to take advantage of previous theoretical proposals that can guide the collection and analysis of data. From this perspective, the relevant use of case studies as a research methodology has been verified in various subjects, including several associated with business management, management and organizational change, and innovation (Biggart, 1977; Brown and Eisenhardt, 1997; Higgins-Desbiolles, 2018; Mccutcheon and Meredith, 1993; Nieto and Pérez, 2000; Van de Ven and Poole, 1990).

As the name indicates, the case study requires a real case. To create this manual, a company from the tourism sector, which usually works with influencers, has been chosen. Specifically, it is an online price comparison company. Thus, the case study is based on the practices carried out by this company in its routine business activities when designing and implementing campaigns with influencers. This business seems to be a good choice because it is a leader in the use of influencers in tourism, as it has applied the technique successfully, and in different markets.

A case study has the following steps (George & Bennett, 2005; Yin, 1994):

Step 1: Study design
Step 2: Development of the study
Step 3: Analysis and conclusions

In the first step, the objective of the study is established. As already mentioned, in this case, this consists of creating a results-oriented influencer marketing manual to help organizations successfully run influencer campaigns. During the second step, the data collection activity is planned and implemented using the different sources considered for the case. These consist mainly of interviews with the influencer marketing campaign manager at the company selected. Specifically, in order to obtain the necessary information to create this manual, an in-depth interview has been carried out, as well as a series of subsequent follow-ups to clarify certain points with the two people responsible for managing these campaigns. The interviews took place in June 2017. The last step analyzes the evidence. There are different criteria to interpret the results of a study. When the case study methodology is cause-explanation, the dynamic aims to find coinciding behavior patterns and associate them with results. An example of this could be the existence of a systematic relationship between variables. To conclude, a summary of the investigation will be prepared and the results presented in a way that can be used as a manual for interested readers.

INFLUENCER MARKETING MANUAL FOR THE TOURISM INDUSTRY

The process that will be followed in this manual is organized into the following phases: (1) Plan, which consists of preparing an overview of the project that will be used as a guide for the entire process; (2) Search for influencers for the campaign; (3) Evaluation, through a more in-depth look at the profiles and their audience; (4) Contacting the influencers, which is not always a simple task, (5) Proposing a detailed campaign project; (6) Execution, detailing the implementation of the plan; and (7) Analysis of the results, assessing them as accurately as possible based on the established Key Performance Indicators (KPIs).

Campaign Planning

Planning social media and influencer activity is extremely important to efficiently running these types of communication campaigns. A starting point is to determine the frequency and number of campaigns to be run each month. This will mainly depend on the budget and the timeframe. If the organization has a budget and timeframe in mind, the next question to answer is: why is the brand using influencer campaigns? Influencers can help companies pursue different objectives. At the same time, each objective has its own KPIs that will determine the way the rest of the process is structured and evaluated. Therefore, it is very important to establish measurable KPIs. These measurable KPIs will ultimately allow brands to determine whether the campaign was successful or not.

Once the brand has established clear campaign objectives (increase the number of tourists, improve the tourist destination's image, increase visibility, etc.), it must specify the type of influencer(s) and social network channel(s) that best suit the main objective. The target audience must be considered in order to execute an effective influence campaign. It should be borne in mind that many social network profiles are ambitious. This means that although the image they project may be what the company is looking for, the real audience is completely the opposite, since they follow it because it represents their ideals. Therefore, it is necessary to concentrate on the audience behind the influencer.

Some of the most popular social networks are Instagram, YouTube, Twitter, Facebook, Snapchat, blogs, messaging apps (WhatsApp, Line, KakaoTalk, etc.), and the so-called "niche social networks," which target a specific segment of the general population. One example of the latter is Untappd, a mobile phone application that allows its users to check in as they drink beer and share these check-ins and their locations with their friends. Another example is Dribbble, a community of designers, which offers a self-promotion platform for graphic design, web design, illustration, photography, and other creative areas.

Finally, in this phase, it is necessary to come up with a rough description of the type of influencer profile the company is seeking. Appendix 1 shows a quick project planner that will make it easier for organizations to successfully plan their influencer campaigns. The steps to follow to select the type of influencer needed are:

1. **Influencer Origin:**
   a. **Influencer:** A person who is well-known because of social networks. Their entire career was developed on social networks, for example, YouTubers or Instagrammers.
   b. **Celebrity:** They have an impact on social media, but their fame comes from other activities outside the online world. This is the case for actors, singers, politicians, sport stars, and TV presenters.

Both are perfectly capable of driving specific behavior in the target audience, but the type of influencer must be carefully chosen depending on the company's objectives. Sometimes, influencers become celebrities if they attain great fame and start to appear on TV shows or similar platforms. The opposite is true when a celebrity starts to gain significant traction on social media. As noted in the introduction, the term influencers is used for people who are social media experts. However, many of these principles are also applicable to celebrities. Generally, celebrities bring an added bonus to branding, as they are usually well-known in society, whereas influencers are generally only "famous" among their followers. On the other hand, influencers tend to have increased interaction with the social media community.

2. **Type of Profile:** It is helpful to consider different types of profiles for the same campaign:
   a. **Lifestyle:** They get word of mouth from their recommendations. People tend to be very involved.
   b. **Fashion:** Their image can work well with branding campaigns.
   c. **Travel:** This profile is good for category-lovers and early adopters.
   d. **Family:** They make trusted recommendations.
   e. **Technology:** This type is appropriate for app campaigns and early adopters.
   f. **Humor:** Very effective for product placement.
   g. **Photography:** Especially useful for content generation.
   h. **Others:** Some examples are Do it Yourself (DIY), designs, BookTubers, fitness, music, and foodies.
3. **Influencer Size:** This is measured through the number of followers. Categories may vary depending on the market:
   a. **Macro-Influencer:** > 1M followers.
   b. **Big Influencers:** 500K-1M followers.

c. **Medium Influencers:** 100K-500K followers.
d. **Small Influencers:** 50K-100K followers.
e. **Micro-Influencers:** <50K followers.

In general, the lower the size, the greater the number of followers who trust the influencer. On the other hand, the bigger the size, the more people who are reached by the influencer. The most balanced profiles are big and medium.

Searching for Influencers

Once a tourism company has a clear idea of the type of influencer it wants to use in its campaign, it needs to find them. Although this may seem like an easy task, it requires the ability to "zero in" to end up with an accurate selection of profiles. If the brand does not have the resources/time to go through the search process, it can always rely on an agency for this part. However, if it prefers to take on the entire process, the next phase is to answer the question: how do we start the search? This will depend on the company's degree of understanding of the market and the influencer panorama in that country/market. Below are a series of recommendations that companies can follow to carry out a good search:

- **Ask About it:** Brands must find the target market and simply ask them who they are following on social media.
- **Google it**: The simpler way to start is googling it. After a bit of research, companies should be able to find a couple of names repeated in different articles. They should try searching for something like "Top influencers in [Market]," "Best YouTuber [Market]," "Unique [Market] Instagram profiles," "Top [Market] fashion bloggers," etc.
- **Use External Tools:** YouTube, and especially Instagram, are not big fans of external tools and almost all of them are blocked. However, there are some useful tools to help marketing campaign managers find new users. SocialBlade is one of them, and it offers the possibility of checking the top 100 most-followed accounts for a country. This information should be treated with care, as the most-followed are not necessarily the most profitable, the most influential, or those that have the greatest reach.
- **Explore Competitors:** What are competitors or even other non-related brands doing? Brands can check competitors' social profiles to identify influencers. They should check tagged photos or top posts for specific hashtags.
- **Friends of influencers:** Once a couple of influencers have been found, brands should check their top friends, as They are usually tagged several times in their posts.
- **Use Tools Provided by the Social Network:** Not all of them provide features, but here are some examples.
    - Instagram.
        - **Suggested Profiles:** Instagram displays a little arrow next to the "Follow" button. When it is tapped, a menu will come up with "other suggested profiles." These suggested profiles tend to have characteristics similar to the original (see Figure 1, in which the arrow appears, pointing down, to the right of the "Follow" button).

        - **Explore:** The explore option in Instagram lets users see recommended profiles. Some may be of interest to brands looking for influencers (see in Figure 2, the option that looks like a magnifying glass).
    - YouTube.
        - **Trending:** YouTube allows users to discover what videos are trending at that moment. This can be extremely helpful for finding new YouTubers or new profiles that are growing fast (see Figure 3, the "Trending" option).
        - **Suggested Videos:** If companies already have a YouTuber in mind, they can check a few videos and pay extra attention to the suggested videos. They usually include similar influencers.
    - **Twitter:** It is important to check trending topics. Twitter is open to collaborating with other tools, so external tools such as "Audiense" can be used to find the perfect influencer.
- **Partners:** If the company has some influencers who are already working with the team, managers can explore different ways to collaborate with them to get additional exposure.
- **Influencer Network:** In line with the above, influencers with whom the brand has a good relationship can serve as guidance and can be a good source of new ideas. Asking them about other profiles, or even using them to get some feedback about specific profiles can be helpful. They will most likely know all the pros and cons of their competitors.
- **Previous Influencers you Have Already Worked With:** Although it is recommended to work with different influencers over time, long-term relationships are always welcome and could be a great opportunity for businesses.

Figure 1. "Explore similar profiles" feature on Instagram

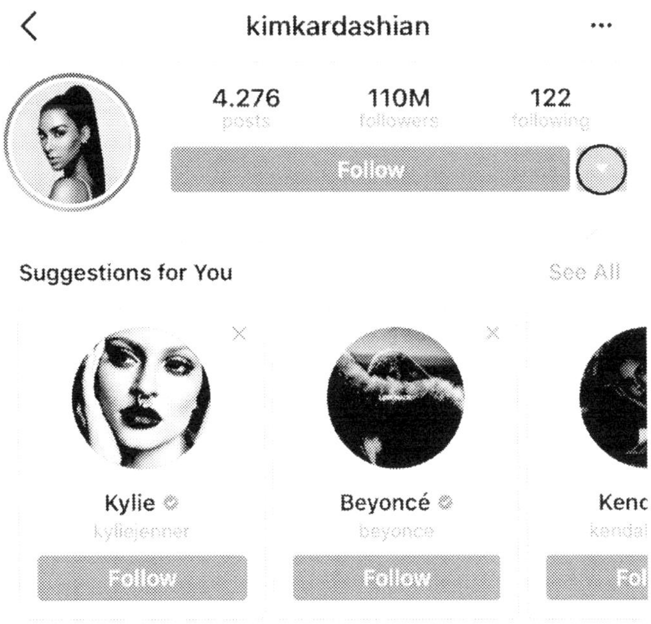

Figure 2. "Explore" feature on Instagram

Figure 3 "Trending" menu in YouTube

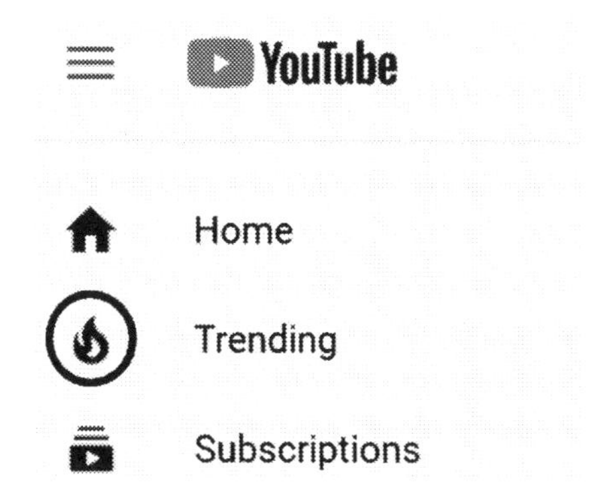

These are just a few ideas; however, it is always a good idea to innovate and look for new ways to find those perfect influencers.

Evaluate

When the company has a large database of influencers, it must filter them to collaborate with those that can best meet its requirements. Not all the profiles the brands have found and selected are equally valuable. Although the degree of influence that the profile exerts on users cannot be measured accurately, there are some metrics that will allow the brand to understand the users' level of implication or commitment to the influencer. This is an important subject to consider when dealing with campaigns with clear calls to action. Sometimes, agencies can provide extra information that can be useful, such as internal metrics, specifically designed tools, or prior experience. However, there is one key metric that is easy to obtain: engagement. Engagement can be defined as a psychological state that occurs because of individuals' interactive and co-creative experiences with specific actors, objects, or events (Hollebeek, Glynn & Brodie, 2014). This state will play a vital role in the evaluation and selection process.

In order to evaluate influencers, a simple formula has been developed:

$$\text{Engagement } \% = (A*1+B*2+C*3) / D$$

This formula assigns different values to different engagement items, (i.e. likes or comments) assuming that each of them requires a higher or lower level of user involvement. The formula is also applicable to different social media networks. In this regard, the proposed formula is based on different user interactions for the main social networks, as shown in Table 1. For the analysis to be representative, it should be based on at least the last five posts, meaning that each of the items must be recorded as the average (i.e.: Likes = Average of likes for the last five posts). Appendix 2 includes clarifications for calculating engagement.

With respect to the number of followers, when conducting the evaluation, companies must be careful of fake followers. Therefore, it is important that they are able to identify so-called "fake followers" on influencers' accounts. Here are some ways to identify them:

- **Spikes in Follower Growth:** SocialBlade provides a history of follower growth. If abrupt spikes are seen regularly, they should be double-checked, as they could be bought followers.
- **Fake Accounts as Followers:** Sometimes, an influencer's followers are fake accounts. Signs of fake accounts are profiles with no photos, few posts and that are "following" many accounts.
- **Fake Comments:** If comments are standard comments such as "cool photo," "nice shot," or just an emoji, there is a chance that they are automated comments. This phenomenon is the result of tools that can post comments automatically.
- **Comment Pods:** Sometimes, a group of users creates a private Instagram group where they share their latest posts and members agree to make comments. These comments look natural and genuine but they are also fake. To identify them, browse some posts. If the commenters are the same, this may indicate that the influencer is using this practice.
- **Inactive Audience:** If engagement goes below 5% on Instagram, or below 50% in YouTube, the users may be real, but they are not interested in the content anymore.

Given that the most popular social networks for influencers today are Instagram, YouTube, and Snapchat, it is worth delving deeper into how these three market leaders are assessed:

- **Instagram:** Since engagement is the core of the evaluation process, brands should aim for a minimum engagement of 6% and, ideally, over 10%. Additionally, they must take into account engagement on branded posts vs. non-branded posts. Other important metrics, such as the consistency of posts, should also be considered. If an influencer publishes regularly, they are likely more valuable than those who only post from time to time. In this sense, it is a good practice to follow an account's activity for some time before deciding to work with it, so as to better understand the influencer's communication style, personality, likes and dislikes. This will also help create a better proposal. Another good indicator are "fan accounts." When an Instagrammer has a huge influence, fans start to create fan clubs or fan accounts. Finally, demographic data or the internal account

Table 1. Importance of user interactions on the main social networks

Social Media	A	B	C	D
Instagram	Likes	Comments	-	Followers
YouTube	Views	Likes	Comments	Subscribers
Twitter	Likes	Retweets	Comments	Followers
Facebook	Likes	Comments	Shares	Fans
Snapchat	Views	Screenshots	-	-

Source: Developed by author

statistics can be very valuable. These metrics are a bit more difficult to obtain, since the company will have to contact the influencer directly and request them. However, if the brand gets this information, they will have a more realistic understanding of the audience. These metrics include:
  - Instagram story views, as accounts that are highly active on Instagram stories also have better, more natural engagement with the audience.
  - Audience age.
  - Country and city of the audience.
  - Reach and impressions per post.
- **YouTube:** Views on this platform are also considered when calculating engagement, but the engagement rate should be higher than on Instagram. Brands should look for a minimum rate of 50%, and, ideally, above 80%; this means that the YouTuber is growing quickly, as there are more people watching them regularly than subscribers. Moreover, YouTube provides statistics similar to Instagram, but provides them in a more accurate way, so that the company can research:
  - Audience age.
  - Audience gender.
  - Combined audience age and gender.
  - Country of origin.
  - Retention rate, that is, the average duration of the view and the average percentage of the videos viewed.
  - Devices.
  - Other information such as traffic source, subscriber evolution, etc.

As some of these data may be personal, companies should be careful in their use to avoid problems with data protection laws.

- **Snapchat:** Many users in this field use apps that hack Snapchat, where they appear in a ghost mode, mainly for three reasons: (1) They can take screenshots of the story without the user knowing; (2) They can watch the story without the user knowing; and (3) They can save pictures and videos without the other user knowing. So, keeping this in mind, and given that most use third-party applications like Snap-Hack, SaveSnap, SnapPad, etc., asking the influencer to send a screenshot of the number of views does not make sense, as it does not reflect the real number or level of engagement. There are three alternative ways to evaluate a Snap influencer:
  - Checking the influencer's Instagram and assessing it as previously stated.
  - Checking YouTube: Many followers or fans create YouTube channels for the influencer and upload all their snaps. Thus, companies can check the comments, views, and subscribers on these channels.
  - Checking Twitter: The company can hashtag the influencer's name and see what is being posted about them. Most users go to Twitter to give their opinion. Thus, the influencer's snaps can be found posted on Twitter with a hashtag, along with some comments. This can show how popular they are, the type of followers they have, and to what extent their followers are influenced by them.

Contacting Influencers

If it has not yet been done in the evaluation phase, the next step is to get in touch with influencers and propose that they work with the company or with the tourism destination. There are several ways to establish communication with influencers. Email works for the first approach, but after this first contact, many of them prefer to switch to phone calls, WhatsApp, Line, Facebook messenger, or other messaging apps. To make this first contact, the influencer's email address can be found in different ways:

- **Instagram:** Many influencers include a contact email in their biography. If this information is not shown there, this may be because they have an "email" button on their Instagram profile. Keep in mind that this button is only available in the Instagram application and will not be displayed when checking Instagram from a PC or web browser (see Figure 4, the "email" option).
- **YouTube:** To contact YouTubers, go to their YouTube profiles and click on the "About" tab. Once there, click on "view email address" and verify that that is not a robot.

If an influencer's email address is not available on YouTube or Instagram, the company can try to check other social media such as Twitter to see if they one in their bio. Also, it should be noted that, even though many influencers manage their collaborations by themselves, there is a large part of the influencer community that relies on agencies or personal managers for this.

Figure 4. Email address or call button in Instagram profile

Proposal

At this stage of the process, it is time to design the creative aspects of the campaign. The proposal can be specifically designed for the audience and for the influencers that best align with the company's objectives, who have been selected previously. This proposal can be as creative as desired. Therefore, only the most basic structure will be covered here to let the rest be guided by marketing creatives. The three core points are: cost, type of impacts, and creativity and concept.

- **Cost:** There are two basic ways of rewarding influencers for their collaboration: fees (price per post) or sponsorship (exchange).
  - **Fees:** This is the traditional method. It consists of a previously agreed-upon fixed price per post. Influencers are open to negotiating the fee when companies purchase more than one post or posts on different platforms. It tends to be more expensive than sponsorship, but it is faster and more time-effective when managing several campaigns. This option is also worth considering when the influencer is not big enough to be worth covering the cost of the product or service that would be given as an exchange. Fees can range from $20 for small profiles to over $20,000 per YouTube video. Of course, fees depend on reach, style, popularity, the influencer's business model, etc. In this regard, celebrities tend to be far more expensive than influencers, even if they have the same number of followers.
  - **Sponsorship:** This consists of giving something away in exchange for social media exposure. Some tourism-related examples would be:
    - **Free Product or Service:** For example, the brand could fully or partially cover an upcoming trip or give away free trials.
    - **Discounts:** Brands can provide long-term discounts, for example, 10% off influencer purchases during the next X months.
    - **Win-Win:** The company can also offer to pay them based on the performance (purchases, visits, etc.) they manage to generate through their audience.
- **Type of Impacts:** One of the most important parts of the entire process is how the influencer will implement the promotion and the impact this will have on the audience. There are several ways they can make an impact, depending on the social media platform. Below are some components of the different types of social media posts as well as some tips of what they should include:
  - Instagram Post:
    - **Caption:** This is the text that accompanies the image. It is important to include a mention and it is good practice to have it displayed on the first line. The company can also include #Hashtags. Another recommendation is to avoid standard captions such as "thanks to [Brand]," which add no value. The more natural, the better.
    - **Tags:** The brand's profile can be tagged on a normal post.
    - **Location:** The influencer can add the location. This is good when promoting specific destinations.
    - **Promoted Feature:** The influencer can tag the brand's Instagram account as a sponsor. This will let the tagged brand automatically access internal data such as reach, impressions, and engagement.
    - **No Links:** There is no way to include clickable content in a normal post caption.

- Instagram stories: These are images or short videos in vertical format, only available for 24 hours.
  - **Image or Video:** The content will be shown full screen, so it needs to be vertical (like the smartphone screen). Only images or a maximum of 13 seconds of video are allowed.
  - **Text or Drawing:** This function is available on Instagram stories.
  - **Mentions and Hashtags:** It is possible to add mentions that link to a profile or clickable hashtags. When adding hashtags, there is a chance that the stories will be displayed in the hashtag feed.
  - **Location:** This option also offers the chance of being featured in the location feed.
  - **Swipe up:** Stories allow the company to add links; simply add them to the stories and users only need to "swipe up" to see them. This feature is only available to a restricted number of profiles, such as verified accounts for profiles with many followers.
- Instagram Live:
  - **Pinned Comments:** Specific comments can be pinned so that they are always shown when watching the live broadcast.
  - **Co-Streaming:** This feature allows live streamers to invite an external user to the stream in split-screen mode.
- YouTube:
  - **Video:** Generally, videos should last less than 10 minutes. However, some influencers prefer to make long videos, about 15 minutes. If that is what they usually do and that is how they became popular, it can be a good practice. The important thing is to let the influencer follow their usual practices.
  - **Description Box:** The brand can add several links and plain text in the description box below the video player.
  - **Linked Content in the Video:** YouTube allows links to old videos or other YouTube videos during playback, but does not allow external links.
- YouTube Live: YouTube also allows creators to do live videos.
- Snapchat:
  - **Image or Video:** As with Instagram stories, the content will be shown full screen and only images or a maximum of 13 seconds of video are allowed.
  - **Location:** This function is only available for some places.
- Tweet:
  - **Text:** In general, a maximum of 280 characters are allowed. Links and photos will not be counted as such.
  - **Photos or Videos:** Can be posted on Twitter.
  - **Links:** Also allowed on Twitter.
- Facebook Post:
  - Facebook allows text as well as multimedia content. Branded posts are also possible, as the influencer can tag the specific publication as a sponsored post.
- Blog Post:
  - These posts allow the influencer to give a more detailed review of the product or service. "Do-follow" links will help improve search engine optimization (SEO) positioning.

- **Creativity and Concept:** Brands must also define what type of post is going to be published. There are different types. The first could be called "inspirational" because it shows the influencer taking advantage of the benefit of using the company's brand. The key aspect of these publications is their naturalness. Thus, the publication should not be seen as being promoted, for example, the influencer could be on a plane, or having breakfast. The second type of posts are promotional posts. In this case, the influencer will make it obvious that it is a collaboration and this allows for a more aggressive style of posts, for example, showing the app on the post, or promoting the benefits of using the brand. Both types are equally effective. Brands should seek to find a balance between the two to get the perfect mix for their country/market. Furthermore, the brand also needs to develop a concept behind the campaign. This will vary depending on the objective of the entire campaign. In this regard, we offer some advice:
    - **Co-Creation:** The company must let the influencer take part in creating the concept. Some ideas may seem crazy, but it must be remembered that they know their audience and they also know what similar companies have proposed so far.
    - Be a "mentor" for the influencer: Influencers might know their audience, but the brand should share its knowledge of its products and current customers, as well as previous campaigns.
    - Influencer campaigns are full of "last-minute issues." The company should not let them ruin the campaign, and must be ready for unexpected changes.

Execution

When everything has been planned and arranged, the last step is to implement the plan. Below is the suggested implementation process:

1. **Agreement:** It is always recommendable to sign a contract or agreement to ensure a fair collaboration on both sides. Some important points must be outlined in these agreements. One is to make it clear what the brand is offering (hotel, fee, plane tickets, meals, dates, etc.). Another important question to be answered is what the influencer will do in return (number of posts, hashtags to include, mentions, etc.). In addition to these issues, other important clauses that must be included are:
    - The post, for example the photo or video, cannot be deleted afterwards.
    - Post dates, if required.
    - Whether the company can use the content in other promotional activities. Or, in other words, how the content can be used.

    Appendix 3 includes a sample agreement that could be used for this purpose.

2. **Payments and Products Delivered:** If the collaboration requires some type of payment (fees, expenses, etc.), the company must be sure to process it on time, or must reach an agreement with the influencer about the payment schedule. If a product needs to be delivered or service needs to be reserved, it is important to book them in accordance with the agreed-upon dates.
3. **Influencer:** Influencers must be provided with the right information to build the perfect post. Influencers should be informed of:

a. Brand guidelines should be written that include best publication practices. They should include, for example, the concept of the campaign and information about the company.
b. If the influencer needs to include a link, the brand must be sure to provide it. Links must be tracked to better evaluate their effects, except in SEO campaigns.
c. The company must also provide influencers with any additional materials, such as logos, images, or whatever the influencer may need to be able to publish the necessary content.

Analysis

Once the campaign has been executed, it is time to measure the results. The way in which the results are evaluated will depend upon the campaign's objective and established target KPIs. The best way to provide an overview of the campaign will be to write a report with all of this information. To do so, the company will need to take a look at the original plan sheet and analyze the results in terms of the previously established KPIs. Table 2 shows some examples for establishing measurable KPIs.

It is vital to compare expected results with those ultimately obtained and, where appropriate, to analyze any deviations. Likely, many tests will be required until the prefect mix is achieved. Due to the unlimited possibilities of KPIs, only the most common ones are listed here:

- **App Downloads:** There are two main ways to verify the impact on downloads:
  - **Organic Downloads:** Checking the overall trend of app downloads is a good way to gauge the impact of an influencer campaign. However, it also has some limitations. For example, spikes in downloads may be caused by other marketing activities. Therefore, it is always important to keep in mind any other marketing campaigns in place and avoid overlapping with them.
  - **Tracked Downloads:** Tracked links allow brands to accurately attribute downloads to a specific campaign. However, there are two important limitations. First, not all social networks allow links to be included, for example, Instagram. Secondly, not all users follow the link.
- **Reach, Impressions, Likes and Comments:** While likes and comments are public and easily accessible for everyone, reach and impressions need to be obtained directly from the account manager. This information about reach and impressions, as with other similar information, is private and should be obtained by requesting it directly from the influencer.

Table 2. Examples of measurable KPIs depending on the objective

Objective	Measurable KPI Examples
App downloads.	Number of apps downloaded.
Branding.	Post reach, impressions, comments, or buzz generated.
Promote specific properties or destinations.	Increase in visits/leads/… to the destination.
Promote specific features (filters, map, etc.).	Changes in user behavior.
Social media channel growth.	Number of followers.
Revenue growth.	Increase in revenue, market share, channel share, etc.
Diversification of the audience.	Change in user demographics.

Source: Developed by author

- **Followers and Demographics:** To accurately measure growth on social media, it is helpful to use a business profile so as to have access to profile statistics and use external tools. The statistics provided by the social network will allow the company to see their current user profile (country, age, gender, etc.). Thus, they can compare that profile to the type of users added during the campaign, or, to the user profile from previous similar campaigns. Regarding external sources, SocialBlade.com lets users have access to a history of follower fluctuations on Instagram and YouTube.
- **Revenue:** This is the indicator that is most difficult to measure accurately. A data team is essential for the internal analysis of a company's results. Tracked links will serve as a first approach, displaying all revenue generated via this link. Unfortunately, as mentioned above, not all users will reach the brand via tracked link, or on the same device used to follow the tracked link. It would be best for brands to discuss with their data team the most suitable way to approach this variable, case by case, and prepare a specific report that excludes other unrelated marketing campaigns as well as organic growth. Once the results have been analyzed, they must be compared to the cost and other marketing activities that pursue the same objective. It is very important to consider the cost, since this in turn allows the company to calculate the Return on Investment (ROI) based on the results obtained. Once the ROI has been evaluated, the company can make decisions, such as whether it wants to continue to run the campaign, or what changes need to be made to improve the results.

CONCLUSION

Traditional approaches to communications have been challenged by social media and influencers. Both are tools with great potential for companies, as they allow them to reach customers in a more natural, friendly way. In fact, influencers offer a series of advantages to tourism companies, given that they describe their own travel experiences, which provides greater credibility. In addition, this type of campaign can reach a very specialized audience at a much lower cost than traditional campaigns. They can also create innovative content, allowing brands to connect with visitors and help to better understand their potential audience, attracting tourists who identify with the values and experiences that the destination offers.

The results of this work show that influencer marketing is a tool as measurable as any other, and can be included in any tourism company's communication strategy to successfully reach their commercial objectives. Indeed, although influencer marketing campaigns may be a challenge for different companies (unknown tool, new techniques, etc.), this study provides a manual that can be used across the tourism industry in order to pursue business' objectives, step by step.

To create this manual, the case study methodology was used. The case chosen was a company from the tourism sector that works with influencers frequently. Specifically, it is an online price comparison company. Based on this company's experience, a series of phases for implementing an influencer marketing campaign were identified. The process begins with the planning phase, in which the objectives are established in such a way that allows for their subsequent evaluation. Then, the search for possible influencers that allow the brand to achieve its objectives in a satisfactory manner is essential. Third, the company must evaluate all of its options to choose the influencer or influencers who will allow them to obtain the best results. Next, they must know how to contact them, which is not always a simple task, in order to subsequently offer them a proposal in which both parties win. Sixth, the campaign must be executed, and, finally, the results analyzed.

For a campaign to be evaluated correctly so that changes can then be made to the campaign, it is essential to be able to measure to what extent the objectives are being achieved. Therefore, the fact that this manual is results-oriented is of great importance. Establishing measurable KPIs leads to a more precise analysis of influencer marketing activities and to a greater efficiency in their use.

As this is a pioneering manual, it should be followed by scientific studies that, through empirical research, provide evidence on key aspects of influencer marketing. Thus, for example, future research could analyze which strategy provides the best results. Likewise, and given the importance of evaluating the campaign, future studies could focus on outlining and clarifying the measures used to adapt evaluation methods to the reality of campaign results. In any event, and given the relative novelty of the subject, this manual is intended to be a first step that will lead to other works in which the phases described here are more fully developed. It is now the companies that must take on the challenge.

ACKNOWLEDGMENT

This work was supported by the University of Málaga, Spain [grant Proyect "Diagnóstico y Posicionamiento de los SMIs" from Jóvenes Investigadores – I Plan Propio de Investigación y Transferencia Plan Propio].

REFERENCES

Armano, D. (2011). *Pillars of the new influence*. Retrieved from https://hbr.org/2011/01/the-six-pillars-of-the-new-inf

Axon Marketing & Communication. (2018). *Influencer marketing trends in the tourism sector for 2018*. Retrieved from https://axonlatam.lpages.co/white-paper-influencer-marketing-trends-in-the-tourism-sector-for-2018/

Barone, L. (2013). *The 5 types of influencers on the Web*. Retrieved from http://smallbiztrends.com/2010/07/-the-5-types-of-influencers-on-the-web.html

Biggart, N. W. (1977). The creative-destructive process of organizational change: The case of the post office author. *Administrative Science Quarterly*, *22*(3), 410–426. doi:10.2307/2392181

Bitner, M. J., Faranda, W. T., Hubbert, A. R., & Zeithaml, V. A. (1997). Customer contributions and roles in service delivery. *International Journal of Service Industry Management*, *8*(3), 193–205. doi:10.1108/09564239710185398

Booth, N., & Matic, J. A. (2011). Mapping and leveraging influencers in social media to shape corporate brand perceptions. *Corporate Communications*, *16*(3), 184–191. doi:10.1108/13563281111156853

Brown, D., & Hayes, N. (2008). *Influencer Marketing. Who really influences your customers?* Oxford, UK: Butterwort.

Brown, S. L., & Eisenhardt, K. M. (1997). The art of continuous change: Linking complexity theory and time-paced evolution in relentlessly shifting organizations. *Administrative Science Quarterly*, *42*(1), 1–34. doi:10.2307/2393807

Casaló, L. V., Flavián, C., Guinalíu, M., & Ekinci, Y. (2015). Do online hotel rating schemes influence booking behaviors? *International Journal of Hospitality Management*, *49*, 28–36. doi:10.1016/j.ijhm.2015.05.005

Castelló-Martínez, A., & del Pino-Romero, C. (2015). La comunicación publicitaria con influencers. *READMARKA UIMA*, *14*(1), 21–50.

Cave, J., & Jenkin, H. (2012). Social media as influencers of tourism decisions made by backpackers and fully independent travellers. *Indian Journal of Applied Hospitality and Tourism Research*, *4*, 69–80.

Chae, J. (2017). Media psychology explaining females' envy toward social media influencers. *Media Psychology*, *21*(2), 246–262. doi:10.1080/15213269.2017.1328312

Chatzigeorgiou, C. (2017). Modelling the impact of social media influencers on behavioural intentions of millennials: The case of tourism in rural areas in Greece. *Journal of Tourism. Heritage & Services Marketing*, *3*(2), 25–29.

Fatanti, M. N., & Suyadnya, I. W. (2015). Beyond User Gaze: How Instagram Creates Tourism Destination Brand? *Procedia: Social and Behavioral Sciences*, *25*, 1089–1095. doi:10.1016/j.sbspro.2015.11.145

Gebauer, H., Johnson, M., & Enquist, B. (2010). Value co-creation as a determinant of success in public transport services. *Managing Service Quality: An International Journal*, *20*(6), 511–530. doi:10.1108/09604521011092866

George, A. L., & Bennett, A. (2005). *Case studies and theory development*. Cambridge, UK: MIT Press.

Gretzel, U. (2006). Consumer generated content -trends and implications for branding. *Ereview of Tourism Research*, *4*(3), 9–11.

Hernández, E., Fuentes, M. L., & Marrero, S. (2012). Una aproximación a la reputación en línea de los establecimientos hoteleros españoles. *Papers de Turisme*, *52*, 63–88.

Higgins-Desbiolles, F. (2018). Event tourism and event imposition: A critical case study from Kangaroo Island, South Australia. *Tourism Management*, *64*, 73–86. doi:10.1016/j.tourman.2017.08.002

Hollebeek, L., Glynn, M., & Brodie, R. (2014). Consumer brand engagement in social media: Conceptualization, scale development and validation. *Journal of Interactive Marketing*, *28*(2), 149–165. doi:10.1016/j.intmar.2013.12.002

Iyiola, O., & Akintunde, O. (2011). Perceptions as influencer of consumer choice behavior: The case of tourism in Nigeria. *Journal of Marketing Development and Competitiveness*, *5*(7), 27–36.

Kanellopoulos, D. N., & Panagopoulos, A. A. (2008). Exploiting tourism destinations' knowledge in an RDF-based P2P network. *Journal of Network and Computer Applications*, *31*(2), 179–200. doi:10.1016/j.jnca.2006.03.003

Kosnik, E. (2018). Production for consumption: Prosumer, citizen-consumer, and ethical consumption in a postgrowth context. *Economic Anthropology*, *5*(1), 124–134. doi:10.1002ea2.12107

Lindh, C., & Rovira, E. R. (2018). New service development and digitalization: Synergies of personal interaction and IT integration. *Services Marketing Quarterly*, *39*(2), 108–123. doi:10.1080/15332969.2018.1436777

Marine-Roig, E. (2017). Measuring destination image through travel reviews in search engines. *Sustainability*, *9*(8), 1–18. doi:10.3390u9081425

Mccutcheon, D. M., & Meredith, J. R. (1993). Conducting case study research in operations management. *Journal of Operations Management*, *11*(3), 239–256. doi:10.1016/0272-6963(93)90002-7

Nielsen. (2015). *Global Trust in Advertising – 2015*. Retrieved from http://www.nielsen.com/us/en/insights/reports/2015/global-trust-in- advertising-2015.html

Nieto, M., & Pérez, W. (2000). The development of theories from the analysis of the organisation: Case studies by the patterns of behaviour. *Management Decision*, *38*(10), 723–734. doi:10.1108/00251740010360588

Özüdoğru, A. G., Ergün, E., Ammari, D., & Görener, A. (2018). How industry 4.0 changes business: A commercial perspective. *International Journal of Commerce and Finance*, *4*(1), 84–95.

Pan, B., MacLaurin, T., & Crotts, J. C. (2007). Travel blogs and the implications for destination marketing. *Journal of Travel Research*, *46*(1), 35–45. doi:10.1177/0047287507302378

Richter. (2014). Prosumption: Evolution, revolution, or eternal return of the same? *Journal of Consumer Culture*, *14*(1), 3-24.

Sammis, K., Lincoln, C., & Pomponi, S. (2016). *Influencer marketing for dummies*. John Wiley & Sons.

Stylidis, D., & Cherifi, B. (2018). Characteristics of destination image: Visitors and non-visitors' images of London. *Tourism Review*, *73*(1), 55–67. doi:10.1108/TR-05-2017-0090

Suciati, P., Maulidiyanti, M., & Lusia, A. (2018). Cultivation effect of tourism Tv program and influencer's instagram account on the intention of traveling. In *The 1st International Conference on Social Sciences University of Muhammadiyah Jakarta, Indonesia, 1–2 November 2017* (pp. 267–278). Indonesia: International Conference on Social Sciences.

Toffler, A. (1980). *The third wave*. New York: Bantam Books.

Van de Ven, A. H., & Poole, M. S. (1990). Methods for studying innovation development in the Minnesota Innovation Research Program. *Organization Science*, *1*(3), 313–335. doi:10.1287/orsc.1.3.313

Vargo, S. L., & Lusch, R. F. (2004). Evolving to a new dominant logic for Marketing. *Journal of Marketing*, *68*(1), 1–17. doi:10.1509/jmkg.68.1.1.24036

Wang, Y., Yu, Q., & Fesenmaier, D. R. (2002). Defining the virtual tourist community: Implications for tourism marketing. *Tourism Management*, *23*(4), 407–417. doi:10.1016/S0261-5177(01)00093-0

Wöber, K. (2006). Domain specific search engines. In D. R. Fesenmaier (Ed.), *Destination recommendation systems: Behavioral foundations and applications*. Wallingford, UK: CABI. doi:10.1079/9780851990231.0205

Word of Mouth Marketing Association-WOMMA. (2013). *Influencer guidebook 2013*. Retrieved from https://es.slideshare.net/svenmulfinger/womma-influencer-guidebook-2013-pdf

Xiang, Z., Wöber, K., & Fesenmaier, D. R. (2008). Representation of the online tourism domain in search engines. *Journal of Travel Research*, *47*(2), 137–150. doi:10.1177/0047287508321193

Xie, C., Bagozzi, R. P., & Troye, S. V. (2008). Trying to prosume: Toward a theory of consumers as co-creators of value. *Journal of the Academy of Marketing Science*, *36*(1), 109–122. doi:10.100711747-007-0060-2

Yin, R. K. (1994). *Case study research : Design and methods*. London, UK: SAGE Publications.

APPENDIX 1

See Figure 5.

Figure 5.

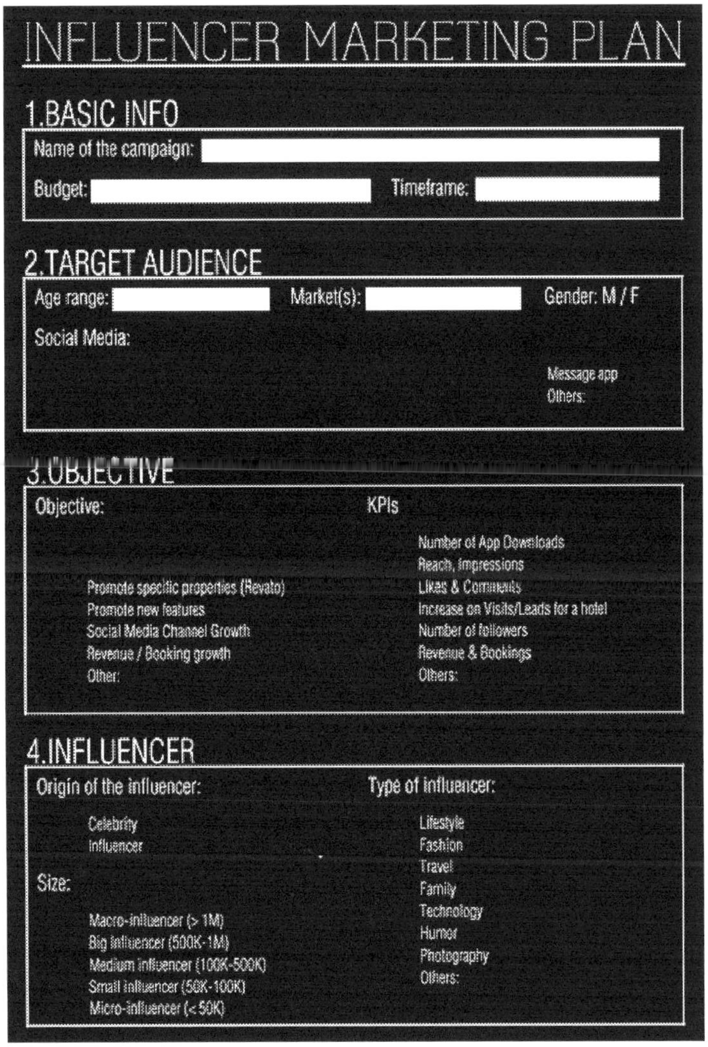

INFLUENCER MARKETING PLAN

1.BASIC INFO

Name of the campaign:

Budget: Timeframe:

2.TARGET AUDIENCE

Age range: Market(s): Gender: M / F

Social Media:

Message app

Others:

3.OBJECTIVE

Objective:

Promote specific properties (Revato)
Promote new features
Social Media Channel Growth
Revenue / Booking growth
Other:

KPIs

Number of App Downloads
Reach, impressions
Likes & Comments
Increase on Visits/Leads for a hotel
Number of followers
Revenue & Bookings
Others:

4.INFLUENCER

Origin of the influencer:

Celebrity
Influencer

Size:

Macro-influencer (> 1M)
Big influencer (500K-1M)
Medium influencer (100K-500K)
Small influencer (50K-100K)
Micro-influencer (< 50K)

Type of influencer:

Lifestyle
Fashion
Travel
Family
Technology
Humor
Photography
Others:

APPENDIX 2

See Figure 6.

Figure 6.

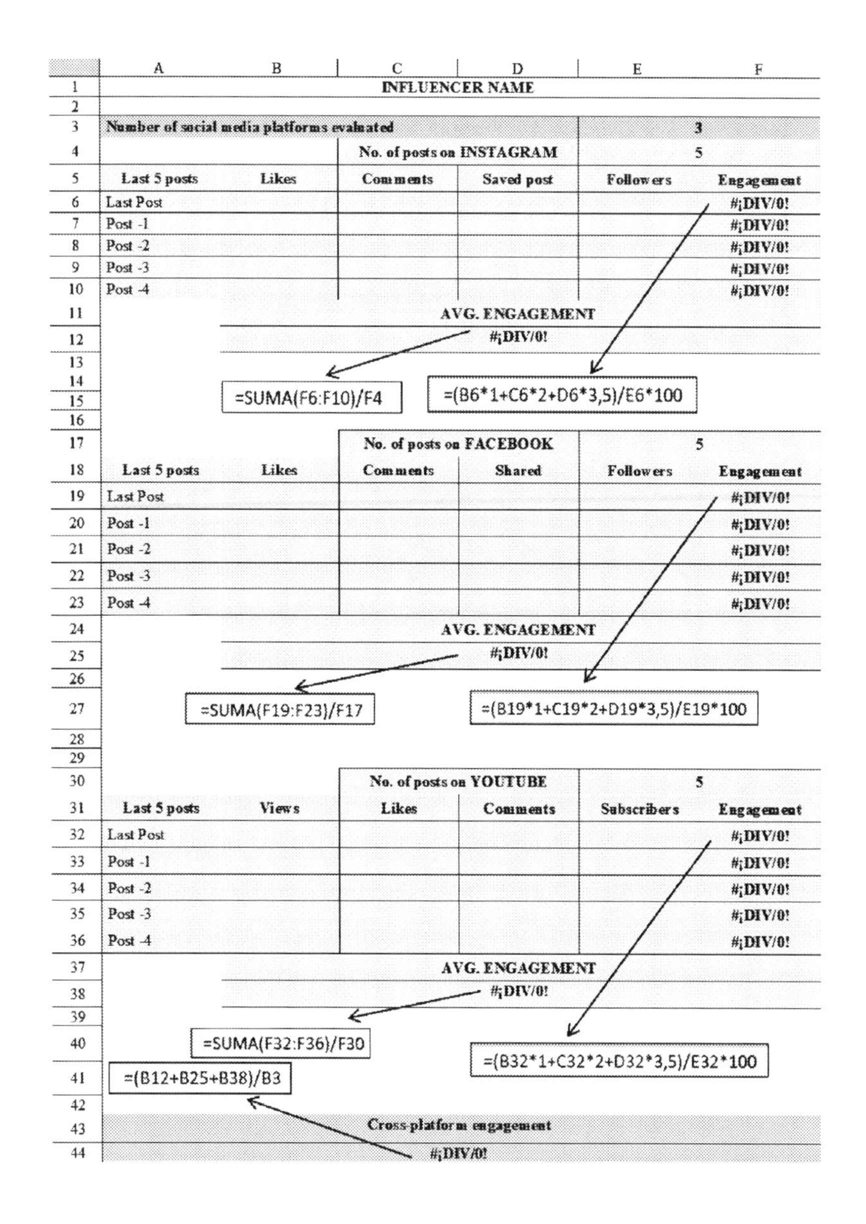

APPENDIX 3

COLLABORATION AGREEMENT

Company Name
Company ID
Company Address
Influencer Name:
Influencer Address:
Influencer ID:
This document sets out the different terms and details of the collaboration, as well as what both parties agree to do:
Concept of the campaign:
The influencer: Influencer name
Impact on different social media platforms, in accordance with the collaboration manual previously shared with the influencer:

- Instagram Posts
- Instagram Stories
- YouTube Videos

The brand:
The brand will provide:

- The amount of $-----, paid by transfer
- Exchange of: Product or Service

Date: MM/DD/YYYY
Signed:
Brand Agency/Manager/Influencer

Chapter 15
The Agricultural Innovation Under Digitalization

Michel J. F. Dubois
Institut Polytechnique UniLaSalle, France

Fatma Fourati-Jamoussi
Institut Polytechnique UniLaSalle, France

Jérôme Dantan
Institut Polytechnique UniLaSalle, France

Davide Rizzo
Institut Polytechnique UniLaSalle, France

Mehdi Jaber
Institut Polytechnique UniLaSalle, France

Loïc Sauvée
Institut Polytechnique UniLaSalle, France

ABSTRACT

This chapter aims to discuss how the rapid evolution of digital technologies is creating opportunities for new agricultural business models. First, it provides an overview of what the authors consider to be part of the digitalization in agriculture. Then it addresses the emergence of a community of practice based upon the data exchange and interconnections across the agricultural sector. New business opportunities are presented first through an overview of emerging start-ups, then discussing how the inventor farmer profile could create opportunities for new business models through the appropriation of technologies, eventually highlighting the limits of some classic farm business models. Finally, the chapter presents an example of farmer-centered open innovation based on the internet of things and discusses the related business model. The conclusion provides some perspectives on the use of agricultural digitalization to increase the share kept by farmers in the value chain of agricultural productions.

DOI: 10.4018/978-1-5225-7262-6.ch015

INTRODUCTION

Agriculture in the 21st century should produce sufficient high-quality food, for more than 9 billion inhabitants, without increasing the surface of production, while limiting ecological impact. The production's increase should be about 70% (World Population Prospect, 2012). Historically, during the second half of 20th century, the intensification of agriculture has been supported by mechanization, plant breeding and chemistry, with some bad consequences over the long term, like pollution and loss of biodiversity.

So, farmers have to decrease the inputs (pesticides, fertilizers, energy, and soil cultivation) and to increase the outputs (production) over the same global surfaces. According to FAO – Food and Agriculture Organization of the United Nations –, the global challenge for agriculture is to produce more with less, in keeping with sustainability (FAO, 2016). Agriculture and farming are inherently linked to food with two aims: enough production and good quality. However, the global increase for both encounter the physical limits of the planet. Waste has to be controlled, water use should be optimized and its quality increased, chemicals and energy inputs should decrease, and biodiversity should be at least maintained. At the same time, it is forecasted an increased innovation in plant breeding, including "orphan crops" (Vanderschuren, 2012), probably a growing amount of livestock and an increased role of urban agriculture. New Business Model (BM) for the food industry are requested and nutrition will be a priority for consumers (Traitler et al. 2018).

This situation leads to a deep paradigmatic change. The soil should no more be used as a simple substrate to which all the necessary elements should be added. It has to be recognized as a global system integrating physical elements, chemical and biochemical elements, and living beings. The mass of living beings in the soil is generally much higher than our domesticated animals which put a hoof on the ground. The interactions of the soil, the plants and the climate define the core of the productive system. The aim of agriculture is to obtain the maximum production, through minimum inputs and so to optimize the functionalities of any element of the system, living or inert. For a farmer, the time is to be spent in thinking about how to organize crop rotations, intercrops, double and even triple plant cultivations, how to maintain the productivity of the soil and the biodiversity of the whole agro-ecosystem, how to sell at the best price a good quality production, how to provide useful energy without buying fossil energy, how to decrease the production costs. The aim of the activity of the farmer cannot be any more a simple producer whose all actions are answers to demands. In order to be adapted to local conditions, it appears that agriculture will become more and more diversified, including both urban agriculture and aquaculture. This new paradigm transforms the farmer job. To be able to increase production by overcoming these constraints, the approach must be globally and radically changed. The agriculture world must leave a simplifying and homogenizing model to move towards a systemic approach in which interactions are increased at different levels. This means being able to produce in a sustainable way, according to the pedo-climatic context and the needs of the territories. Agriculture should be both more and more precise and flexible in a context of global growth (Traitler et al. 2018).

Generally speaking, digitization is the process of producing information into a digital format. It produces a collection of a finite number of signs taken from a countable set of valid signs. Digitization "is of crucial importance to data processing, storage and transmission, because it allows information of all kinds in all formats to be carried with the same efficiency and also intermingled" (McQuail, 2000). Digitizing means the conversion of analog source material into a numerical format through which is provided a discrete representation; hence it is a reduced or restricted point of view about the use of digital tools.

Herein, the authors use the definition of digitalization according to Business Dictionary. The concept of "digitalization" is much more general than digitization; it includes also the new activities which need digitalized data that means robotics and connected objects, any sensors able to transmit stable information, any tool for the transmission, the treatment or the sharing of data and any type of innovation made possible by digitalization. Digitalization can be considered as a powerful process to produce memory, traceability and previsions.

Can digitalization be a good answer to the challenges of future agriculture? Can digitalization offer new opportunities for agricultural development? Is digitalization a new set of tools for extracting more value from agriculture to upstream and downstream actors of agriculture, in the food chains? Or can it be an opportunity for the farmers to obtain more value from their production? And how can digitalization transform the BM of the farmer in order to create added value?

The bias and the commitment of this text is to analyze the activity of a farmer as if he ran a conventional business. His strategy would be to optimize his revenues by taking into account all its markets and sales and also all its costs. Its choices correspond to a well-identified business model. Instead of looking for subsidies, he demands the payment of the services and products he brings. He uses digitalization as a new way to increase revenue and behaves strategically (mutualization, sales of services, etc.). This proposal is therefore a liberal based model.

First the authors propose hereinafter a simple classification and a description of the digitalization in agriculture. The second part, shows how digitalization of agriculture is an answer to agro-ecological evolution and to the emergence of communities of practice. The third part, describes the dynamics of digitalization in agriculture insofar and some perspective for the twenty coming years. The fourth part is focused on BM in agriculture and the possibility to set up innovative BMs in agriculture, as a fast pace of agricultural digitalization will need innovation BM. The fifth part presents an experiment which offers the possibility to test new models. The conclusion provides some challenges and recommendations for the farmers to keep a part of the added value within agriculture.

DIGITALIZATION IN AGRICULTURE: WHAT IS ALL ABOUT?

Computers, robotics, connected agriculture, Big Data and Artificial Intelligence happen in accordance with the global economic and social change and not only as an answer to the challenges of food production. Both the absolute and relative growths of urban population, increasing cost of labor, decreasing interest to work in the field, or computers and smartphones as companions of everyday life are some robust worldwide trends. The result is a large production of data, so large that Big Data technologies could be useful and Artificial Intelligence (including deep learning) could become a necessity even more than an opportunity.

Nevertheless, the global unfolding of such new techniques can be also an important answer to the challenges of food and agriculture production entangled with the properly agricultural solutions for "producing more with less" (FAO, 2016). Digitalization occurs when agriculture undergoes a need for more precision associated with many constraints on labor and input consumption; so, precision agriculture can be understood as a positive technical answer and could be an opportunity for farmers. However, one can ask the question: under which conditions? We try to propose one of them below.

Precision farming is not a static concept, but an evolutionary process. It is difficult to identify the exact dates when emergence of each cluster of inventions that could take place in the agricultural sector. In most cases, the innovations are made from and through the existing technologies adoption from other fields. We must also take into account the improvement of all these technologies that are constantly evolving. Nevertheless, it is clear enough that the geolocalization began in the 1980s; the agronomic information gathering and data flows processing exist through the development of both Geographic Information Systems (GIS) and sensors for harvesting, soil, climate, plants, during the years 1990-2000; controlled traffic farming, and agricultural production management software around 2000-2010. Data acquisition systems, sharing, integration and valuation (Big data), decision-support tools and artificial intelligence are arising since 2010. The digitalization evolution of these technologies contributes to optimizing and anticipating agricultural production. All this is supported by the mechanical evolution of agricultural equipment.

Computer and Digital Tools

For many years, as soon as micro computers were becoming available, most of the farmers in industrialized countries have been using classical tools of data management, including accounting software and many computerized management tools. In the most advanced countries, for greenhouse agricultural production, computer use has been oriented towards the control of lightening, sprinkling, watering and temperature control. From the outset, drip irrigation was computer-controlled. The most technical agricultural production, in market gardening, horticulture, greenhouse production, and also irrigated production, requires the use of automatisms or computer control that generates data.

Market Data

Farmers can access market information according to commercial data sources, directly from web platforms. It is a source for inspired decisions to sow, harvest or sell. With the global computerization, each decision provides data, useful for traceability, for analyzing the effects of such decisions and for modelling of expected future actions. The market data can shape a completely new customer-oriented agriculture.

Sensors

Many farmer's actions are typically limited by agronomic constraints including weather conditions. These conditions can be monitored by different sensors, on the moving equipment (flying, rolling, and walking) or set up in the field. The sensors concern different light wave lengths (including infrared and ultraviolet), level of humidity, temperature, ions concentration in the soil, soluble fertilizers, density of some living organisms; they can be associated with image recognition software. More and more sensors are now used at different scales, in such a way that the entire technical itinerary can be further analyzed and compared to other ones.

Robotics

The first real robots seem to appear in animal breeding (John et al. 2016), and the reason is easily understandable. An animal is a valuable concentrate, and this is probably the historical reason for the development of livestock and herds. Breeding requires a lot of human labor and constant monitoring of the herd.

The converging elements can be resumed as following: labor costs, monitoring and accuracy, improved productivity, and also animal welfare and quality of life for workers. So, it is precisely in the production with the greatest added value that robots have been first put in place.

Generally speaking, the development of IT (Information Technologies), robotics and digital technology began within high value-added professions. The fall in the costs of these technologies (Moore's law) has led to a gradual widening of uses, with the following chronology of computerization: military industry, banking-insurance, pharmacy, aeronautics, automobile, food industry, then agriculture (livestock followed by vegetable farming with high added value and then large field crop farms).

For crops, the highest added values are in vineyards, market gardening, horticulture, and then fruit groves. When highly mechanized, large field crops can also have good added value per worker. New regulations to limit herbicides and phytosanitary products lead to opportunities for land rolling and flying robots. In particular, the will to avoid herbicides leads to robotic weeding, the first uses of which concern vineyards and market gardening. The robots' latest generation carry more and more sensors.

GPS, Imagery, Mapping

Field crops have entered the digital age thanks to global techniques used in other markets (GPS, image processing). Cognitive processes derived from knowledge of plant physiology and agronomy have been integrated into the system to provide precise tools for monitoring and anticipation.

Smartphone

Farmers have adopted smartphones (Pongnumkul et al. 2015) very quickly as a support for many applications in monitoring production and acquiring daily information. Smartphones become terminals allowing access to all the more or less autonomous tools of the farm. It can also be used as a sensor (imaging).

Internet of Things (IoT)

Today, all tools can be interconnected, allowing farmers to follow in real time what is happening on their farms. They can now even design new tools that they will be able to control. Connected objects, which produce a large amount of data, can feed Artificial Intelligence and interconnected robots can evolve towards a more and more adapted behavior through detecting and monitoring of the actions.

Data Processing, Big Data, Artificial Intelligence and Deep Learning

All these techniques produce data from different origins, in different formats and in such quantities that no classical statistic can be used for translation or interpretation. It is the domain of Big Data.

As the conditions on a farm are all the time changing, Artificial Intelligence and Deep Learning will be very quickly useful. As the cost of these techniques should decrease, it could be within farmers' grasp. Big Data, Artificial Intelligence and Deep Learning could be done at the level of the farm and at the level of groups of farms before being directed at integrating operations across larger scales. These treated Data could have high values.

IMPLEMENTATION OF DIGITALIZATION AT FARM LEVEL: EMERGENCE OF NEW COMMUNITIES OF PRACTICE

The global digitalization of economies around the world has its counterpart in agriculture. Precision farming cannot be developed on a large scale without digital tools. Mapping of farms is now possible thanks to GPS and drones with multispectral sensors. The geospatial information provided by the satellite has an important application in so-called "precision agriculture", improving crop management. Drones could also be important for a similar aim. Specific physic-chemical sensors provide the measurement of edaphic and meteorological conditions. Sensors can also drive precision mechanics and new technological tools (autonomous tractors, robots). Collected data can be transferred and processed at the level of the farm or a collective of farmers for decision-making aids. All information coming from the field can be connected to the analysis of marketing and consumption data. Web-based platforms can be used for information exchange and compare performance between the farmers. The challenge of information connection networks is important for farmers because it is their autonomy that is at stake. They must be able to process the information themselves before scaling up and being able to value the information.

Technologies are allowing agriculture to be more precise and smart. The development of precision agriculture and its support, the digitalization of agriculture, is leading to the emergence of many new approaches that combine agronomic knowledge and agricultural know-how with computer skills (Caroux et al., 2018). The emergence of such trends is seen by the publication of recommendations or work by actors and specialists of the agriculture domain, by the creation of many startups since 2010 and by the development of new applications. Indeed, all around the world, there are hundreds of start-ups counting on the digitalization of agriculture.

The animators of the Digital & Agriculture Network ACTA in France, analyze the problematics related to the explosion and the capture of Big Agricultural Data (Brun and Haezebrouck, 2017). They propose through ten recommendations, three key areas of improvement for data management and use in agriculture: "Innovate": collaboration and technicality, "Fluidize": both accessible and shared data and "Reassure": a transparent and respectful use. *Farmers Business Network SM*, an independent network of thousands of America's most advanced farmers is already sharing data for production improvement. Technologies related to agriculture data are able to transform the profession of farmers and also that of companies in the sector: cooperative or private agricultural distribution companies, food industrial firms (traceability for crisis prevention), consumer relations (personalized marketing), etc. Agricultural social networks providing valuable data is a possible consequence of the emergence of agricultural "Big Data".

Precision agriculture is made of techniques that allow farmers to deal specifically with those parts of a field that require special or different attention; this precision could reach the level of one plant.

Digitalization involves networking and connecting farmers: with each other, with their suppliers, their advisers, their customers, and also with the connected objects providers. Connected objects are mobile or transportable objects like sensors in the field, robots, drones, smartphones, or even autonomous tractors.

All these connected objects could send their information to treatment centers at different spatial scales. Farmers should be aware that information control concerning their own data is a really an important issue. The precision required for the monitoring of agricultural production will require many precise actions that can only be performed by more or less mobile, interconnected and interacting objects.

All the data produced already by agriculture require large processing capacities. The generalization of precision agriculture, which means connected objects and "internet of things", will require the implementation of Big Data tools. However, these developments need first to be able to qualify the data that means to choose the useful data for agriculture and the first best level is the farm.

Connected agriculture can be described as a community of interconnected farmers suppliers and customers through digital operators (solution providers, agri-equipment and robotics players). Connected agriculture allows for a high accuracy of actions and monitoring of production. Interconnection of data coming from connected agriculture, from the different markets of the food chain and from the demands of consumers could provide a lot of new solutions to solve the global social demands.

Current solutions, such as service providers where data is retrieved by reselling companies either as Decision Support Systems (DSS) or to other companies for marketing / commercial purposes, are unsatisfactory to farmers. (1) Most of the proposed solutions are generic solutions, they do not necessarily correspond to specific cases; (2) customized solutions realized by businesses, (often SMEs) would be too expensive (3) the collaborative aspect between farmers involving knowledge sharing is then low: indeed, providers can centralize both data and "black box" decision tools, without great collaboration between farmers of the same regions (Information collected through interviews with French farmers).

According to Stratus Ag Research survey conducted in 2016, less than half interrogated farmers are satisfied with their current methods for analyzing and interpreting their agronomic data to make decisions.

In addition, farmers are increasing their awareness and concerns about the access to and the use of their farm data (American Farm Bureau Federation, 2016) and the related major shift in role and power relations. Finally, a report prepared for the European Parliament warns "As a result of these asymmetries, farmers' own particular needs and rights may be ignored, and inequalities are at risk of growing due to data-driven insights, rather than be reduced" (Kritikos, 2017, p. 41).

THE NEW BUSINESS OPPORTUNITIES OF DIGITALIZATION IN AGRICULTURE

Emergence of Start-Up in Agriculture

The development of precision agriculture and its support, the digitalization of agriculture, is leading to the emergence of many new professions that combine agronomic knowledge and agricultural know-how with computer skills. The emergence of these new professions can be seen by the creation of the many Startups since 2010. A lot of them are founded by young graduates from the agricultural world.

For example, among the fifty or so start-ups of the French Agtech in March 2017, we can list: Agriconomie, a web platform specialized in agricultural supplies; Agrifind, a web platform that allows experienced farmers to maximize the value of their know-how and/or to help other farmers who need reliable and operational information; Airinov, with the first multi-spectral sensor carried by a drone to measure crop growth; CarbonBee that uses specific sensors to discover blight at an early stage; ComparateurAgricole.com, a web marketplace where farmers sell their grain at the best price; Diimotion, with the PiX, a "printer" for phytosanitary products; Ekylibre, which offers open source tools to enable more efficient and simpler management of farms. Inalve, which proposes the production of a microalgae-based plant meal for animal feed; Karnott, the connected notebook, allowing to share agricultural machines; Les Grappes, a community platform for the direct purchase of wine from winemakers; MiiMOSA, a crowdfunding platform dedicated to agriculture and food; Monpotager.com, a concept of kitchen garden connected between the producer and the consumer; Myfood, sells connected greenhouses to those who wish to produce their own food. Naïo, sells agricultural robots helping farmers to weed, hoe and harvest; NeXXtep, offers smart and connected objects for farms (traceability, supervision, security); Piloter sa ferme, the robot consulting platform in agriculture dedicated to risk management; La Ruche qui dit Oui !, a short distribution channel through direct exchanges between local producers and consumer communities; Tibot technologies, automates some difficult and repetitive tasks in poultry farming; Vitirover, the robot which maintains the inter-rows of vineyards. YourMachine.com, a digital platform of localization / mutualization of agricultural equipment; Weather Measures, an expert in precision meteorology; Weenat, with ultra-local data sensors transmitted in real time to farmers.

In addition to start-ups, groups are developing new applications in many areas. All stages of agricultural activity are now transformed by the use of digital tools.

Early Adopters or Inventor Farmers

Why should farmers be inventors? How can they be? It seems that there are several modalities of technical invention in agriculture. Caroux et al. (2018) categorize innovators according to two ordinates: more or less machines, and more or less strong suitability for the dominant model. However, whatever are the general representations and objectives of innovative farmers, their main and common quality is to seek concrete solutions to local and recurrent problems.

Cost and lack of knowledge are widely identified as the two main obstacles that farmers have to overcome to involve into new technologies (Doye et al. 2000; Reichardt et al. 2009; Pignatti et al. 2015). The approach we will propose in the last part of this study is to put farmers at the heart of innovation, for instance by observing their propensity to innovation based on low cost and do-it-yourself tools. The aim is to correct a common research and development approach that considers farmers as simple end-users. The main limit of such an approach is to fail at embracing the complexity of their decision-making (Douthwaite and Hoffecker 2017).

The presentation hereafter aims to trace the evolution of agriculture in terms of BMs that have been designed and developed for about 60 years to understand how the farm operates in its environment. Then, a new type of platform, such as "AgriLab®", a collaborative innovation center for agriculture based in Beauvais (northern France), will be described, with the questions: can this new technical proposal lead to a new type of successful BM? Can farmers be inventors directly involved in the future of agriculture? Is the digitalization of agriculture able to foster a new kind of development? Is there a new type of development able to answer to the challenges of today's agriculture? The authors will try to clarify these questions and to propose a global model and possible tests in the conclusion.

AN OPPORTUNITY FOR BUSINESS MODEL CHANGE: APPROPRIATION OF TECHNOLOGIES BY FARMERS

Business Model Concept in Agriculture as a Conceptual Revolution

For a very long time, during the history of agriculture, farmers were producers. Farming was more a way of life, the basic way of life of the population, than a job. They were obliged to produce enough to feed their family and also to return a tithe that means to feed more than their own family. In some historical cases, reduced by serfdom or slavery to meet only survival needs, they could achieve a much higher production, added value that was confiscated.

Jobs and businesses were invented about the same time as agriculture and increased with metallurgy, development of handicraft and trade. Since antiquity commercial activities have developed in more or less luxurious delicacies: bakery, sausages, dairy products, sweet and alcoholic products, drinks, sweets, ice creams ... We know the history of Thales who was able to cause a shortage of oil and an increase in prices by buying a sufficient number of oil presses in time.

The customer-oriented products were not pure farmer's products. And even the grain trade was a state activity or wealthy merchant's activity.

The famous "crisis" of the tulip in the Netherlands, which occurred in 1637, can be conceived as an indicator of the beginning of a new type of agriculture, market and business oriented. It is significant that it began in horticulture, in the richest European country. Many new species were cultivated and an economy of plant breeding appeared. In northern Europe, and more particularly in the United Provinces, horticulture and gardening became a real market. Dutch gardeners cultivated roses, lilies, iris, peonies, columbines, wallflowers, carnations, anemones, snapdragons, hyacinths, jasmines, lilacs and of course tulips. Flanders and United Provinces entered in a new technical and marketing revolution in agriculture, from a producer stance to a merchant stance.

In parallel, the movement of enclosures began in the United Kingdom at the end of the sixteenth century. Between 1604 and 1914, thousands of official enclosure acts were passed, covering millions of acres. Community-grown open fields and pastures were converted by wealthy landowners into pastures for sheep in the aim to answer to an expanding wool trade. In more rural countries, the agricultural economy remained a productive economy.

According to the historian Patrick Verley (1985, p. 204), "historiography has long focused on the phenomenon of enclosures and its social consequences, however they do not constitute an agricultural revolution, they are only a prerequisite, which does not lead automatically to a progress in production and productivity". To transform the farmers from producers to entrepreneurs and managers is a long way. It is probably not a coincidence that the most efficient and most digitalized European agriculture is in the Nederland.

The concept of the farmer's BM is not yet obvious. Even in the most advanced countries, many barriers do exist when farmers are led to consider the possible innovations of agricultural BMs. A start-up has a high probability to fail. So, the farmer cannot take this type of risk. . "The farmers were quite comfortable talking about their existing farming operations with the help of the Canvas (its blocks and barriers). However, farmers had more difficulties in creating a new BM that required innovation. Their focus was nearly always on their own farms, and not on meeting customers' needs" (Sivertsson and Tell, 2015).

The history of Dutch agriculture, very early focused to meet customers' needs could give an idea of the difficulties to build a BM in agriculture. Any business is focused first to the customers' needs (Amit and Zott, 2012). However most of the farmers, on many European farms, are focused on production, on what is done on the farm. It could be useful to understand what the implicit BMs are in standard agriculture and to have a review of the existing proposed BMs up to now.

Is There an Implicit Business Model of European Farms?

It is difficult, and it could have no sense, to define a single BM for agriculture. Indeed, there are different types of production for which the relative costs of land, investment in material or productive living organisms, or even labor costs are different. For example, viticulture, arboriculture, field crops, market gardening, floriculture, breeding for milk production, production of eggs or broilers; there BM is as different as the markets they serve.

It is true that a large part of agricultural production releases a small margin (as any row materials production), contrary to high value products/services. Nevertheless, there are a few counter-examples. The range of the food and beverage products remains very broad, from commodity products to highly specific technical products even to luxury products; for example, how can you compare the wheat food chain and Champagne?

Generally speaking, the design of a BM in agriculture poses specific problems compared to anything known in industry or services. Indeed, a fundamental trait of the implicit agricultural BMs, as it was installed in the developed countries, since the 50s of the XXth century, is to associate the features of a heavy production industry of intermediate goods - which has long required large structures - to those of micro-enterprises which cannot individually control their markets.

About the production of high yielding cereals, in France, structural costs are close to 75% of total expenses, those corresponding to material investments (land + equipment) often exceed 40% (ARVALIS, 2005; Longchamp and Pagès, 2012). In viticulture or arboriculture, the investment for plant growing until production begins may require from 3 to 10 years.

An important share of agricultural products has a short shelf-life and consequently a significant degradation of the potential value over time. It should be sold quickly, even if it means selling off. Under these conditions, the end customer loses its importance; the priority concerns of quick sale if not immediate. In some cases, the answer is to be found in additional investments (for example, conservation of apples or potatoes in cold rooms). The exception concerns mainly cereals which nevertheless know the other two problems (structural costs and return on investment).

A farm manager is very often, if not generally, under the pressure of a high debt ratios with a return on investment that is constituted by the overall capital valuation of the farm. It can be compared to insurance companies, drinking water suppliers or steel industry. It is therefore the conventional risks of a low-differentiated intermediate goods company that must invest continuously in order to lower its production costs. Such managers have acquired the necessary training that farmers have generally not received.

Apart from the most advanced agriculture, as in the Netherlands, whose historical component we have seen, in agriculture we continue to calculate with reference to the surface of the exploitation (ARVALIS, 2005), that is to say by hectare and not according to the quantities produced and the value of the production per unit produced. Farmers are able to track their production costs per unit of economic output are still few, and of course fewer master any marginal analysis.

As a result, obtaining accurate data on cost allocations per unit of economic output is generally difficult in agriculture.

A farm is mostly a microenterprise because even a farm of 3000 ha or more, in field crops, is a small business, with no more than six full time employees. In comparison a 30-ha farm of vegetable production can be equivalent or even larger in turnover, and so is rarely described. Agricultural enterprise with a staffing level near 100 full-time employees exists only in countries like Brazil, Argentina, and Malaysia or in the "Black Sea countries". There are concerns in France about emergence of 1000 cow' farms without realizing that compared to any sector, it is only a small office. The majority of farms in Europe are personal enterprises whose business transfer is that of inter-family and intergenerational exchanges of the property.

Patrimonial security prevails over all other considerations. There is probably a fundamental contradiction between the agricultural legal status and the BM. The separation between land and ownership of productive material could reduce the inconsistency between BM and heritage issues, while it can make more visible the other components that slow down the choice of a new BM.

This leads on almost all agricultural markets to the fact that there is still a large number of farmers producing the same products, with the result that there is a very great difficulty of differentiation, with a market definition that can only be obtained through large mutualization of the production. Consequently, farmers are focused on their production tools much more than on the satisfaction of their customers.

It is under those terms that agricultural activity has to deal with a variability that most other economic activities have been able to protect themselves against climate variability. It leads to more or less chaotic succession of over- and underproduction. The current globalization context leads to competition between multiple agricultures whose cost structures, pedo-climatic conditions and climatic variability are very diverse; this also does not simplify the construction of suitable BM, apart from the focus on lowering costs and the race for productivity. Historical series, from longer terms perspective, show that agriculture needs regulation, stock control and subsidies in order to maintain a satisfactory level of production at a reasonable price and good quality and to be able to quickly compensate for local under-production due to climatic accidents, and/or for global overproduction that causes price collapses and local failures. Can a new type of BM be installed within an agricultural undertaking that would be closer to standard industry and service valuations? Is this transformation conceivable? Moreover, if the value brought by the farmer also includes the maintenance of the landscape, we should define who pays for it. It's an old debate (Beuret, 2002; Raymond et al. 2015; Kissinger et al. 2013). Can and should we incorporate subsidies into the BM? According to the BM approach, subsidies should be considered as a revenue which should be received according to a production or a service. Is it just a solidarity between the haves and the have nots, or a payment for a fundamental service to the society?

BUSINESS MODELS REVIEW AND ITS APPLICATION IN AGRICULTURE

Business Models Review

Different notions of BM have been studied in management research, including Canvas BM (Osterwalder, 2004; Teece, 2010), BM Innovation (Chesbrough, 2007a; Amit and Zott, 2012; Foss and Saebi, 2015; 2017), BM for sustainable innovation (Boons and Lüdeke-Freund, 2013) and sustainable BM (Bocken et al. 2014; Biloslavo et al. 2018). These authors have tried to clarify the origin of these concepts and the future research questions using some theories (such as Innovation and Entrepreneurship theories) and research streams like open innovation and sustainability (Foss and Saebi, 2018).

Biloslavo et al. (2018) summarized the evolution of 20 definitions and components of BMs between 1998s and 2016. Timmers (1998) defined a BM as "an architecture for the product, service and information flows, including a description of the various business actors and their roles; and a description of the potential benefits for the various business actors; and the description of the sources of revenues" (p. 4). Between 2000 and 2005 the authors, like Hedman and Kalling (2003), focused the BM definitions on connecting different actors with strategic dimensions (core strategy, strategic resources, etc.). Since 2008, the BM definitions included other components: customer value proposition (Johnson et al. 2008); value creation, delivery and capture (Teece, 2010). In order to create value, the BM has been studied also in innovation and technology management domains. Chesbrough (2003) considered on one side the BM as "a subject of innovation". On the other side, he showed that open innovation can contribute to create value by sharing information and knowledge between innovators. In this case, the BM is seen as "the open BM" (Chesbrough, 2007b) or"the BM Innovation" (Mitchell and Coles, 2003; Chesbrough, 2007a, 2010; Foss and Saebi, 2017).

In the last years, some authors have addressed the BM in the Internet of Things (IoT) industry. Metallo et al. (2018) present the main literature about IoT-oriented BM from two perspectives: technical and managerial. The first one, IoT technology is considered as "a platform-based ecosystem". The platform's architecture, governance and environmental dynamics can influence the co-created value within the ecosystem (Tiwana et al. 2010). The second one, the ecosystem included the connections between BMs designed by the firms and the external environment for IoT technology (Hui, 2014; Westerlund, 2014; Metallo et al. 2018).

Biloslavo et al. (2018) showed that BM definitions exclude "natural and social aspects of organizational environment". Hence, they proposed a new model named 'Value Triangle' that includes three main components (p.755): the value proposition; the value co-creation and co-delivery system and the value capture system (Bocken et al. 2014).

The purpose of this chapter is to focus on new BM for open innovation in four French cases of farmers. So, the authors consider that the farmer, as an actor, can change the type of BM when his/her status changes. The authors tried to identify the BMs of farmers before and after digitalization and how these farmers are building their innovative BMs via IoT technology and are promoting sustainability.

Digitalization is a real challenge in agriculture; it is generally described that digitalization changes the efficient BM, and the emergence of new type of network companies is really convincing. However, as agriculture is still to produce the basis of food, as it has been for millennia, how could digitalization change that? Before trying to answer this question, we have to study what could have been the implicit BMs to explain the agriculture activity.

Business Models Application in Agriculture

The BMs research focused on large companies or technology innovations and pays little attention to the farmers' case. Sivertsson and Tell (2015) identified some barriers, linked to human, contextual or governmental parameters, existing when Swedish farmers tried to innovate in their BM.

The authors tried to adapt the BM canvas, as proposed by Osterwalder and Pigneur (2010), to four farm group models, i.e "four major ways of conceiving and modelling the farm" that can be thought as trials to understand the farm in the evolution of agriculture, according to Laurent et al. (2003). The article by Laurent et al is premonitory by its anticipation of a multifunctional model that is becoming the agricultural reality of today. Through a selective literature review the authors highlighted four models to address farm multi-functionality. The fourth model looks like preliminaries of how to conceive multi-functionality afresh.

Based on the analysis of three research programs that are relevant to three major developments of the world of agriculture that are currently under way, this article described that agriculture is entering a new world that will have to be tackled in a totally new way (anticipation of digitalization). Recall that the year 2003 is the launch of the BlackBerry 7200, the first smartphone sold on a large scale that almost no family farmer could hear about. The market for such a product was concerning geeks or managers in large or international companies. The same year, organic agriculture, in its first period of growth, reached in France 400 000 ha, about 1.3% of agricultural land, the OGM surface reached nearly 80 million ha in the world (about 5% of cultivated land) after a regular growth during seven years and conservation agriculture began also a large growth and could have reached at least 50 million ha, most of them in the Americas and Australia.

For the last twenty years, agriculture has been changing at an accelerated pace, and it seems that the acceleration is going on, all around the world. This chapter presents a French experiment, although the change it described could be easily extended to European and worldwide agriculture.

Here the authors propose a simplified presentation of these four "farmer" groups of BMs based on the categorization by Laurent et al. (2003), according to the canvas proposed by Osterwalder and Pigneur (2010). Usually, the value proposition is considered as a value for the customer. We choose here to define this value as a pillar for the farmer. Even when he/her produces a labelled product (organic farming, protected designation of origin, quality labels, etc.), the labelling is mostly collective and not the property of the farmer.

This group of models (Table 1), described here schematically, according to BM Canvas, is based on the micro-economic theory of the firm with the explicit aim of providing support for farm management and contributing to the establishment of a scientific organization in agriculture. It has been developed first in the United-States, before the Second World War, and at the beginning of the fifties until the end of the nineteenth of the XXth century in Europe. This BM also attempts to describe the actual functioning of farms, with an aim in the background to analyze the functioning of the sector and to assess the impact of agricultural policies.

According to this type of model (Table 1), the farm is described as a business from a normative perspective. It seems difficult to give some precise empirical content to this type of model. This group of BM has become more complex through the integration of the utility function of the farmer and taking into account the perception of risk in farmers' behavior. Nevertheless, it does not lead to a new concept of farming that simultaneously accounts for its environmental, social and productive functions.

In this model (Table 1), most of the value added is transferred to partners, downstream or upstream. For the farmer, the value he or she appropriates corresponds first of all to a patrimonial model: the value of the land estimated through its yield potential.

This second group of models (Table 2) claims an approach to farming as part of a larger economic and social dimension, the regulation of which does not depend only on market forces. The point of view is more structural. The farm is integrated in a set of farms more or less similar, which leads to business combination and divisions of labor. As for the previous model, work is conducted in a normative perspective. It is a model that meets the work of rural sociology. It puts the farm more into perspective which takes into account all the agricultural transformations. However, like the previous model, the articulation of environmental issues with economic and social dimensions remains limited.

This group of models (Table 2) incorporates the role of functions or material mutualization and therefore of cooperatives, which are at the same time partners, customers and suppliers. It is possible that the mutualization is a way of preserving the relatively small farms, maintained by the existence of cooperatives, whose activity is to pool a number of functions that are those of conventional enterprises. These cooperatives play an entrepreneurial role by mastering the logistics, the storage and in some cases the marketing and the transformation and thus position themselves as competitors of the private companies.

These models began to be developed during the 1970s and continued to expand until the late 1990s.

This third group of models (Table 3) stems, among other things, from the fact that farmers do not automatically adopt agronomic research proposals. They do not seem to behave like "rational actors" seeking to maximize profit or income. In these models, the point of view of the farmers on their actions and decisions is much more taken into account and even will be the most recommended. These models use the theory of the general system and postulates a limited rationality of the actors. Real agricultural practices are integrated and lead to the development of expert systems, their effectiveness in the face of hazards are evaluated. They open up the possibilities of understanding the multifunctional nature of farms, although they put non-agricultural activities out of the scope of analysis.

These models began to be designed and developed during the late 1970s and continued to expand until the late 1990s.

This fourth group of models (Table 4) analyzes farming by asking first and foremost the nature of farmers' organizational choices and their efficiency. Modeling is no longer built through the production function. The farm is considered as an organization that coordinates various activities. Like the first two sets of models, since the measure of efficiency is central, the designers' posture seems initially normative. The aim is to evaluate the economic efficiency of organizational choices, from the minimization of transaction costs and organizational costs.

There are crossing points between the third group and this one. The economic theories of the organizations are mobilized, the outsourcing choices are taken into account. Several different analytical approaches are used. The problems of natural hazards, including climatic hazards are taken into account. The diversification choices are studied, according to several approaches, including pluri-activity. Some approaches understand farming as an information and communication system.

This group of models (Table 4) is the most disparate, different theoretical fields are mobilized, the empirical work applied to farms being relatively rare. Finally, it turns out that this model group brings together different lines of research that come out of standard micro-economic models. This is not, despite the initial appearance, a normative stance; it is a research that tries to capture the new characteristics of farms in a context of change.

Synthetically, the first group of models tries to describe the articulation between the economic and agro-technical dimension. The second adds the social dimension, the third considers the strategic approach and the fourth anticipates the emergence of information systems, flow optimization and information processing. We can see the contribution of transdisciplinary and interdisciplinary in the agricultural world with the evolution of approaches. Implicitly, these models are bent on the agro-ecological and territorial integration, they lack both an integrated conceptualization and the right adapted tools (Berthet, 2014).

The Transformation of Agricultural Activities and the Need of Digitalization

What emerges from the article by Laurent et al (2003) is that, since that date, both society's expectations of agriculture and the practices of farmers have changed. In 2017, all researchers studying the evolution of agriculture are convinced, even if the models of the farm of tomorrow remain uncertain.

Ten years after that article, Marraccini et al (2013) confirmed that farmers are still, and increasingly asked, to fulfill several functionalities and services (basic food production, energy save and production, waste recycling, landscapes management, economic vitality of rural communities' maintenance, protection of sensitive biotopes, biodiversity blooming, etc.). Farmers also seek a balance of their activities either by pluri-activity or by diversification. All of this combined, it appears a completely new model that leads to reconsider the analysis of the operation at varying spatial scale, depending on the point of view considered and usually largely exceeding the farm (Munroe et al. 2014).

The requirement of traceability of the sectors associated with landscape and biotope conservation, and also Carbon Dioxide (CO_2) caption and storage, generates new constraints far exceeding the scope of the farm. Furthermore, the need for precision farming to save inputs while optimizing production can also provide the means for increased traceability of actions. Combining precision, digital tracking and traceable action are technically similar goals, thanks to digitalization. It is of course necessary to define the economics of such developments.

Farmers diversify their production and develop pluri-activity. It is important for any type of statistics about agriculture that the farms managed by such farmers be accounted for in agricultural activities. It is even more important that they are also digitized in a similar way.

Any study or benchmark corresponding to scale levels beyond the activity of the farm must be able to be based on the activities of all farms included in the referenced higher scale.

The references of agricultural production methods have been transformed: organic farming, conservation agriculture, agroforestry, sustainability requirements, peri-urban and urban agricultures. In terms of quality, we have attended the multiplication of quality labels and protected designation of origin (mostly in Europe). If production optimization tools in each cultural context and for each cultural itinerary are digitally monitored, it will be possible to follow the production and guarantee its traceability. We can imagine that this will provide tools for monitoring and productivity improvement. The whole question will be to have the right sensors corresponding to the criteria that will have to be followed.

At the same time, a still growing surge of connected objects, Big Data, Artificial Intelligence and robots is taking place. The number of smartphones in the world has reached 2.3 billion, farmers becoming one of the most equipped social groups (Beza et al. 2017). The number of connected objects has exceeded the number of human beings and its exponential growth continues. The number of industrial robots increases by nearly 700,000 per year, and now they reach agriculture. Connected objects means connected tools; any type of existing tool or new tools or new sensors could offer the possibility for collecting and monitoring precise information.

Faced with the complexity of agricultural activity, now at the crossroads of agro food chains and controlled regions that can be revitalized by the bio-economy, it is much more than just a simple economic model that is to be accomplished. We have to build the concepts and the tools for the various and integrated models that will emerge. It appears that digitalization could offer the possibility to accumulate precise data for two different and complementary aims: traceability and improvement of efficiency. Both of these goals should be pursued far beyond the agricultural parcel and farm levels.

Internet of Things Business Models: A New Opportunity for Farmers?

The new situation in the development of a multifunctional agriculture and in the enlargement of farmer's activities has some heavy consequences. First, this situation could result in the need to separate the farmer from the farm, which could lead to a change in farm businesses law. This may lead to another revolution, as this modification will have to be made taking into account other modifications of the current general context.

Second, it leads to asking who are the customers of the farmers who participate to functionalities which go beyond the perimeter of their own farm? (Hansen et al. 2014). It can be about biodiversity, even relatively local, protection of the grounds and water, landscape maintenance, CO_2 capture, etc. Indeed, it seems hard to imagine that a farm manager would invest time, skills and money in activities whose benefits revert to different collectives and at different levels of scale and yet this is taking place (Bühler and Raymond, 2012; Raymond et al. 2015; Primdahl et al. 2013). He must be paid and his service should be evaluated. The evaluation of farmers' services requires various tools (Digitalization) that will indicate, on the one hand the actions actually undertaken and on the other hand the measured effects (Everard, 2011). It will be necessary to move also towards outcomes guarantee. And it will require data to build the relationship between means and outcomes. So, the farmer could be paid as his service is evaluated, and the payers could be satisfied to go beyond the only guarantee of means.

All data provided by connected objects must be shaped and analyzed by cognitive models derived from scientific knowledge to be transformed into measurement tools and decision support tools. Only then can data make sense. As noted by Corentin Cheron (in Caroux et al. 2018, p. 103), the raw data collected must be analyzed and transformed locally because it is first a question of defining relevant interpretation and acting directly. It is after this first stage that the aggregations at different levels can become interesting. In the same parcel, and on the same farm, data comes from various connected objects that contribute to new significant data. While it is important to maintain raw databases for traceability purposes, it may be important to know what is useful for integrations at larger scales.

Used as results' proofs but also as tools for improvement, data may belong only to those who produced it. However, this may not always be the case. For the study and the follow-up of functionalities which can only be followed beyond the farm, it seems logical that data can be the property of the bodies funding the data collection at the larger scale.

We can conclude that, even if it is still a question of research, it now seems possible to design and implement specific BM built from IoT. It can be a real opportunity for farmers to obtain adequate financing for their production of non-marketable features.

AN EXAMPLE OF FARMING DIGITALIZATION: THE FIRST AGRILAB® BOOTCAMP

Farming digitalization is considered to require high skills in new technologies. Workshops and open labs can facilitate the appropriation of such technologies by farmers. In this perspective, UniLaSalle organized its first bootcamp – an intensive participative workshop – about the IoT in smart farming (UniLaSalle, 2017).

Goal and Organization of the First Bootcamp

This first bootcamp aimed to provide an open learning environment hindered on the farmers' needs to enhance their decision-making by connected sensors. It lasted two days and a half (24-26 November 2017), with the participation of six voluntary farmers (members of UNEAL, NORIAP and Agora cooperatives), 39 master degree students (mostly in agriculture and agroindustry), ten experts in digital technologies and agronomy, and some observers. The perspective was to start a free and open knowledge base shared under Creative Commons license (AgriLab 2017). Experiences like Open Source Ecology (http://opensourceecology.org/) and its French homologue Atelier Paysan (https://www.latelierpaysan.org/) show the interest for boosting grassroots innovation path drawing upon free access to others' experiences.

The bootcamp was coordinated by AgriLab®, a collaborative innovation center for agriculture based in Beauvais (northern France), and part of the sustainable development UniLaSalle program, whose main aim is to promote digital innovation and open source involving all stakeholders. The bootcamp was backed by the Chaire "Agro-Machinisme & Nouvelles Technologies" together with by the Region Hauts-de-France through the INS'Pir regional program for digital and social innovation. In addition, the Wolfram Company and RS Components provided their technological expertise.

Farmers were prompted to address their main current needs for farming management, namely by exploring the use of connected sensors to improve the monitoring of physical environmental variables and for decision-making support. They can choose among the latest cheap and open source technological "building blocks": Grove sensors, Arduino and Raspberry Pi, gateways and LoRa radios, a 3D printer, etc. In addition, they had access to a cloud data repository (Wolfram DataBin, InfluxDB time-series database) and to Wolfram technology data analysis software (Mathematica and Wolfram Cloud).

The first half-day started with the introductory conferences dealing with data, IoT and project examples. In particular, CongDuc Pham (LIUPPA, University of Pau, France) provided a wide array of examples concerning the use of low-cost antenna technology for low-power wide-area network IoT in rural applications (Pham et al. 2017) issued from the WaziUp H2020 research project (http://www.waziup.eu). Afterwards, four farmers formed each a group with the students, starting to define their work issues through a design thinking approach.

On the second day, each farmer group kicked-off the connected sensor prototype by addressing the physical variables that they wish to monitor. In the afternoon, the participants split into learning thematic sessions about the key prototype components: sensors and electronics, 3D printing and packaging, digital network interfaces, data collection and processing. In particular, the data session focused on the farmers' expected human-machine interface and the decision support system. The third day the groups continued and completed the prototyping. Finally, in the afternoon, each group presented its concept prototype to the other participants and a conclusive debate allowed to wrap-up the main workshop outcomes and perspectives.

Main Bootcamp Outcomes

The four projects shared the same general workflow architecture (Figure 1), except for the specifications concerning the sensors for the physical data collection and for the desired DSS (Figure 2).

The implied physical variables may be temperature, CO_2, humidity, distance, etc. Sensors are directly connected to an Arduino Printed Circuit Board, which acquires data and then sends them to a database via a network, which may be a wired network or a wireless network (e.g. Wifi, Lora), depending on the needs. Databases and data mining processes may either be installed locally or in the cloud.

Four open source prototypes of connected sensors/DSS were designed based upon the farmers' needs (Table 5) and tested in simulated environment. Each project consisted of two main steps. First, the specification that includes the definition of the context, and of the input and output variables. Second, the implementation, which consists in sensor choice & packaging, connectivity and data collection and processing.

All the participants to the bootcamp agreed that it is relatively easy to design and implement the most appropriate connected solutions suited to the farmers' needs. Through this process, we show that farmers should be involved in the agriculture digitalization, and even more that they may participate to a problem-solving approach to broaden their skills. In addition, relating this to the students' training, by project learning, can help to raise their confidence in their capabilities to acquire new knowledge and new know-how and to develop new solutions.

Such an open innovative model farmer-oriented can benefit from the involvement of students and experts in agronomy and in digitalization technologies. In the end, this approach to innovation leads to a large range of new BMs, among which the creation of a start-up becomes just one of the multiple possibilities through which the involved actors can create a new activity. For instance, the farmer-oriented concept prototypes can be adopted by established industries for the development of improved solutions.

TOWARDS A NEW TYPE OF MODEL: SUSTAINABLE INTERNET OF THINGS OPEN INNOVATION MODEL

The fifth BM should take into account different levels of analysis:

1. Pure open innovation with open data, and open software. It means the emergence of an economy of functionality. The utility of data acquires value and can be rapidly expanded at different scales.
2. Control on some data which could be useful for different levels. The data acquired by farmers whose activity is completely digitalized can be aggregated at different levels that are not all hierarchical. This may concern territorial levels, and also institutional levels. Local innovation can come from a recovery of innovations from various locations worldwide (open software travels very fast ...). These local innovations can then expand in an ascending way, by aggregation.
3. Partially closed innovation for specific tools that can be crucial for competitiveness. Some process innovations can lead to open innovations in terms of products. The sale of the process can be implemented in different ways (intellectual or industrial property, production of specific parts, etc.
4. Multiple levels of customers. As the farmers are producing a lot of pluri-level functionalities, their customers are at all the scales mentioned above. All these features correspond to various societal demands that may seem contradictory and who's only practical application will be to disentangle these possible contradictions (protect the soil, capture CO_2, protect aquifers and arable crops, increase biodiversity, protect endangered species, improve taste quality and sanitary quality, increase

the volumes produced and their diversity, etc. These various supplies of new services and products are addressed to private structures, territorial authorities, geographical areas, national or regional institutions, rural areas or agglomerations, cooperatives, food chain operators, etc.

All these activities have a cost for farmers, and at the same time nobody is able to conceive the integration of all these demands addressed to them neither to compute their real values. How to monetize the immaterial produced by agriculture? This question is studied in the field of ecosystem goods and services.

To capitalize farmers' knowledge and skills, accumulated knowledge, techniques and processes, they must be shared by farmers and at the same time be financed, because no farmer will invest without hope of return. How can farmers sell this knowledge and know-how? To answer, we must find a solution to the weakness of farmers' power due to the low relative weight of each farm.

It is conceivable that this increase in power goes through the construction of cooperatives. However, the loss of farmers' power in cooperatives, as soon as they develop, shows that the farmer's BM needs to be reconsidered.

This fifth model (Table 6) describes a change of attitude for farmers. The land holdings are no longer a source of value, yet they are the whole complex of digitalized knowledge and know-how. Indeed, the technical mastery of this complex creates value that relies upon the knowledge of its digitalized features.

The agro-ecological knowledge is also a value, a place-based value that depends though on the data usage skills, which is to say by digitalization.

Given the widespread context of knowledge, know-how digitalization, and the identification, monitoring and tracking of real practices, the data control master is the final owner. Yet, this ownership is its value, perhaps the basis of negotiation, which is to say of monetary exchanges, in a multi-scale framework. This market of agricultural knowledge is not necessarily a traditional market since the status of buyers depends on the scale at which these data are valued. These buyers can be enterprises, territorial or political institutions, and the value can be fluctuating and dependent on political choices at different levels.

Farmers will therefore have data whose value will depend on the level of data integration. They will have to share innovation with the peers. They will have to foster solutions even in the case of open innovation and to participate in monitoring and data integration platforms. As everything is traceable, it will be possible to move from collective valuation to individual valuation. In the cases of very advanced data integration, it is conceivable that the value of data and the sharing of value will derive from the collective recognition that it is a common good.

CONCLUSION AND PERSPECTIVE

Digitalization is finally a general concept that encompasses all innovations currently under development and using digital tools. For example, the evolution of Artificial Intelligence has the consequence that they will be quickly portable and that anyone could have these tools, including farmers. As the use of Artificial Intelligence will be relatively easy (which is not true of its conception), one can conceive very profound changes of professional relations and thus of BM. Farmers who will master the use of tools with Artificial Intelligence and Deep Learning could have the means to analyze data, to create value and to preserve his or her autonomy.

Among the large trends in food and agriculture, digitalization appears to be a dominant one, as it is far more than a tool. It can adapt together to large production in a changing environment and to specific and diversified products. Evermore, digitalization could be the necessary tool to meet the agro-ecological challenges and the general needs for a more efficient agriculture. The innovator farmer could find some new ways to better food security, better optimization of inputs in agriculture, and methods to be better paid for their actions.

The presentations of this chapter focus on a new innovative BM's type. It could offer to the farmers the possibility to be empowered through mastering the use of all digital based tools. Special cases of this BM model can be built as part of partnerships between farmers, engineering school, students and experts in digitalization technologies. Nevertheless, this global BM leads to a large range of new BMs. The Start-Up approach is one possibility through which a new activity can be created. In addition, this experience interests some project sponsors for the development of new activities. It means that this new type of BM integrates the possibility of the emergence of multiple SMEs, with local and specific activities (Chiappini and Toccaceli, 2013), most of them controlled in part by farmers, who will act in the agricultural sector. Large groups may also have interests in these SMEs, although there is no guarantee that a large industrial group will have the agility to be present at the multiple interfaces generated by digitalization.

Digitalization arrives in agriculture later than in any other economic activity, because the added value is small. In fact, the tools that will develop in agriculture would often be reframing or reconstructing technologies already tested elsewhere.

It can therefore be inferred that the digitalization of agriculture can revolutionize agricultural practices. We believe that one of the keys to the success of such a revolution is farmer' mastery of the technological innovations. Open IoT is a very significant way to achieve this. Indeed, thanks to AgriLab® approach, farmers will be enabled to develop their own "custom" solution; proposed solutions are not expensive due to the low cost of the equipment; conception cost can be low and integrated in apprenticeship training; and finally, open IoT favors collaboration between farmers, scientists, experts and students.

The AgriLab® project falls within Sustainable Development approach and aims to promote digital innovation and open source involving all stakeholders. Through this process, it can be shown that the farmers can be involved in the agriculture digitalization, and even more that they can participate in a problem-solving approach to broaden their skills. The students' training, by project learning, serve to raise their confidence in their capabilities to acquire new knowledge and new know-how and to realize new solutions. All of this should be built upon and improved for future experiments or farmer's project.

This new and open innovative BM involves farmers, engineering schools, students and experts in agronomy and Information and Communication Technology (ICT). Nevertheless, this fifth model leads to a large range of specific BMs. What are the barriers to open the 5th BM via IoT technologies in order to promote sustainability? A benchmarking with different types of FabLabs, in agriculture and in other domains, could be very useful.

In the process of digitalization of agriculture, three different axes appear:

1. An agronomic axis: low-cost sensors, mobiles or fixes, are expected to enable farmers to better monitor the environment, thus to facilitate the implementation of precision farming. This possibility to follow in real time many parameters leads to the possibility of optimizing and simplifying the technical cultural itineraries. For instance, the main outcome of the AgriLab® bootcamp was to improve the farmers understanding of sensor technology and data management. They were eventu-

ally helped to clarify their expectations from the new technology providers and advisors. More in general, from the agronomic point of view, we can stress that digitalization will require farmers to ask more and more precise questions, then to consider how the novel monitoring tools will enable new practices to finally build models allowing anticipation, follow-up and decision-making in complex situations.

2. An economic axis: digitalization could provide added value; this can explain why farmers, or their children, adopt such new technologies. The implementation of AgriLab® platform can play a part on the adoption and the dissemination of technology by farmers, students and experts in digitalization technologies. For the farmers, the challenge is to be able to own, adopt and adapt the 5th BM.
3. A technological axis: proposed indicators are still descriptive data. It will be necessary to develop predictive analyzes (Artificial Intelligence and Deep Learning), which will require bigger data sets, sometimes involving data collection over several years. It could be necessary to validate the different sensors, may be to define some other, to define some others, and to think about the needs of sensors according to the scales of integration of the data.

We can first mention the fears and desires of farmers, and the necessity of a culture of innovation and entrepreneurship. As a result, the level of education and information of farmers will become crucial. For this, AgriLab® will play also as training and learning center.

The digitalization of agriculture can have profound consequences. The traceability of actions and their results can be built by integrating pedo-climatic factors, agro-ecological choices and societal criteria (protection of the environment, pollution, biodiversity). The new drivers of digital agriculture could depend on a multi-scale societal demand policy.

Depending on political choices defining the role of agricultural activity in each agricultural region or sub-region, as well as in peri-urban or urban agriculture, it could become possible to ensure stable incomes that will no longer be subsidies but service payments to the community. This has already begun, however, without digitalization, the traceability of actions and effects is far from complete. Today, it is trivial to attribute without evidence the effect of farmers' actions on environmental or food security issues. One can imagine that with the general digitalization of agriculture it could become possible to carry out such follow-ups.

The digitalization of agriculture can also change the status of farmers who could become experts for the actual evaluation of any new variety, even before the variety is registered. Digitalization and complete crop monitoring can turn any field into a potential field trial. Farmers could be able to sell the data of a well traced production in real conditions to the breeder of the variety. The plant breeder will realize the integration of the data.

These challenges require farmers to expand their knowledge to be able to master these innovations such as digital machine control, embedded sensors, big data management, etc. Thanks to the lowering cost and miniaturization of advanced technologies, farmers are pushed and eager to shift from intuitive to fact-based farming practices. Any decision in agricultural production could be controlled and accounted for at the intra-field level (Bencini et al. 2012; Aqeel-ur-Rehman et al. 2014). The increased data collection and monitoring capacities can be a good answer to the need for a better use of natural resources to reduce farming trade-offs, thus meeting the society expectations for sustainable development. Yet, the fast-increasing amount of harvested data remains largely unexploited because the first users - farmers - are still poorly involved in the development of processed information relevant for their decision making and often poorly equipped.

Innovation in training and educational programs, as well as new forms of knowledge transfer could be a solution. Digitalization can lead to innovation processes that would change the way of farming, with a focus on the impacts and expectations for the agricultural knowledge and information systems. Indeed, digitalization, through web platforms, could allow experienced farmers to maximize the value of their know-how, help other farmers who need reliable and operational information, and also for all farmers to acquire knowledge and know-how for the success of their business. Such farmers could obtain some value from their knowledge and from the data they could be able to produce.

Digitalization can therefore become a general tool for diversifying agricultural services to businesses and institutions. The greater the number of services and their financial weight, the less farmers will be sensitive to market fluctuations.

ACKNOWLEDGMENT

The authors acknowledge the fundamental work done by the farmers and students who participated to the first AgriLab® bootcamp. We wish to thanks Bernard De Franssu - AgriLab® co-manager and director of the Sustainable Development Department at UniLaSalle, Narges Bahi-Jaber - associate professor at UniLaSalle, Nathalie Schnuriger - associate professor at UniLaSalle, Carolina Ugarte - associate professor at UniLaSalle, Pedro Fonseca (SUEZ) and Dorian Birraux (Wolfram inc.) for their contribution to the design and realization of the first bootcamp. The authors thank William E Edmonds, Research Lecturer, at UniLaSalle for the reading and revision of the final version.

Table 1. 1st model: Farm as "a micro-economic unity"

<table>
<tr><td rowspan="2">Partners
Upstream/ downstream agencies</td><td>Activities
Crop and livestock production</td><td rowspan="2">Value for the farmer
Family patrimony</td><td>Customer Relationship
Logistic
Stock</td><td rowspan="2">Customer segments
Companies (private, trading)</td></tr>
<tr><td>Resources
Scarce, soil</td><td>Channels
B to B</td></tr>
<tr><td colspan="2">Costs
Production unit</td><td colspan="3">Revenues
Agricultural raw materials</td></tr>
</table>

Table 2. 2nd model: Farm as "a component of social system"

<table>
<tr><td rowspan="2">Partners
Upstream/
downstream
agencies
Other farmers
Financial
institutions</td><td>Activities
Crop and/or livestock production
Specialization
Division of labor</td><td rowspan="2">Value for the farmer
Family patrimony
Investment in equipment.
Improved productivity.</td><td>Customer Relationship
Logistic
Stock</td><td rowspan="2">Customer segments
Cooperatives
Companies
(private, trading)</td></tr>
<tr><td>Resources
Soil
Labor force</td><td>Channels
B to B</td></tr>
<tr><td colspan="2">Costs
Set of production units</td><td colspan="3">Revenues
Return on investment</td></tr>
</table>

Table 3. 3rd model: Farm as "a controlled system"

<table>
<tr><td rowspan="2">Partners
Upstream/ downstream agencies
Other farms
Financial institutions
Development organizations
Public institutions</td><td>Activities
Different production "workshops"
Various marketing channels
Marketing
Logistics
Social function</td><td rowspan="2">Value for the farmer
Family patrimony
Investment in equipment.
Improved productivity.
Variety of the production flows
Efficiency</td><td>Customer Relationship
Logistic
Stock
Products for consumption</td><td rowspan="2">Customer segments
Companies
Cooperatives
Consumers</td></tr>
<tr><td>Resources
Soil
Labor force
Vision of the situation</td><td>Channels
B to B/ Distribution
B to C</td></tr>
<tr><td colspan="2">Costs
All factors contributing directly to production including some processing and sales.</td><td colspan="3">Revenues
Sales/ return on investment, sales, processing
Subsidies</td></tr>
</table>

Table 4. 4th model: Farm as "a complex organization"

<table>
<tr><td rowspan="2">Partners
Upstream/ downstream agencies
Other farms
Financial institutions
Development organizations
Public institutions
Consulting</td><td>Activities
Production
Marketing
Logistics
Information and communication systems</td><td rowspan="2">Value for the farmer
Family patrimony
Investment in equipment (hardware and software).
Improved productivity.
Variety of the production flows
Efficiency
Life quality
…</td><td>Customer Relationship
Logistic
ERP
Web services
…</td><td rowspan="2">Customer segments
Companies
Cooperatives
Consumers
…</td></tr>
<tr><td>Resources
Natural
Information
Other activity</td><td>Channels
B to B
Distribution
B to C</td></tr>
<tr><td colspan="2">Costs
All factors contributing directly to production and sales, including software, robots, internet connection, distribution</td><td colspan="3">Revenues
Sales/ return on investment/ transformation</td></tr>
</table>

Table 5. Overview of the outcomes from the first AgriLab® bootcamp

Project name	Issue	Specifications		Digitalization perspectives
		Physical data	Output	
iPatate	Quality monitoring of potato stocks	Temperature CO2 Humidity phyto-hormones	Intervention of the operator	Remote management of separated stocks
SiloTeam	Monitoring of the filling level of poultry food storage silos	Height of food remaining in the silo	Exact tonnage of food remaining in the silos	Automation of the supply management
VegData	Early rot detecting in salads	Air, foliar and soil temperature and humidity	Raising soil moisture to warn the farmer when the moisture reaches a threshold that requires irrigation	Creation of a territorial network with neighbors and test of an artificial intelligence approach
Decisio	Soil moisture monitoring to identify best time for: (i) sowing flax; (ii) harvesting potatoes	Soil moisture and temperature Moisture and air temperature Foliar development Rainfall	Concerning flax: knowing the best period for sowing. Concerning potatoes: predicting the stoppage of vegetation growth and thus the harvest	Simplification of monitoring through multipurpose sensors.

Figure 1. Workflow architecture of the bootcamp prototyping activity (source: adapted from Dantan et al. 2018)

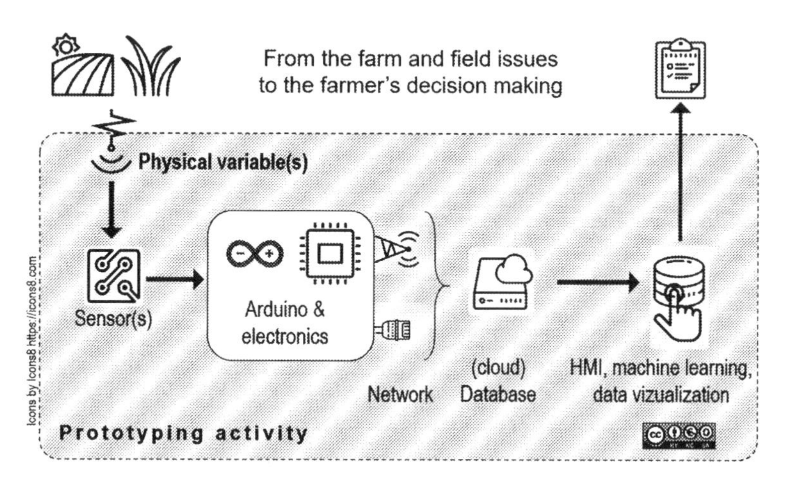

Figure 2. Decisio sensor-to-DSS package (source: adapted from Dantan et al. 2018)

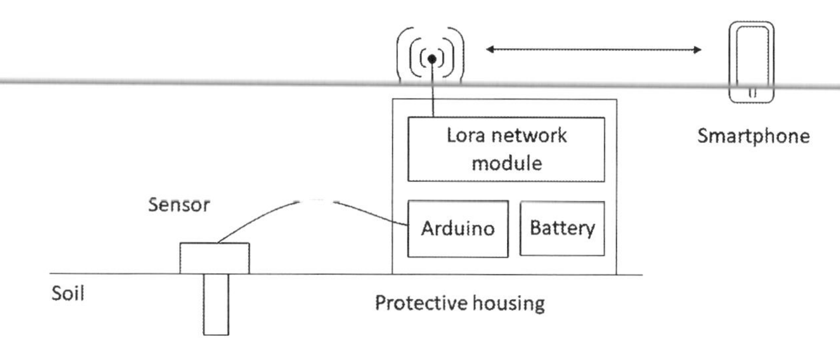

Table 6. 5th model: Pluri-activity Farm as "a connected organization"

Partners	Activities / Resources	Value for the farmer	Customer Relationship / Channels	Customer segments
Partners Upstream downstream agencies Other farms Financial institutions Development organizations Public and territorial institutions « AgriLab® » type, FabLab Teaching and research institutions Startup	**Activities** Diversified (Tourism, renewable energies, environmental features, material innovation…)	**Value for the farmer** Processing Data Information and knowledge sharing The knowledge of the soil, the climate, his agronomic choices, etc. Technical itinerary, markets, …,	**Customer Relationship** Comparative Websites CRM solutions	**Customer segments** Cooperatives « Locavores » Companies Startup (distribution, sales, platforms) Consumers
	Resources Mechanical - mechatronics Data Hardware and software All FabLab, AgriLab® included.		**Channels** B to B Distribution B to C Short B to C Direct business on farm	
Costs All factors contributing directly to production and sales, including software, robots, internet connection, distribution, etc.			**Revenues** Sales/ return on investment/ transformation Valuation at several levels of scale (sale corresponding to territorial requirements) Data Hacking (Collaborative Platforms) Grants considered not as aid but as compensation for "intangible property sales"	

REFERENCES

AgriLab. (2017). *Open IoT in Smart Farming*. Retrieved from https://web.archive.org/web/20180415191211/http://agrilab.unilasalle.fr/projets/projects/open-iot-in-smartfarming/wiki

Agtech. (n.d.). Retrieved from https://www.frenchweb.fr/agtech-40-start-up-francaises-qui-font-passer-lagriculture-a-lheure-du-digital/281907

Amit, R., & Zott, C. (2012). Creating value through Business Model innovation. *MIT Sloan Management Review*, *53*, 41–49.

Aqeel-ur-Rehman, Abbasi, A. Z., Islam, N., & Shaikh, Z. A. (2014). A review of wireless sensors and networks' applications in agriculture. *Computer Standards & Interfaces*, *36*(2), 263–270. doi:10.1016/j.csi.2011.03.004

Artificial Intelligence. (n.d.). Retrieved from https://www.aiforhumanity.fr/

ARVALIS. (2005). *Livre blanc réduction des charges*. ARVALIS.

Bencini, L., Maddio, S., Collodi, G., Di Palma, D., Manes, G., & Manes, A. (2012). Development of Wireless Sensor Networks for Agricultural Monitoring. In Smart Sensing Technology for Agriculture and Environmental Monitoring (pp. 157-86). Springer. doi:10.1007/978-3-642-27638-5_9

Berthet, E. (2014). *Concevoir l'écosystème, un défi pour l'agriculture*. Paris: Presses des Mines.

Beuret, J. E. (2002). À qui appartient le paysage? To whom does the landscape belong? *Nature Sciences Sociétés*, *10*(2), 47–53. doi:10.1016/S1240-1307(02)80070-X

Beza, E., Steinke, J., Etten, J. V., Reidsma, P., Fadda, C., Mittra, S., ... Kooistra, L. (2017). What are the prospects for citizen science in agriculture? Evidence from three continents on motivation and mobile telephone use of resource-poor farmers. *PLoS One*, *12*(5), e0175700. doi:10.1371/journal.pone.0175700 PMID:28472823

Biloslavo, R., Bagnoli, C., & Edgar, D. (2018). An eco-critical perspective on business models: The value triangle as an approach to closing the sustainability gap. *Journal of Cleaner Production*, *174*, 746–762. doi:10.1016/j.jclepro.2017.10.281

Bocken, N. M. P., Short, S. W., Rana, P., & Evans, S. (2014). A literature and practice review to develop sustainable Business Model archetypes. *Journal of Cleaner Production*, *65*(15), 42–56. doi:10.1016/j.jclepro.2013.11.039

Boons, F., & Lüdeke-Freund, F. (2013). Business Models for sustainable innovation: State of the art and steps towards a research agenda. *Journal of Cleaner Production*, *45*, 9–19. doi:10.1016/j.jclepro.2012.07.007

Brun, F., & Haezebrouck, T. P. (2017). AgTech – Digital Agriculture - Current development in France, Big Data, a multiscale solution for a sustainable agriculture. *Business Dictionary*. Retrieved from http://www.businessdictionary.com/definition/digitalization.html

Bühler, È. A., & Raymond, R. (2012). Pratiques agricoles et gestion territoriale de la biodiversité en contexte de grandes cultures. *Revue Geographique des Pyrenees et du Sud-Ouest*, 65–78. doi:10.4000oe.120

Caroux, D., Dubois, M. J. F., & Sauvée, L. (2018). *Evolution agro-techniques contemporaine II. Transformations de l'agro-machinisme: fonction, puissance, information, invention*. U. T. B. M. Belfort, Ed.

Chesbrough, H. W. (2003). *Open innovation: The new imperative for creating and profiting from technology*. Boston: Harvard Business School Press.

Chesbrough, H. W. (2007a). Business Model innovation: It's not just about technology anymore. *Strategy and Leadership*, *35*(6), 12–17. doi:10.1108/10878570710833714

Chesbrough, H. W. (2007b). Why companies should have open Business Models. *MIT Sloan Management Review*, *48*(2), 22–28.

Chesbrough, H. W. (2010). Business Model innovation: Opportunities and barriers. *Long Range Planning*, *43*(2), 354–363. doi:10.1016/j.lrp.2009.07.010

Chiappini, S., & Toccaceli, D. (2013). The relevance of district contexts in the utilization of rural development policies: experience from Italy. *International Agricultural Policy*, 4. Retrieved from https://ageconsearch.umn.edu/bitstream/190606/2/3.pdf

Dantan, J., Rizzo, D., Fourati-Jamoussi, F., Dubois, M. J. F., & Jaber, M. (2018). *Farmer-oriented innovation: outcomes from a first bootcamp*. Paris: FRA; bit.ly/agid2018

Douthwaite, B., & Hoffecker, E. (2017). Towards a complexity-aware theory of change for participatory research programs working within agricultural innovation systems. *Agricultural Systems*, *155*, 88–102. doi:10.1016/j.agsy.2017.04.002

Doye, D., Jolly, R., Hornbaker, R., Cross, T., King, R. P., Lazarus, W. F., & Yeboah, A. (2000). Case studies of farmers' use of information systems. *Review of Agricultural Economics*, *22*(2), 566–585. doi:10.1111/1058-7195.00039

Everard, M. (2011). *Common ground: the sharing of land and landscapes for sustainability*. Zed Books.

FAO. (2016). *Produce more with Less*. Available from: http://www.fao.org/ag/save-and-grow/fr/accueil/index.html

Foss, N. J., & Saebi, T. (2015). Business Models and Business Model innovation: bringing organization into the field. In N. J. Foss & T. Saebi (Eds.), *Business Model Innovation: the Organizational Dimension* (pp. 1–23). Oxford, UK: Oxford University Press. doi:10.1093/acprof:oso/9780198701873.003.0001

Foss, N. J., & Saebi, T. (2017). Fifteen years of research on Business Model innovation: How far have we come, and where should we go? *Journal of Management*, *43*(1), 200–227. doi:10.1177/0149206316675927

Foss, N. J., & Saebi, T. (2018). Business Models and Business Model innovation: Between wicked and paradigmatic problems. *Long Range Planning*, *51*(1), 9–21. doi:10.1016/j.lrp.2017.07.006

Hansen, R., & Pauleit, S. (2014). From Multifunctionality to Multiple Ecosystem Services? A Conceptual Framework for Multifunctionality in Green Infrastructure Planning for Urban Areas. *Ambio*, *43*(4), 516–529. doi:10.100713280-014-0510-2 PMID:24740622

Hedman, J., & Kalling, T. (2003). The Business Model concept: Theoretical underpinnings and empirical illustrations. *European Journal of Information Systems*, *12*(1), 49–59. doi:10.1057/palgrave.ejis.3000446

Hui, G. (2014). How the Internet of Things changes Business Models. *Harvard Business Review*, 29.

John, A. J., Clark, C. E. F., Freeman, M. J., Kerrisk, K. L., Garcia S. C., & Halachmi I. (2016). Review: Milking robot utilization, a successful precision livestock farming evolution. *Animal*, *10*(9), 1484-1492.

Johnson, M. W., Christensen, C. M., & Kagermann, H. (2008). Reinventing your Business Model. *Harvard Business Review*, *86*, 50–59. PMID:18681297

Kissinger, G., Brasser, A., & Gross, L. (2013). *Reducing Risk: Landscape Approaches to Sustainable Sourcing. In EcoAgriculture Partners, on behalf of the Landscapes for People*. Washington, DC: Food and Nature Initiative.

Kritikos, M. (2017). *Precision agriculture in Europe. Legal, social and ethical considerations*. PE 603.207. European Parliamentary Research Service. Retrieved from http://www.lafermedigitale.fr/

Laurent, C., Maxime, F., Mazé, A., & Tichit, M. (2003). Multifonctionnalité de l'agriculture et modèles de l'exploitation agricole. *Economie Rurale*, *273/274*, 134-152. doi:10.3406/ecoru.2003.5395

Longchamp, J. Y., & Pagès, B. (2012). Charges de mécanisation et structure d'exploitation. *Document de travail - Commission des comptes de l'agriculture de la nation*, 7. Retrieved from http://agreste.agriculture.gouv.fr/IMG/pdf/doctravail70712-2.pdf

Marraccini, E., Rapey, H., Galli, M., Lardon, S., & Bonari, E. (2013). Assessing the Potential of Farming Regions to Fulfill Agro-Environmental Functions: A Case Study in Tuscany (Italy). *Environmental Management*, *51*(3), 759–776. doi:10.100700267-012-9997-0 PMID:23263567

McQuail, D. (2000). *McQuail's Mass Communication Theory* (4th ed.). London: Sage.

Metallo, C., Agrifoglio, R., Schiavone, F., & Mueller, J. (2018). (in press). Understanding Business Model in the Internet of Things industry. *Technological Forecasting and Social Change*. doi:10.1016/j.techfore.2018.01.020

Mitchel, D., & Coles, C. (2003). The ultimate competitive advantage of continuing Business Model innovation. *The Journal of Business Strategy*, *24*(5), 15–21. doi:10.1108/02756660310504924

Munroe, D. K., McSweeney, K., Olson, J. L., & Mansfield, B. (2014). Using economic geography to reinvigorate land-change science. *Geoforum*, *52*, 12–21. doi:10.1016/j.geoforum.2013.12.005

Osterwalder, A. (2004). *The Business Model ontology: A proposition in a design science approach* (Dissertation). University of Lausanne, Switzerland.

Osterwalder, A., & Pigneur, Y. (2010). *Business Model generation: a handbook for visionaries, game changers, and challengers*. John Wiley & Sons.

Pham, C., Ferrero, F., Diop, M., Lizzi, L., Dieng, O., & Thiaré, O. (2017). Low-cost Antenna Technology for LPWAN IoT in Rural Applications. In *Proceedings of the 7th IEEE International Workshop on Advances in Sensors and Interfaces (IWASI'17)*. IEEE. Retrieved from http://cpham.perso.univ-pau.fr/Paper/IWASI17.pdf

Pignatti, E., Carli, G., & Canavari, M. (2015). What really matters? A qualitative analysis on the adoption of innovations in agriculture. *Agrárinformatika. Agrárinformatika Folyóirat*, *6*(4), 73–84. doi:10.17700/jai.2015.6.4.212

Pongnumkul, S., Chaovalit, P., & Surasvadi, N. (2015). Applications of Smartphone-Based Sensors in Agriculture: A Systematic Review of Research. *Journal of Sensors*. doi:10.1155/2015/195308

Primdahl, J., Kristensen, L. S., & Busck, A. G. (2013). The Farmer and Landscape Management: Different Roles, Different Policy Approaches: The Farmer as a Landscape Manager. *Geography Compass*, *7*(4), 300–314. doi:10.1111/gec3.12040

Raymond, C. M., Bieling, C., Fagerholm, N., Martin-Lopez, B., & Plieninger, T. (2015). The farmer as a landscape steward: Comparing local understandings of landscape stewardship, landscape values, and land management actions. *Ambio*, 1–12. doi:10.100713280-015-0694-0 PMID:26346276

Reichardt, M., Jürgens, C., Klöble, U., Hüter, J., & Moser, K. (2009). Dissemination of Precision Farming in Germany: Acceptance, Adoption, Obstacles, Knowledge Transfer and Training Activities. *Precision Agriculture*, *10*(6), 525–545. doi:10.100711119-009-9112-6

Sivertsson, O., & Tell, J. (2015). Barriers to Business Model innovation in Swedish agriculture. *Sustainability*, *7*(2), 1957–1969. doi:10.3390u7021957

Teece, D. J. (2010). Business Models, business strategy and innovation. *Long Range Planning*, *43*(2), 172–194. doi:10.1016/j.lrp.2009.07.003

Timmers, P. (1998). Business Models of electronic markets. *Electronic Markets*, *8*(2), 3 8. doi:10.1080/10196789800000016

Tiwana, A., Konsynski, B., & Bush, A. A. (2010). Research commentary - Platform evolution: Coevolution of platform architecture, governance, and environmental dynamics? *Information Systems Research*, *21*(4), 675–687. doi:10.1287/isre.1100.0323

Traitler, H., Dubois, M. J. F., Heikes, K., Petiard, V., & Zilberman, D. (2018). *Megatrends in Food and Agriculture: Technology, Water Use and Nutrition*. John Wiley & Sons Ltd.

UniLaSalle. (2017). *Agriculture connectée : un bootcamp sur l'internet des objets à UniLaSalle*. Retrieved from https://web.archive.org/web/20180415163809/https://www.unilasalle.fr/lasalleactus/agriculture-connectee-bootcamp-linternet-objets-lagriculture-a-unilasalle/

Vanderschuren, H. (2012). Can genomics boost productivity of orphan crops? *Nature Biotechnology*, *30*(12), 1172–1175. doi:10.1038/nbt.2440 PMID:23222781

Verley, P. (1985). La révolution industrielle. Paris: Gallimard.

Westerlund, M., Leminen, S., & Rajahonka, M. (2014). Designing Business Models for the Internet of Things. *Technology Innovation Management Review*, *4*(7), 5–14. doi:10.22215/timreview/807

World Population Prospect. (2012). Available from: https://esa.un.org/unpd/wpp/publications/files/key_findings_wpp_2015.pdf

Compilation of References

Abbas, A., Zhang, L., & Khan, S. U. (2014). A literature review on the state-of-the-art in patent analysis. *World Patent Information*, *37*, 3–13. doi:10.1016/j.wpi.2013.12.006

Aboelmaged, M. G. (2010). Predicting e-procurement adoption in a developing country: An empirical integration of technology acceptance model and theory of planned behavior. *Industrial Management & Data Systems*, *110*(3), 392–414. doi:10.1108/02635571011030042

Abolhassan, F. (2016). *The Drivers of Digital Transformation*. Springer International Publishing.

Abolhassan, F. (2017). Pursuing Digital Transformation Driven by the Cloud. In *The Drivers of Digital Transformation* (pp. 1–11). Cham: Springer. doi:10.1007/978-3-319-31824-0_1

Abrell, T., Pihlajamaa, M., Kanto, L., Brocke, J., & Uebernickel, F. (2016). The Role of Users and Customers in Digital Innovation: Insights from B2B Manufacturing Firms. *Information & Management*, *53*(3), 324–335. doi:10.1016/j.im.2015.12.005

Accenture. (2014). Big data analytics in supply chain: Hype or here to stay? Retrieved from https://acnprod.accenture.com/_acnmedia/Accenture/Conversion-Assets/DotCom/Documents/Global/PDF/Dualpub_2/Accenture-Global-Operations-Megatrends-Study-Big-Data-Analytics.pdf#zoom=50

Accenture. (n.d.). *The Future of Digital Advertising*. Retrieved from https://www.accenture.com/us-en/~/media/Accenture/next-gen/pulse-of-media/pdf/Accenture-Future-Of-Advertising-POV.pdf

Acs, Z. J., Estrin, S., Mickiewicz, T., & Szerb, L. (2017a). *Entrepreneurship, Institutions and Productivity Growth: A Puzzle*. Available at SSRN: https://ssrn.com/abstract=3060982

Acs, Z. J., Autio, E., & Szerb, L. (2014). National Systems of Entrepreneurship: Measurement Issues and Policy Implications. *Research Policy*, *43*(1), 476–494. doi:10.1016/j.respol.2013.08.016

Acs, Z. J., Stam, E., Audretsch, D. B., & O'Connor, A. (2017b). The lineages of the entrepreneurial ecosystem approach. *Small Business Economics*, *49*(1), 1–10. doi:10.100711187-017-9864-8

Acs, Z. J., Szerb, L., & Lloyd, A. (2017c). Entrepreneurship and the Future of Global Prosperity. In *Global Entrepreneurship and Development Index 2017* (pp. 11–27). Cham: Springer. doi:10.1007/978-3-319-65903-9_2

Addo-Tenkorang, R., & Helo, P. T. (2016). Big data applications in operations/supply-chain management: A literature review. *Computers & Industrial Engineering*, *101*, 528–543. doi:10.1016/j.cie.2016.09.023

Agarwal, L., Shrivastava, N., Jaiswal, S., & Panjwani, S. (2013). Do not embarrass: re-examining user concerns for online tracking and advertising. In *Proceedings of the Ninth Symposium on Usable Privacy and Security (SOUPS '13)*. ACM. 10.1145/2501604.2501612

Agarwal, R., Gao, G., DesRoches, C., & Jha, A. K. (2010). Research commentary—The digital transformation of healthcare: Current status and the road ahead. *Information Systems Research*, *21*(4), 796–809. doi:10.1287/isre.1100.0327

Aglietta, M. & Orléan, A. (Eds.) (1998). *La monnaie souveraine*. Paris: Odile Jacob.

AgriLab. (2017). *Open IoT in Smart Farming*. Retrieved from https://web.archive.org/web/20180415191211/http://agrilab.unilasalle.fr/projets/projects/open-iot-in-smartfarming/wiki

Agtech. (n.d.). Retrieved from https://www.frenchweb.fr/agtech-40-start-up-francaises-qui-font-passer-lagriculture-a-lheure-du-digital/281907

Aharonson, B. S., & Schilling, M. A. (2016). Mapping the technological landscape: Measuring technology distance, technological footprints, and technology evolution. *Research Policy*, *45*(1), 81–96. doi:10.1016/j.respol.2015.08.001

Ahlgren, P., Jarneving, B., & Rousseau, R. (2003). Requirements for a cocitation similarity measure, with special reference to Pearson's correlation coefficient. *Journal of the Association for Information Science and Technology*, *54*(6), 550–560.

Ahluwalia, P., & Varshney, U. (2009). Composite quality of service and decision making perspectives in wireless networks. *Decision Support Systems*, *46*(2), 542–551. doi:10.1016/j.dss.2008.10.003

Ahuja, G., & Katila, R. (2001). Technological acquisitions and the innovation performance of acquiring firms: A longitudinal study. *Strategic Management Journal*, *22*(3), 197–220. doi:10.1002mj.157

Aiden, E., & Michel, J. B. (2014). The Predictive Power of Big Data. *Newsweek*. Retrieved from http://www.newsweek.com/predictive-power-big-data-225125

Ajzen, I. (1991). The Theory of Planned Behavior. *Organizational behavior and human decision*, *50*(2), 179-211.

Ajzen, I. (1991). The theory of planned behavior. *Organizational Behavior and Human Decision Processes*, *50*(2), 179–211. doi:10.1016/0749-5978(91)90020-T

Akkermans, H., Bogerd, P., Yucesan, E., & Van Wassenhove, L. (2003). The impact of ERP on supply chain management: Exploratory findings from a European Delphi Study. *European Journal of Operational Research*, *146*(2), 284–294. doi:10.1016/S0377-2217(02)00550-7

Akter, S., Wamba, S. F., Gunasekaran, A., Dubey, R., & Childe, S. J. (2016). How to improve firm performance using big data analytics capability and business strategy alignment? *International Journal of Production Economics*, *182*, 113–131. doi:10.1016/j.ijpe.2016.08.018

Akturan, U., & Tezcan, N. (2012). Mobile banking adoption of the youth market: Perceptions and intentions. *Marketing Intelligence & Planning*, *30*(4), 444–459. doi:10.1108/02634501211231928

Alavi, M., & Leidner, D. E. (2001). Review: Knowledge management and knowledge management systems. *Management Information Systems Quarterly*, *25*(1), 107–136. doi:10.2307/3250961

Albani, A., & Dietz, J. L. G. (2009). Current trends in modeling inter-organisational cooperation. *Journal of Enterprise Information Management*, *22*(3), 275–297. doi:10.1108/17410390910949724

AlBar, A. M., & Hoque, M. R. (2017). Factors affecting cloud ERP adoption in Saudi Arabia: An empirical study. *Information Development*. doi:10.1177/0266666917735677

Aldrich, H. (2014). The democratization of entrepreneurship? Hackers, makerspaces, and crowdfunding. In Annual Meeting of the Academy of Management, Philadelphia, PA, October 27.

Aldrich, H. E. (2014). The democratization of entrepreneurship? Hackers, makerspaces, and crowdfunding. *Annual Meeting of the Academy of Management*.

Alharbi, F., Atkins, A., Stanier, C., & Al-Buti, H. A. (2016). Strategic Value of Cloud Computing in Healthcare organisations using the Balanced Scorecard Approach: A case study from A Saudi Hospital. *Procedia Computer Science*, *98*, 332–339. doi:10.1016/j.procs.2016.09.050

Alharthi, A., Krotov, V., & Bowman, M. (2017). *Addressing barriers to big data*. Business Horizons.

Allen, K. R. (2015). *Launching new ventures: an entrepreneurial approach*. Cengage Learning.

Al-Mubaraki, H. M., Muhammad, A. H., & Busler, M. (2014). *Innovation and entrepreneurship: Powerful tools for a modern knowledge-based economy*. Springer.

Al-Mudimigh, A., Zairi, M., & Al-Mashari, M. (2001). ERP software implementation: An integrative framework. *European Journal of Information Systems*, *10*(4), 216–226. doi:10.1057/palgrave.ejis.3000406

Aloulou, W. (2016). Understanding entrepreneurship through the chaos and complexity perspective. In *Şefika Şule Erçetin and Hüseyin Bağcı, Handbook of Research on Chaos and Complexity Theory in the Social Sciences*. IGI Global.

Alvesson, M., & Karreman, D. (2000). Varieties of discourse: On the study of organizations through discourse analysis. *Human Relations*, *53*(9), 1125–1149. doi:10.1177/0018726700539002

Amadi-Echendu, J. E. (2007). Thinking styles of technical knowledge workers in the systems of innovation paradigm. *Technological Forecasting and Social Change*, *74*(8), 1204–1214. doi:10.1016/j.techfore.2006.09.002

Amit, R., & Zott, C. (2012). Creating value through Business Model innovation. *MIT Sloan Management Review*, *53*, 41–49.

Amshoff, B., Dülme, C., Echterfeld, J., & Gausemeier, J. (2015). Business Model Patterns for Disruptive Technologies. *International Journal of Innovation Management*, *19*(3). doi:10.1142/S1363919615400022

Anand, A., Wamba, S. F., & Gnanzou, D. (2013). A Literature Review on Business Process Management, Business Process Reengineering, and Business Process Innovation. In *Workshop on Enterprise and Organizational Modeling and Simulation* (pp. 1-23). Springer.

Anaya, L., Dulaimi, M., & Abdallah, S. (2015). An investigation into the role of enterprise information systems in enabling business innovation. *Business Process Management Journal*, *21*(4), 771–790. doi:10.1108/BPMJ-11-2014-0108

Andal-Ancion, A., Cartwright, P. A., & Yip, G. S. (2003). The digital transformation of traditional business. *MIT Sloan Management Review*, *44*(4), 34–41.

Angalakudati, M., Balwani, S., Calzada, J., Chatterjee, B., Perakis, G., Raad, N., & Uichanco, J. (2014). Business analytics for flexible resource allocation under random emergencies. *Management Science*, *60*(6), 1552–1573. doi:10.1287/mnsc.2014.1919

Angst, C. M., Wowak, K. D., Handley, S. M., & Kelley, K. (2017). Antecedents of Information Systems Sourcing Strategies in US Hospitals: A Longitudinal Study. *Management Information Systems Quarterly*, *41*(4), 1129–1152. doi:10.25300/MISQ/2017/41.4.06

Anussornnitisarn, P., & Nof, S. Y. (2003). E-work: The challenge of the next generation ERP systems. *Production Planning and Control*, *14*(8), 753–765. doi:10.1080/0953728031000164793l

Aqeel-ur-Rehman, Abbasi, A. Z., Islam, N., & Shaikh, Z. A. (2014). A review of wireless sensors and networks' applications in agriculture. *Computer Standards & Interfaces*, *36*(2), 263–270. doi:10.1016/j.csi.2011.03.004

Archer, S., Hull, L., Soukup, T., Mayer, E., Athanasiou, T., Sevdalis, N., & Darzi, A. (2017). Development of a theoretical framework of factors affecting patient safety incident reporting: a theoretical review of the literature. *BMJ open*, *7*(12). Digital entrepreneurship. Retrieved from https://www.slideshare.net/GiovanniConigliaro/digital-entrepreneurship-67720006

Archibugi, D., & Planta, M. (1996). Measuring technological change through patents and innovation surveys. *Technovation*, *16*(9), 451519–468. doi:10.1016/0166-4972(96)00031-4

Armano, D. (2011). *Pillars of the new influence*. Retrieved from https://hbr.org/2011/01/the-six-pillars-of-the-new-inf

Armitage, C. J., Armitage, C. J., Conner, M., Loach, J., & Willetts, D. (1999). Different perceptions of control: Applying an extended theory of planned behavior to legal and illegal drug use. *Basic and Applied Social Psychology*, *21*(4), 301–316. doi:10.1207/S15324834BASP2104_4

Arnesen, S. (2013). Is a Cloud ERP Solution Right for You? *Strategic Finance*, (February), 45-50.

Arthur, W. B. (2011). The second economy. *The McKinsey Quarterly*, *4*, 90–99.

Artificial Intelligence. (n.d.). Retrieved from https://www.aiforhumanity.fr/

ARVALIS. (2005). *Livre blanc réduction des charges*. ARVALIS.

Athreye, S., & Keeble, D. (2000). Technological convergence, globalisation and ownership in the UK computer industry. *Technovation*, *20*(5), 227–245. doi:10.1016/S0166-4972(99)00135-2

Autio, E. (2017). *Digitalisation, Ecosystems, Entrepreneurship and Policy. Perspectives into Topical Issues Is Society and Ways to Support Political Decision Making. Government's Analysis, Research and Assessment Activities Policy Brief 20/2017*. Helsinki: Prime Minister's Office.

Autio, E., Nambisan, S., Thomas, L. D., & Wright, M. (2017). Digital affordances, spatial affordances, and the genesis of entrepreneurial ecosystems. *Strategic Entrepreneurship Journal*, *12*(1), 72–95. doi:10.1002ej.1266

Avison, D., Jones, J., Powell, P., & Wilson, D. (2004). Using and validating the strategic alignment model. *The Journal of Strategic Information Systems*, *13*(3), 223–246. doi:10.1016/j.jsis.2004.08.002

Awa, H. O., Ojiabo, O. U., & Emecheta, B. C. (2015). Integrating TAM, TPB and TOE frameworks and expanding their characteristic constructs for e-commerce adoption by SMEs. *Journal of Science & Technology Policy Management*, *6*(1), 76–94. doi:10.1108/JSTPM-04-2014-0012

Axon Marketing & Communication. (2018). *Influencer marketing trends in the tourism sector for 2018*. Retrieved from https://axonlatam.lpages.co/white-paper-influencer-marketing-trends-in-the-tourism-sector-for-2018/

Bagchi, S., Kanungo, S., & Dasgupta, S. (2003). Modeling use of enterprise resource planning systems: A path analytic study. *European Journal of Information Systems*, *12*(2), 142–158. doi:10.1057/palgrave.ejis.3000453

Ballings, M., & Van den Poel, D. (2015). CRM in social media: Predicting increases in Facebook usage frequency. *European Journal of Operational Research*, *244*(1), 248–260. doi:10.1016/j.ejor.2015.01.001

Bamoriya, D., & Singh, P. (2012). Mobile banking in India: Barriers in adoption and service preferences. *Journal of Management*, *5*(1), 1–7.

Banker, R. D., Bardhan, I. R., Chang, H., & Lin, S. (2006). Plant Information Systems, Manufacturing Capabilities, and Plant Performance. *Management Information Systems Quarterly*, *30*(2), 315–337. doi:10.2307/25148733

Banker, R. D., Chang, H., & Kao, Y. (2010). Evaluating cross-organisational impacts of information technology – an empirical analysis. *European Journal of Information Systems*, *19*(2), 153–167. doi:10.1057/ejis.2010.9

Bansal, A., & Bagadia, P. (2018). The Effect of Financial Risk Tolerance on Adoption of Mobile Banking in India: A Study of Mobile Banking Users. *IUP Journal of Bank Management*, *17*(1), 50–76.

Baran, R. J., & Galka, R. J. (2017). *Customer Relationship Management: the foundation of contemporary marketing strategy* (2nd ed.). Routledge.

Barney, J. B. (1991). Firm resources and sustained competitive advantage. *Journal of Management*, *17*(1), 99–120. doi:10.1177/014920639101700108

Barone, L. (2013). *The 5 types of influencers on the Web*. Retrieved from http://smallbiztrends.com/2010/07/-the-5-types-of-influencers-on-the-web.html

Barton, D., & Court, D. (2012). Making advanced analytics work for you. *Harvard Business Review*, *90*(10), 78. PMID:23074867

Barton, D., & Court, D. (2012). Making advanced analytics work for you. *Harvard Business Review*, *90*(10), 78–83. PMID:23074867

Bartosova, D. (2011). The future of the media professions: Current issues in media management practice. *International Journal on Media Management*, *13*(3), 195–203. doi:10.1080/14241277.2011.576963

Bass, T., & Mabry, R. (2004) Enterprise architecture reference models: a shared vision for Service-Oriented Architectures. In *Proceedings of the IEEE MILCOM* (pp. 1-8).

Bayles, M. J. (2002). Designing online banner advertisements: Should we animate? *Proceedings of CHI*, 363–366.

Beatty, R. C., & Williams, C. D. (2006). ERP II: Best practices for successfully implementing an ERP upgrade. *Communications of the ACM*, *49*(3), 105–109. doi:10.1145/1118178.1118184

Bechthold, L., Fischer, V., Hainzlmaier, A., Hugenroth, D., Ivanova, L., Kroth, K., & Sitzmann, V. (2015). 3D printing: A qualitative assessment of applications, recent trends and the technology's future potential. *Studien zum deutschen Innovations system.*

Beelders, T., & Bergh, L. (2014). Age as differentiator in online advertising gaze patterns. In *Proceedings of the Southern African Institute for Computer Scientist and Information Technologists Annual Conference 2014 on SAICSIT 2014 Empowered by Technology (SAICSIT '14)*. ACM. doi:10.1145/2664591.2664595

Bencini, L., Maddio, S., Collodi, G., Di Palma, D., Manes, G., & Manes, A. (2012). Development of Wireless Sensor Networks for Agricultural Monitoring. In Smart Sensing Technology for Agriculture and Environmental Monitoring (pp. 157-86). Springer. doi:10.1007/978-3-642-27638-5_9

Benlian, A., & Hess, T. (2011). Comparing the relative importance of evaluation criteria in proprietary and open-source enterprise application software selection – a conjoint study of ERP and Office systems. *Information Systems Journal*, *21*(6), 503–525. doi:10.1111/j.1365-2575.2010.00357.x

Benway, J., & Lane, D. (1998). Banner blindness: Web searchers often miss 'obvious' links. *Internetworking*, *1*, 3.

Beretta, S. (2002). Unleashing the integration potential of ERP systems. *Business Process Management Journal*, *8*(3), 254–277. doi:10.1108/14637150210428961

Berger, E. S. C., & Kuckertz, A. (2016). The Challenge of Dealing with Complexity in Entrepreneurship, Innovation and Technology Research – An Introduction. In E. S. C. Berger & A. Kuckertz (Eds.), *Complexity in entrepreneurship, innovation and technology research – Applications of emergent and neglected methods* (pp. 1–9). Cham, Switzerland: Springer International Publishing.

Berger, P., & Luckman, T. (1971). *The Social Construction of Reality*. Harmondsworth: Penguin.

Berman, S. J. (2012). Digital transformation: Opportunities to create new business models. *Strategy and Leadership*, *40*(2), 16–24. doi:10.1108/10878571211209314

Berry, L. L., & Parasuraman, A. (1993). Building a new academic field-The case of services marketing. *Journal of Retailing*, *69*(1), 13–60. doi:10.1016/S0022-4359(05)80003-X

Berthet, E. (2014). *Concevoir l'écosystème, un défi pour l'agriculture*. Paris: Presses des Mines.

Berthon, B., Hintermann, F., & Berjoan, S. (2014). *The promise of digital entrepreneurs*. Accenture.

Bessen, J., Ford, J., & Meurer, M. M. (2012). The private and social costs of patent trolls: Do nonpracticing entities benefit society by facilitating markets for technology? *Regulation*, *34*(4), 26–35.

Beuret, J. E. (2002). À qui appartient le paysage? To whom does the landscape belong? *Nature Sciences Sociétés*, *10*(2), 47–53. doi:10.1016/S1240-1307(02)80070-X

Beza, E., Steinke, J., Etten, J. V., Reidsma, P., Fadda, C., Mittra, S., ... Kooistra, L. (2017). What are the prospects for citizen science in agriculture? Evidence from three continents on motivation and mobile telephone use of resource-poor farmers. *PLoS One*, *12*(5), e0175700. doi:10.1371/journal.pone.0175700 PMID:28472823

Bharadwaj, A., El Sawy, O. A., Pavlou, P. A., & Venkatraman, N. (2013). Digital Business Strategy: Toward a Next Generation of Insights. *Management Information Systems Quarterly*, *37*(2), 471–482. doi:10.25300/MISQ/2013/37:2.3

Bhardwaj, G., Agrawal, A., & Tyagi, R. (2015). Combination therapies or standalone interventions? Innovation Options for pharmaceutical firms fighting cancer. *International Journal of Innovation Management Vol.*, *19*(03), 1540003. doi:10.1142/S1363919615400034

Bidar, R., Fard, M. B., Salman, Y. B., Tunga, M. A., & Cheng, H. I. (2014). Factors affecting the adoption of mobile banking: sample of Turkey. *16th International Conference on Advanced Communication Technology (ICACT)*. 10.1109/ICACT.2014.6779165

Biggart, N. W. (1977). The creative-destructive process of organizational change: The case of the post office author. *Administrative Science Quarterly*, *22*(3), 410–426. doi:10.2307/2392181

Biloslavo, R., Bagnoli, C., & Edgar, D. (2018). An eco-critical perspective on business models: The value triangle as an approach to closing the sustainability gap. *Journal of Cleaner Production*, *174*, 746–762. doi:10.1016/j.jclepro.2017.10.281

Binder, M., & Clegg, B. T. (2006). A conceptual framework for enterprise management. *International Journal of Production Research*, *44*(18/19), 3813–3829. doi:10.1080/00207540600786673

Binder, M., & Clegg, B. T. (2007). Enterprise management: A new frontier for organisations. *International Journal of Production Economics*, *106*(2), 406–430. doi:10.1016/j.ijpe.2006.07.006

Bingi, P., Sharma, M., & Godla, J. (1999). Critical issues affecting an ERP implementation. *Information Systems Management*, *16*(3), 7–14. doi:10.1201/1078/43197.16.3.19990601/31310.2

Birch, D. G. W. (1999). Mobile financial services: The Internet isn't the only digital channel to consumers. *Journal of Internet Banking and Commerce*, *4*(1), 20–29.

Bitner, M. J., Faranda, W. T., Hubbert, A. R., & Zeithaml, V. A. (1997). Customer contributions and roles in service delivery. *International Journal of Service Industry Management*, *8*(3), 193–205. doi:10.1108/09564239710185398

Blackstone, J. H. Jr. & Cox, J.F. (2005) APICS Dictionary (11th ed.). Chicago, IL: APICS: The association for Operations Management.

Bland, S., & Conner, B. (2015). Mapping out the additive manufacturing landscape. *Metal Powder Report, 70*(3), 115–119. doi:10.1016/j.mprp.2014.12.052

Blascheck, T., Kurzhals, K., & Raschke, M. (2014). State-of-the-art of visualization for eye tracking data. *Proc EuroVis.*

Bock, R., Iansiti M., & Lakhani, K. (2017, January 31). What the companies on the right side of the digital business divide have in common. *Harvard Business Review Digital Articles.*

Bocken, N. M. P., Short, S. W., Rana, P., & Evans, S. (2014). A literature and practice review to develop sustainable Business Model archetypes. *Journal of Cleaner Production, 65*(15), 42–56. doi:10.1016/j.jclepro.2013.11.039

Bond, B., Genovese, Y., Miklovic, D., Wood, N., Zrimsek, B., & Rayner, N. (2000) ERP is dead - long live ERPII, Retrieved from www.pentaprise.de/cms_showpdf.php?pdfname=infoc_report

Boons, F., & Lüdeke-Freund, F. (2013). Business Models for sustainable innovation: State of the art and steps towards a research agenda. *Journal of Cleaner Production, 45*, 9–19. doi:10.1016/j.jclepro.2012.07.007

Booth, N., & Matic, J. A. (2011). Mapping and leveraging influencers in social media to shape corporate brand perceptions. *Corporate Communications, 16*(3), 184–191. doi:10.1108/13563281111156853

Borés, C., Saurina, C., & Torres, R. (2003). Technological convergence: A strategic perspective. *Technovation, 23*(1), 1–13. doi:10.1016/S0166-4972(01)00094-3

Borys, M., & Plechawska-Wójcik, M. (2017). Eye-tracking metrics in perception and visual attention research. *European Journal of Medical Technologies, 3*(16), 11–23. Retrieved from http://www.medical-technologies.eu/upload/2_eye-tracking_metrics_in_perception_-_borys.pdf

Bose, I., & Chen, X. (2009). Quantitative models for direct marketing: A review from systems perspective. *European Journal of Operational Research, 195*(1), 1–16. doi:10.1016/j.ejor.2008.04.006

Bose, I., Pal, R., & Ye, A. (2008). ERP and SCM systems integration: The case of a valve manufacturer in China. *Information & Management, 45*(4), 233–241. doi:10.1016/j.im.2008.02.006

Bott, J., & Milkau, U. (2014). Mobile wallets and current accounts: Friends or foes? *Journal of Payments Strategy and Systems, 8*(3), 6–19.

Bounfour, A. (2016). *Digital Futures, Digital Transformation*. Cham: Springer. doi:10.1007/978-3-319-23279-9

Bower, J. L., & Christensen, C. M. (2015). Disruptive technologies: Catching the wave. *Harvard Business Review, 12*, 350–365.

Boyd, D., & Ellison, N. (2008). Social Network Sites: Definition, History, and Scholarship. *Journal of Computer-Mediated Communication, 13*(1), 210–230. doi:10.1111/j.1083-6101.2007.00393.x

Brajnik, G., & Gabrielli, S. (2010). A review of online advertising effects on the user experience. *International Journal of Human-Computer Interaction, 26*(10), 971–997. doi:10.1080/10447318.2010.502100

Breschi, S., Lissoni, F., & Malerba, F. (2003). Knowledge-relatedness in firm technological diversification. *Research Policy, 32*(1), 69–87. doi:10.1016/S0048-7333(02)00004-5

Brettel, M., Friederichsen, N., Keller, M. and Rosenberg, M., How virtualisation, decentralisation and network building change the manufacturing landscape: an industry 4.0 perspective". *International Journal of Mechanical, Aerospace, Industrial, Mechatronic and Manufacturing Engineering*, *8*(1) 37-44.

Brown, C. V., & Magill, S. L. (1994). *Alignment* of the IS functions with the enterprise: Toward a model of antecedents. *Management Information Systems Quarterly*, *18*(4), 371–403. doi:10.2307/249521

Brown, D., & Hayes, N. (2008). *Influencer Marketing. Who really influences your customers?* Oxford, UK: Butterwort.

Brown, S. L., & Eisenhardt, K. M. (1997). The art of continuous change: Linking complexity theory and time-paced evolution in relentlessly shifting organizations. *Administrative Science Quarterly*, *42*(1), 1–34. doi:10.2307/2393807

Brun, F., & Haezebrouck, T. P. (2017). AgTech – Digital Agriculture - Current development in France, Big Data, a multiscale solution for a sustainable agriculture. *Business Dictionary*. Retrieved from http://www.businessdictionary.com/definition/digitalization.html

Brunswicker, S., Bertino, E., & Matei, S. (2015). Big data for open digital innovation–a research roadmap. *Big Data Research*, *2*(2), 53–58. doi:10.1016/j.bdr.2015.01.008

Brunswick, G. (2014). A Chronology Of The Definition Of Marketing. *Journal of Business & Economics Research*, *12*(2), 105–114. doi:10.19030/jber.v12i2.8523

Bruyat, C., & Julien, P. A. (2001). Defining the field of research in entrepreneurship. *Journal of Business Venturing*, *16*(2), 165–180.

Brynjolfsson, E., & McAfee, A. (2014). *The Second Machine Age: Work, Progress, and Prosperity in a Time of Brilliant Technologies*. New York: WW Norton & Company.

Brynjolfsson, E., & McAfee, A. (2014). *The second machine age: Work, progress, and prosperity in a time of brilliant technologies*. WW Norton & Company.

Bühler, È. A., & Raymond, R. (2012). Pratiques agricoles et gestion territoriale de la biodiversité en contexte de grandes cultures. *Revue Geographique des Pyrenees et du Sud-Ouest*, 65–78. doi:10.4000oe.120

Burk, D. L. (2013). Patent reform in the US: Lessons learned. *Regulation*, *35*(4), 20–25.

Buscher, G., Cutrell, E., & Morris, M. R. (2009). What Do You See When You're Surfing? Using EyeTracking to Predict Salient Regions of Web Pages. *Proceedings of the 27th International Conference on Human Factors in Computing Systems*, 21–30.

Busch, O. (2016). *Programmatic Advertising: The Successful Transformation to Automated, Data-Driven Marketing in Real-Time*. Cham, Switzerland: Springer International. doi:10.1007/978-3-319-25023-6

Buterin, V. (2014), A next generation smart contract and decentralized platform (White Paper). *Ethereum*. Retrieved from https://cryptorating.eu/whitepapers/Ethereum/Ethereum_white_paper.pdf

Bylinskii, Z., & Borkin, M. A. (2015). *Eye fixation metrics for large scale analysis of information visualizations*. ETVIS Work. Eye Track. Vis.

Byrne, J. A., & Brandt, R. (1993, February 8). The virtual corporation. *Business Week*, 36-41.

Calcagno, M. (2008). An investigation into analyzing patents by chemical structure using Thomson's Derwent World Patent Index codes. *World Patent Information*, *30*(3), 188–198. doi:10.1016/j.wpi.2007.10.007

Calder, B., Malthouse, E., & Schaedel, U. (2009). An Experimental Study of the Relationship between Online Engagement and Advertising Effectiveness. *Journal of Interactive Marketing*, *23*(4), 321–331. doi:10.1016/j.intmar.2009.07.002

Camagni, R. (2017). Technological change, uncertainty and innovation systems: Toward a dynamic theory of economic space. In R. Capello (Ed.), *Seminal Studies in Regional and Urban Economics* (pp. 65–97). doi:10.1007/978-3-319-57807-1_4

Cambra-Fierro, J., Florin, J., Perez, L., & Whitelock, J. (2011). Inter-firm market orientation as antecedent of knowledge transfer, innovation and value creation in networks. *Management Decision*, *49*(3), 444–467. doi:10.1108/00251741111120798

Cambridge English Dictionary. (n.d.). *Text ad meaning*. Retrieved from https://dictionary.cambridge.org/dictionary/english/text-ad

Candido, G., Barata, J., Colombo, A. W., & Jammes, F. (2009). SOA in reconfigurable supply chain: A research roadmap. *Engineering Applications of Artificial Intelligence*, *22*(6), 939–949. doi:10.1016/j.engappai.2008.10.020

Cao, M., & Zhang, Q. (2011). Supply chain collaboration: Impact on collaborative advantage and firm performance. *Journal of Operations Management*, *29*(3), 163–180. doi:10.1016/j.jom.2010.12.008

Capdevila, I. (2015). Co-working spaces and the localized dynamics of innovation in Barcelona. *International Journal of Innovation Management*, *19*(03), 1540004. doi:10.1142/S1363919615400046

Carayannis, E. G., Grigoroudis, E., Campbell, D. F., Meissner, D., & Stamati, D. (2018). The ecosystem as helix: An exploratory theory-building study of regional co-opetitive entrepreneurial ecosystems as Quadruple/Quintuple Helix Innovation Models. *R & D Management*, *48*(1), 148–162. doi:10.1111/radm.12300

Caroux, D., Dubois, M. J. F., & Sauvée, L. (2018). *Evolution agro-techniques contemporaine II. Transformations de l'agro-machinisme: fonction, puissance, information, invention*. U. T. B. M. Belfort, Ed.

Carvalho, L. C. (2017). Entrepreneurial Ecosystems: Lisbon as a smart start-up city. In L. C. Carvalho (Ed.), *Handbook of Research on Entrepreneurial Development and Innovation Within Smart Cities*. IGI Global. doi:10.4018/978-1-5225-1978-2.ch001

Casaló, L. V., Flavián, C., Guinalíu, M., & Ekinci, Y. (2015). Do online hotel rating schemes influence booking behaviors? *International Journal of Hospitality Management*, *49*, 28–36. doi:10.1016/j.ijhm.2015.05.005

Castelló-Martínez, A., & del Pino-Romero, C. (2015). La comunicación publicitaria con influencers. *READMARKA UIMA*, *14*(1), 21–50.

Castriotta, M., & Di Guardo, M. C. (2016). Disentangling the automotive technology structure: A patent co-citation analysis. *Scientometrics*, *107*(2), 819–837. doi:10.100711192-016-1862-0

Caudron, J., & Van Peteghem, D. (2014). *Digital transformation. A model to master digital disruption*. Duval Union Consulting.

Cave, J., & Jenkin, H. (2012). Social media as influencers of tourism decisions made by backpackers and fully independent travellers. *Indian Journal of Applied Hospitality and Tourism Research*, *4*, 69–80.

Caviggioli, F. (2016). Technology fusion: Identification and analysis of the drivers of technology convergence using patent data. *Technovation*, *55*, 22–32. doi:10.1016/j.technovation.2016.04.003

Chae, B. (2015). Insights from hashtag #supplychain and Twitter Analytics: Considering Twitter and Twitter data for supply chain practice and research. *International Journal of Production Economics*, *165*, 247–259. doi:10.1016/j.ijpe.2014.12.037

Chae, B. K., Olson, D., & Sheu, C. (2013). The impact of supply chain analytics on operational performance: A resource-based view. *International Journal of Production Research*, *52*(16), 4695–4710. doi:10.1080/00207543.2013.861616

Chae, B. K., Yang, C., Olson, D., & Sheu, C. (2014). The impact of advanced analytics and data accuracy on operational performance: A contingent resource based theory (RBT) perspective. *Decision Support Systems*, *59*, 119–126. doi:10.1016/j.dss.2013.10.012

Chae, J. (2017). Media psychology explaining females' envy toward social media influencers. *Media Psychology*, *21*(2), 246–262. doi:10.1080/15213269.2017.1328312

Chandra, Y., & Leenders, M. (2012). User Innovation and Entrepreneurship in the Virtual World: A Study of Second Life Residents. *Technovation*, *32*(7-8), 464–476.

Chartered Association of Business Schools (C-ABS). (n.d.). Academic Journal Guide, 2014 and 2018. Retrieved from https://charteredabs.org/academic-journal-guide-2018/

Chatzigeorgiou, C. (2017). Modelling the impact of social media influencers on behavioural intentions of millennials: The case of tourism in rural areas in Greece. *Journal of Tourism. Heritage & Services Marketing*, *3*(2), 25–29.

Chau, P. Y. K. (1995). Factors Used in the Selection of Packaged Software in Small Businesses: Views of Owners and Managers. *Information & Management*, *29*(2), 71–78. doi:10.1016/0378-7206(95)00016-P

Chen, Ph., & Zhang, Ch.-Y. (2014). Data-intensive applications, challenges, techniques and technologies: A survey on Big Data. Information Sciences, 275(10), 314-347.

Chen, C. (2013). Perceived risk, usage frequency of mobile banking services. *Managing Service Quality*, *23*(5), 410–436. doi:10.1108/MSQ-10-2012-0137

Chen, C. L. P., & Zhang, C. Y. (2014). Data-intensive applications, challenges, techniques and technologies: A survey on big data. *Information Sciences*, *275*, 314–347. doi:10.1016/j.ins.2014.01.015

Chen, D., Doumeingts, G., & Vernadat, F. (2008). Architectures for enterprise integration and interoperability: Past, present and future. *Computers in Industry*, *59*(7), 647–659. doi:10.1016/j.compind.2007.12.016

Chengalur-Smith, I., Duchessi, P., & Gil-Garcia, J. R. (2012). Information sharing and business systems leveraging in supply chain: An empirical investigation of one web-based application. *Information & Management*, *49*(1), 58–67. doi:10.1016/j.im.2011.12.001

Cheng, Y., Tao, F., Xu, L., & Zhao, D. (2016). Advanced manufacturing systems: Supply-demand matching of manufacturing resource based on complex networks and Internet of Things. *Enterprise Information Systems*, 1751–7575.

Chen, I. J. (2001). Planning for ERP systems: Analysis and future trend. *Business Process Management Journal*, *7*(5), 374–386. doi:10.1108/14637150110406768

Chen, X. W., & Lin, A. X. (2014). Big data deep learning: Challenges and perspectives. *IEEE Access: Practical Innovations, Open Solutions*, *2*, 514–526. doi:10.1109/ACCESS.2014.2325029

Chen, Y. S. (2011). Using patent analysis to explore corporate growth. *Scientometrics*, *88*(2), 433–448. doi:10.100711192-011-0396-8

Chen, Z. Y., Fan, Z. P., & Sun, M. (2015). Behavior-aware user response modeling in social media: Learning from diverse heterogeneous data. *European Journal of Operational Research*, *241*(2), 422–434. doi:10.1016/j.ejor.2014.09.008

Chesbrough, H. W. (2003). *Open innovation: The new imperative for creating and profiting from technology*. Boston: Harvard Business School Press.

Chesbrough, H. W. (2007a). Business Model innovation: It's not just about technology anymore. *Strategy and Leadership*, *35*(6), 12–17. doi:10.1108/10878570710833714

Chesbrough, H. W. (2007b). Why companies should have open Business Models. *MIT Sloan Management Review*, *48*(2), 22–28.

Chesbrough, H. W. (2010). Business Model innovation: Opportunities and barriers. *Long Range Planning*, *43*(2), 354–363. doi:10.1016/j.lrp.2009.07.010

Chiappini, S., & Toccaceli, D. (2013). The relevance of district contexts in the utilization of rural development policies: experience from Italy. *International Agricultural Policy*, 4. Retrieved from https://ageconsearch.umn.edu/bitstream/190606/2/3.pdf

Chi, L., Ravichandran, T., & Andrevski, G. (2010). Information technology, network structure, and competitive action. *Information Systems Research*, *21*(3), 543–570. doi:10.1287/isre.1100.0296

Chinta, R., & Sussan, F. (2018). A triple-helix ecosystem for entrepreneurship: a case review. In *Entrepreneurial Ecosystems* (pp. 67–80). Cham: Springer. doi:10.1007/978-3-319-63531-6_4

Choi, J. Y., Das, S., Theodore, N. D., Kim, I., Honsberg, C., Choi, H. W., & Alford, T. L. (2015). Advances in 2D/3D printing of functional nanomaterials and their applications. *ECS Journal of Solid State Science and Technology*, *4*(4), 3001–P3009. doi:10.1149/2.0011504jss

Choi, T. Y., & Wu, Z. (2009). Triads in supply networks: Theorizing buyer-supplier-supplier relationships. *The Journal of Supply Chain Management*, *45*(1), 8–25. doi:10.1111/j.1745-493X.2009.03151.x

Chongwatpol, J. (2015). Integration of RFID and business analytics for trade show exhibitors. *European Journal of Operational Research*, *244*(2), 662–673. doi:10.1016/j.ejor.2015.01.054

Chou, D. C. (2014). Cloud Computing: A Value Creation Model. *Computer Standards & Interfaces*. doi:10.1016/j.csi.2014.10.001

Clegg, B., & Wan, Y. (2013). ERP systems and enterprise management trends: A contingency model for the enterprization of operations. *International Journal of Operations & Production Management*, *33*(11/12), 1458–1489. doi:10.1108/IJOPM-07-2010-0201

Coelho, F., & Easingwood, C. (2003). Multiple channel structures in financial services: A framework. *Journal of Financial Services Marketing*, *8*(1), 22–34. doi:10.1057/palgrave.fsm.4770104

Cohendet, P., & Pénin, J. (2011). Patents to exclude vs. include: Rethinking the management of intellectual property rights in a knowledge-based economy. *Technology Innovation Management Review*, *1*(3), 12–17. doi:10.22215/timreview/502

CoinDesk. (2016). CoinDesk's State of Bitcoin.

Coltman, T., Tallon, P., Sharma, R., & Queiroz, M. (2015). Strategic IT alignment: Twentyfive years on. *Journal of Information Technology*, *30*(2), 91–100. doi:10.1057/jit.2014.35

Columbus, L. (2015). Ten ways big data is revolutionizing supply chain management. *Forbes*. Retrieved from http://www.forbes.com/sites/louiscolumbus/2015/07/13/ten-ways-big-data-is-revolutionizing-supply-chain-management/

Condea, C., Thiesse, F., & Fleisch, E. (2012). RFID-enabled shelf replenishment with backroom monitoring in retail stores. *Decision Support Systems*, *52*(4), 839–849. doi:10.1016/j.dss.2011.11.018

Conversy, S., Hurter, C., & Chatty, S. (2010). A descriptive model of visual scanning. Proc. 3rd BELIV'10 Work. BEyond time errors Nov. Eval. methods. *Information Visualization*, *2010*, 35–42.

Costa, C., & Turvani, M. (2013). *New industrial spaces as sustainable communities: the case of digital incubators. ERSA conference papers*. European Regional Science Association.

Cozmei, C., & Caloian, F. (2012). Additive Manufacturing Flickering at the Beginning of Existence. *Procedia Economics and Finance*, *3*, 457–462. doi:10.1016/S2212-5671(12)00180-3

Crabbe, M., Standing, C., Standing, S., & Karjaluoto, H. (2009). An adoption model for mobile banking in Ghana. *International Journal of Mobile Communications*, *7*(5), 515–543. doi:10.1504/IJMC.2009.024391

Cruz, P., Neto, L. N. F., Munoz-Gallego, P., & Laukkanen, T. (2010). Mobile banking rollout in emerging markets: Evidence from Brazil. *International Journal of Bank Marketing*, *28*(5), 342–371. doi:10.1108/02652321011064881

Cuc, J., Paredes, M., & Ventura, R. (2018). Online platform business models for value co-creation within a digital entrepreneurial ecosystem in B2B settings. In *2018 CBIM International Conference* (p. 150). Academic Press.

Curley, M., & Salmelin, B. (2017). Digital Disruption. In *Open Innovation 2.0: The New Mode of Digital Innovation for Prosperity and Sustainability* (pp. 15–25). Cham: Springer.

Curran, C. S. (2013). *The Anticipation of Converging Industries*. London: Springer. doi:10.1007/978-1-4471-5170-8

Curran, C. S., Bröring, S., & Leker, J. (2010). Anticipating converging industries using publicly available data. *Technological Forecasting and Social Change*, *77*(3), 385–395. doi:10.1016/j.techfore.2009.10.002

Curran, C. S., & Leker, J. (2011). Patent indicators for monitoring convergence–examples from NFF and ICT. *Technological Forecasting and Social Change*, *78*(2), 256–273. doi:10.1016/j.techfore.2010.06.021

Cusumano, M. A. (2017). The sharing economy meets reality. *Communications of the ACM*, *61*(1), 26–28. doi:10.1145/3163905

Cutting, S. T., Meitzen, M. E., Wagner, B. P., Backley, C. W., Crum, C. L., & Switzky, B. (2015). Implications of 3D printing for the United States Postal Service. In Postal and Delivery Innovation in the Digital Economy (pp. 43-54). Springer International Publishing.

D'Aveni, R. A., Dagnino, G. B., & Smith, K. G. (2010). The age of temporary advantage. *Strategic Management Journal*, *31*(13), 1371–1385. doi:10.1002mj.897

Daidj, N. (2015). *Developing Strategic Business Models and Competitive Advantage in the Digital Sector*. Hershey, PA: IGI Global. doi:10.4018/978-1-4666-6513-2

Danaher, B., Huang, Y., Smith, M. D., & Telang, R. (2014). An empirical analysis of digital music bundling strategies. *Management Science*, *60*(6), 1413–1433. doi:10.1287/mnsc.2014.1958

Daniel, E. M., & White, A. (2005). The future of inter-organisational system linkages: Findings of an international delphi study. *European Journal of Information Systems*, *14*(2), 188–203. doi:10.1057/palgrave.ejis.3000529

Danowski, J. A., & Choi, J. H. (1998). Convergence in the information industries. Telecommunications, broadcasting and data processing 1981-1996. In H. Sawhney & G.A. Barnett (Eds), Progress in Communication Sciences (pp. 125-150). Stamford, CT: Ablex Publishing

Dantan, J., Rizzo, D., Fourati-Jamoussi, F., Dubois, M. J. F., & Jaber, M. (2018). *Farmer-oriented innovation: outcomes from a first bootcamp*. Paris: FRA; bit.ly/agid2018

Davenport, T. H. (2006). Competing on Analytics. *Harvard Business Review*, *84*(1), 98–107. PMID:16447373

Davenport, T. H. (2013). Analytics 3.0. *Harvard Business Review*, *91*(12), 64–72.

Davidson, E., & Vaast, E. 2010.Digital Entrepreneurship and its Sociometrical Enactment. *Paper presented at 43rd Hawaii International Conference on System Sciences (HICSS)*, January 5-8.

Davidson, E., & Vaast, E. (2010). Digital entrepreneurship and its sociomaterial enactment. In *Proceedings of the 43rd Hawaii International Conference on System Sciences*. IEEE Computing Society.

Davis, E. W., & Spekman, R. E. (2004) *Extended Enterprise: Gaining Competitive Advantage through Collaborative Supply Chains*. New York, NY: Financial Times Prentice-Hall.

Davis, F. D. (1989). Perceived Usefulness, Perceived Ease of Use, and User Acceptance of Information Technology. *Management Information Systems Quarterly*, *13*(3), 319–340. doi:10.2307/249008

Davison, R. M., Powell, P., & Trauth, E. M. (2012). ISJ inaugural edition. *Information Systems Journal*, *22*(4), 257–260. doi:10.1111/j.1365-2575.2012.00417.x

De Rassenfosse, G., Dernis, H., Guellec, D., Picci, L., & de la Potterie, B. V. P. (2013). The worldwide count of priority patents: A new indicator of inventive activity. *Research Policy*, *42*(3), 720–737. doi:10.1016/j.respol.2012.11.002

de Reuver, M., Sørensen, C., & Basole, R. C. (2018). The digital platform: A research agenda. *Journal of Information Technology*, *33*(2), 124–135. doi:10.105741265-016-0033-3

Deep, A., Guttridge, P., Dani, S., & Burns, N. (2008). Investigating factors affecting ERP selection in made-to-order SME sector. *Journal of Manufacturing Technology Management*, *19*(4), 430–446. doi:10.1108/17410380810869905

Delacroix, E., Parguel, B., & Benoit-Moreau, F. (2018). Digital subsistence entrepreneurs on Facebook. *Technological Forecasting and Social Change*. doi:10.1016/j.techfore.2018.06.018

Dell. (2014). Big Data Use Cases. Retrieved from https://fr.slideshare.net/Dell/big-data-use-cases-36019892

Deloitte, L. L. P. (2016). Blockchain Enigma. Paradox. Opportunity. Retrieved from https://www2.deloitte.com/content/dam/Deloitte/uk/Documents/Innovation/deloitte-uk-blockchain-full-report.pdf

Demirkan, H., Spohrer, J. C., & Welser, J. J. (2016). Digital innovation and strategic transformation. *IT Professional*, *18*(6), 14–18. doi:10.1109/MITP.2016.115

Denison, E. (2017). Driving prosperity in the digital era. The role of digital talent, innovation and entrepreneurship. Retrieved from https://www2.deloitte.com/de/de/pages/about-deloitte/articles/prosperity-digital-era.html

Di Guardo, M. C., Galvagno, M., & Cabiddu, F. (2012). Analysing the intellectual structure of e-service research. [IJESMA]. *International Journal of E-Services and Mobile Applications*, *4*(2), 19–36. doi:10.4018/jesma.2012040102

Di Guardo, M. C., & Harrigan, K. R. (2016). Shaping the path to inventive activity: The role of past experience in R&D alliances. *The Journal of Technology Transfer*, *41*(2), 250–269. doi:10.100710961-015-9409-8

Di Guardo, M. C., Harrigan, K. R., & Marku, E. (2018). M&A and diversification strategies: What effect on quality of inventive activity? *The Journal of Management and Governance*, 1–24. doi:10.100710997-018-9437-5

Di Stefano, G., Gambardella, A., & Verona, G. (2012). Technology push and demand pull perspectives in innovation studies: Current findings and future research directions. *Research Policy*, *41*(8), 1283–1295. doi:10.1016/j.respol.2012.03.021

Digital Skills Academy. (2017). Ten Traits of Successful Digital Entrepreneurs. Retrieved from https://digitalskillsacademy.com/blog/10-traits-of-successful-digital-entrepreneurs

Djamasbi, S., Siegel, M., & Tullis, T. (2011). Visual Hierarchy and Viewing Behavior: An Eye Tracking Study. In J. A. Jacko (Ed.), Lecture Notes in Computer Science: Vol. 6761. *Human-Computer Interaction. Design and Development Approaches. HCI 2011*. Berlin: Springer.

Dobrzykowski, D. D., Leuschner, R., Hong, P. C., & Roh, J. J. (2015). Examining Absorptive Capacity in Supply Chains: Linking Responsive Strategy and Firm Performance. *The Journal of Supply Chain Management*, *51*(4), 3–28. doi:10.1111/jscm.12085

Dörner, K., & Edelman, D. (2015). *What 'digital' really means. McKinsey & Company: Insights & Publications*. Retrieved from http://www.mckinsey.com/insights/high_tech_telecoms_internet/what_digital_really_means?cid=digital-eml-alt-mip-mck-oth-1507

dos Santos, D. A. G., Zen, A. C., & Schmidt, V. K. (2017). Entrepreneurship ecosystems and the stimulus to the creation of innovative business: A case in the App industry in Brazil. *Journal of Research in Business, Economics and Management*, *8*(5), 1537–1543.

DoubleClick Creative Solutions. (n.d.). *How rich media works: Rich media creative types*. Retrieved from https://support.google.com/richmedia/answer/2417545?hl=en

Douthwaite, B., & Hoffecker, E. (2017). Towards a complexity-aware theory of change for participatory research programs working within agricultural innovation systems. *Agricultural Systems*, *155*, 88–102. doi:10.1016/j.agsy.2017.04.002

Downes, L., & Nunes, P. F. (2013). Big-bang disruption. *Harvard Business Review*, *91*(3), 44–56.

Doye, D., Jolly, R., Hornbaker, R., Cross, T., King, R. P., Lazarus, W. F., & Yeboah, A. (2000). Case studies of farmers' use of information systems. *Review of Agricultural Economics*, *22*(2), 566–585. doi:10.1111/1058-7195.00039

Drnevich, P. L., & Croson, D. C. (2013). Information Technology and Business-Level Strategy: Toward an Integrated Theoretical Perspective. *Management Information Systems Quarterly*, *37*(2), 483–509. doi:10.25300/MISQ/2013/37.2.08

Dubois, M., Bobillier Chaumn, M. E., & Retour, D. (2011). The impact of development of customer online banking skills on customer adviser skills. *New Technology, Work and Employment*, *26*(2), 156–173. doi:10.1111/j.1468-005X.2011.00266.x

Durif, F., Geay, B., & Graf, R. (2013). Do key account managers focus too much on commercial performance? A cognitive mapping application. *Journal of Business Research*, *66*(9), 1559–1567. doi:10.1016/j.jbusres.2012.09.019

Durkin, M. G., & Howcroft, J. B. (2003). Relationship marketing in the banking sector: The impact of new technologies. *Marketing Intelligence & Planning*, *21*(1), 61–71. doi:10.1108/02634500310458162

Duta, D., & Bose, I. (2015). Managing a big data project: The case of Ramco Cements Limited. *International Journal of Production Economics*, *165*, 293–306. doi:10.1016/j.ijpe.2014.12.032

Dutta, D. & Bose, I. (2015). Managing a Big Data Project: The Case of Ramco Cements Limited. *International Journal of Production Economics: Manufacturing Systems, Strategy & Design*, 1-51.

Ebben, W. (2013). Will smartphones contribute to making payments easier and more efficient? *Journal of Payments Strategy and Systems*, *7*(1), 11–17.

Ecer, F. (2018). An integrated fuzzy ahp and aras model to evaluate mobile banking services. *Technological and Economic Development of Economy*, *24*(2), 670–695. doi:10.3846/20294913.2016.1255275

Eckartz, S., Daneva, M., Wieringa, R., & Hillegersberg, J. V. (2009) Cross-organisational ERP management: How to create a successful business case? In *SAC'09 Proceedings of the 2009 ACM Symposium on Applied Computing*. Honolulu, HI.

Edmond, R. (2015). Five business opportunities surrounding 3-D printing. *The Channel Co*. Retrieved from http://www.itbestofbreed.com/slide-shows/five-business-opportunities-surrounding-3-d-printing/page/0/1

Eisenhardt, K. M., & Martin, J. A. (2000). Dynamic capabilities: What are they? *Strategic Management Journal*, *21*(10/11), 1105–1121. doi:10.1002/1097-0266(200010/11)21:10/11<1105::AID-SMJ133>3.0.CO;2-E

Ekanem, I. (2005). 'Bootstrapping': The Investment Decision-Making Process in Small Firms. *The British Accounting Review*, *37*(3), 299–318. doi:10.1016/j.bar.2005.04.004

Ekufu, T. K. (2012). *Predicting cloud computing technology adoption by organizations: an empirical integration of technology acceptance model and theory of planned behavior* [Doctoral Dissertation]. Capella University, MN.

Engelsman, E. C., & van Raan, A. F. (1994). A patent-based cartography of technology. *Research Policy*, *23*(1), 1–26. doi:10.1016/0048-7333(94)90024-8

Entrepreneur.com. (October 20, 2016). 7 Ways Entrepreneurs Drive Economic Development. Retrieved from https://www.entrepreneur.com/article/283616

Ericson, J. (2001). What the heck is ERPII? Retrieved from http://www.line56.com/articles/default.asp?ArticleID=2851

Ettlie, J., & Pavlou, P. A. (2006). Technology-Based New Product Development Partnerships. *Decision Sciences*, *37*(2), 117–148. doi:10.1111/j.1540-5915.2006.00119.x

European Commission. (2003). Commission recommendation of 6 May 2003 concerning the definition of micro, small and medium sized enterprises. *Official Journal of the European Union, L, 124*(1422), 36–41.

European Commission. (2014). *Fuelling digital entrepreneurship in Europe, Background paper*. Retrieved from http://ec.europa.eu/geninfo/query/resultaction.jsp?QueryText=EU+vision%2C+strategy+and+actions&query_source=GROWTH&swlang=en&x=18&y=8

European Commission. (2015). *Digital Transformation of European Industry and Enterprises; A report of the Strategic Policy Forum on Digital Entrepreneurship*. Available from: http://ec.europa.eu/DocsRoom/documents/9462/attachments/1/translations/en/renditions/native

European Commission. (2017). *Enterprise and industry directorate-general, "strategic policy forum on digital entrepreneurship."* Available from: https://ec.europa.eu/growth/industry/policy/digital-transformation/strategic-policy-forum-digital-entrepreneurship_en

Evans, D. S., & Schmalensee, R. (2016). *Matchmakers: the new economics of multisided platforms*. Harvard Business Review Press.

Everard, M. (2011). *Common ground: the sharing of land and landscapes for sustainability*. Zed Books.

FAO. (2016). *Produce more with Less*. Available from: http://www.fao.org/ag/save-and-grow/fr/accueil/index.html

Faraday, P. (2000). Visually Critiquing Web Pages. *Proceedings of the 6th Conference on Human Factors and the Web*, 1–13.

Farquhar, J. D., & Rowley, J. (1998). Enhancing the customer experience: Contribution from information technology. *Management Decision*, *36*(5), 350–357. doi:10.1108/00251749810220568

Fast Company. (2014). The world's Top 10 most innovative companies in big data. Retrieved from http://www.fastcompany.com/most-innovative-companies/2014/industry/big-data

Fatanti, M. N., & Suyadnya, I. W. (2015). Beyond User Gaze: How Instagram Creates Tourism Destination Brand? *Procedia: Social and Behavioral Sciences*, *25*, 1089–1095. doi:10.1016/j.sbspro.2015.11.145

Fauscette, M. (2013). ERP in the Cloud and the Modern Business (White paper). *International Data Corporation (IDC)*. Retrieved from http://resources.idgenterprise.com/original

Fawcett, S. E., Fawcett, A. M., Watson, B. J., & Magnan, G. M. (2012). Peeking inside the black box: Toward an understanding of supply chain collaboration dynamics. *The Journal of Supply Chain Management*, *48*(1), 44–72. doi:10.1111/j.1745-493X.2011.03241.x

Fawcett, S. E., & Waller, M. A. (2014). Supply chain game changers—mega, nano, and virtual trends—and forces that impede supply chain design (i.e., building a winning team). *Journal of Business Logistics*, *35*(3), 157–164. doi:10.1111/jbl.12058

Fernandes, V. (2012). (Re)discovering the PLS approach in management science. *M@n@gement*, *15*(1), 101-123.

Fessenden, T. (2017). *The Most Hated Online Advertising Techniques*. Retrieved from https://www.nngroup.com/articles/most-hated-advertising-techniques/

Fessenden, T. (2018). *Scrolling and Attention*. Retrieved from https://www.nngroup.com/articles/scrolling-and-attention/

Fichman, R. G., Dos Santos, B. L., & Zheng, Z. E. (2014). Digital innovation as a fundamental and powerful concept in the information systems curriculum. *Management Information Systems Quarterly*, *38*(2), 329–343. doi:10.25300/MISQ/2014/38.2.01

Fischer, T., & Henkel, J. (2012). Patent Trolls on Markets for Technology– An Empirical Analysis of NPEs' Patent Acquisitions. *Research Policy*, *41*(9), 1519–1533. doi:10.1016/j.respol.2012.05.002

Fishbein, M., & Ajzen, I. (2010). Predicting and changing behavior: The reasoned action approach. New York: Psychology Press.

Fitts, P.M., Jones, R.E., & Milton, J.L. (2005). Eye movements of aircraft pilots during instrument-landing approaches. *Ergon Psychol Mech Model*, 3-56.

Fitzgerald, M., & Kruschwitz, D. B., & Welch, M. (2014). Embracing digital technology: A new strategic imperative. *MIT Sloan Management Review*, *55*(2), 1–12.

Fitzgerald, M., Kruschwitz, N., Bonnet, D., & Welch, M. (2014). Embracing digital technology: A new strategic imperative. *MIT Sloan Management Review*, *55*(2), 1.

Fleming, L. (2001). Recombinant uncertainty in technological search. *Management Science*, *47*(1), 117–132. doi:10.1287/mnsc.47.1.117.10671

Forrester, D. (2017). Digital knowledge manager: 5 Skills You Need to Succeed at the Newest Marketing Role. Retrieved from https://www.entrepreneur.com/article/299178

Foss, N. J., & Saebi, T. (2015). Business Models and Business Model innovation: bringing organization into the field. In N. J. Foss & T. Saebi (Eds.), *Business Model Innovation: the Organizational Dimension* (pp. 1–23). Oxford, UK: Oxford University Press. doi:10.1093/acprof:oso/9780198701873.003.0001

Foss, N. J., & Saebi, T. (2017). Fifteen years of research on Business Model innovation: How far have we come, and where should we go? *Journal of Management*, *43*(1), 200–227. doi:10.1177/0149206316675927

Foss, N. J., & Saebi, T. (2018). Business Models and Business Model innovation: Between wicked and paradigmatic problems. *Long Range Planning*, *51*(1), 9–21. doi:10.1016/j.lrp.2017.07.006

Fosso Wamba, S., & Akter, S. (2015). Big data analytics for supply chain management: A literature review and research agenda. In *The 11th International Workshop on Enterprise & Organizational Modeling And Simulation* (EOMAS 2015), Stockholm, Sweden, June 8–9.

Fox, S., & Stucker, B. (2009). *Digiproneurship: New types of physical products and sustainable employment from digital product entrepreneurship*. VTT Working Papers. Finland, VTT Technical Research Centre of Finland: 36.

Gable, G. (1998). Large package software: A neglected technology? *Journal of Global Information Management*, *6*(3), 3–4.

Gambardella, A., & Torrisi, S. (1998). Does technological convergence imply convergence in markets? Evidence from the electronics industry. *Research Policy*, *27*(5), 445–463. doi:10.1016/S0048-7333(98)00062-6

Gandomi, A., & Haider, M. (2015). Beyond the hype: Big data concepts, methods, and analytics. *International Journal of Information Management*, *35*(2), 137–144. doi:10.1016/j.ijinfomgt.2014.10.007

Gangwar, H., Date, H., & Ramaswamy, R. (2015). Understanding determinants of cloud computing adoption using an integrated TAM-TOE model. *Journal of Enterprise Information Management*, *28*(1), 107–130. doi:10.1108/JEIM-08-2013-0065

Gartner. (2017). Big data. Retrieved from http://www.gartner.com/it-glossary/big-data

Gartner. (2017). *Digitalization*. Retrieved from https://www.gartner.com/it-glossary/digitalization

Gartner, W. B., Bird, B. J., & Starr, J. A. (1992). Acting as if: Differentiating entrepreneurial from organizational behavior. *Entrepreneurship Theory and Practice*, *16*(3), 13–32. doi:10.1177/104225879201600302

Gasparin, M., Micheli, R., & Campana, M. (2015). Competing with networks: a case study on the 3D printing. In *Proceeding 1st International Competitiveness Management conference, EIASM*, Copenhagen

Gebauer, H., Johnson, M., & Enquist, B. (2010). Value co-creation as a determinant of success in public transport services. *Managing Service Quality: An International Journal*, *20*(6), 511–530. doi:10.1108/09604521011092866

Gebhart, M., Giessler, P., & Abeck, S. (2016). Challenges of the Digital Transformation in Software Engineering. *ICSEA*, *2016*, 149.

Gefen, D., Straub, D., & Boudreau, M. (2000). Structural equation modeling techniques and regression: Guidelines for research practice. *Communications of AIS*, *4*(7), 2–76.

Gens, F. (2016). *IDC FutureScape: worldwide IT industry 2017 predictions*. Framingham, UK: International Data Corporation.

George, A. L., & Bennett, A. (2005). *Case studies and theory development*. Cambridge, UK: MIT Press.

Geradin, D., Layne-Farrar, A., & Padilla, A. J. (2011). Elves or trolls? The role of nonpracticing patent owners in the innovation economy. *Industrial and Corporate Change*, *21*(1), 73–94. doi:10.1093/icc/dtr031

Gerow, J. E., Grover, V., Thatcher, J. B., & Roth, P. L. (2014). Looking toward the future of IT-business strategic alignment through the past: A meta-analysis. *Management Information Systems Quarterly*, *38*(4), 1059–1085. doi:10.25300/MISQ/2014/38.4.10

Gerow, J. E., Thatcher, J. B., & Grover, V. (2015). Six Types of IT-Business Strategic Alignment: An investigation of the constructs and their measurement. *European Journal of Information Systems*, *24*(5), 465–491. doi:10.1057/ejis.2014.6

Ghezzi, A. (2018). Digital startups and the adoption and implementation of Lean Startup Approaches: Effectuation, Bricolage and Opportunity Creation in practice. *Technological Forecasting and Social Change*. doi:10.1016/j.techfore.2018.09.017

Ghezzi, A., & Cavallo, A. (2018). Agile business model innovation in digital entrepreneurship: Lean Startup approaches. *Journal of Business Research*. doi:10.1016/j.jbusres.2018.06.013

Giones, F., & Brem, A. (2017). Digital Technology Entrepreneurship: A Definition and Research Agenda. *Technology Innovation Management Review*, *7*(5).

Giones, F., & Brem, A. 2017. Digital Technology Entrepreneurship: A Definition and Research Agenda. *Technology Innovation Management Review*, 7(5), 44–51. Retrieved from http://timreview.ca/article/1076

Glaser, B. G., & Strauss, A. L. (1967). *The Discovery of Grounded Theory: Strategies for Qualitative Research*. New York, NY: Aldine.

Gobble, M. M. (2018). Digitalization, Digitization, and Innovation. *Research Technology Management*, *61*(4), 56–59. doi:10.1080/08956308.2018.1471280

Goldberg, J. H., & Kotval, X. P. (1999). Computer interface evaluation using eye movements: Methods and constructs. *International Journal of Industrial Ergonomics*, *24*(6), 631–645. doi:10.1016/S0169-8141(98)00068-7

Goldkind, L., & Wolf, L. (2015). A digital environment approach: Four technologies that will disrupt social work practice. *Social Work*, *60*(1), 85–87. doi:10.1093wwu045 PMID:25643579

Goodhue, D. L., & Thompson, R. L. (1995). Task-technology fit and individual performance. *Management Information Systems Quarterly*, *19*(2), 213–236. doi:10.2307/249689

Gretzel, U. (2006). Consumer generated content -trends and implications for branding. *Ereview of Tourism Research*, *4*(3), 9–11.

Grilo, A., Romero, D., & Cunningham, S. (2016), Digital Entrepreneurship: Creating and Doing Business in the Digital Era, Call for papers, Technological Forecasting and Social Change, Retrieved December 21, 2017, from https://www.journals.elsevier.com/technological-forecasting-and-social-change/call-for-papers/digital-entrepreneurship-creating-and-doing-business-in-the

Groves, W., Collins, J., Gini, M., & Ketter, W. (2014). Agent-assisted supply chain management: Analysis and lessons learned. *Decision Support Systems*, *57*, 274–284. doi:10.1016/j.dss.2013.09.006

Gueydier, P., Pujos, A., & Redien-Collot, R. (2018), Blockchain, the Challenge of Trust, Optic Humana Technologia, http://optictechnology.org/index.php/fr/news-fr/144-revolution-technologique-white-papers-fr

Gu, J. C., Lee, S. C., & Suh, Y. H. (2009). Determinants of behavioral intention to mobile banking. *Expert Systems with Applications*, *36*(9), 11605–11616. doi:10.1016/j.eswa.2009.03.024

Gupta, S., Haejung, Y., Xu, H., & Kim, H. W. (2017). An exploratory study on mobile banking adoption in Indian metropolitan and urban areas: A scenario-based experiment. *Information Technology for Development*, *23*(1), 127–152. doi:10.1080/02681102.2016.1233855

Gupta, S., Misra, S. C., Kock, N., & Roubaud, D. (2018). Organizational, technological and extrinsic factors in the implementation of cloud ERP in SMEs. *Journal of Organizational Change Management*, *31*(1), 83–102. doi:10.1108/JOCM-06-2017-0230

Hacklin, F. (2007). *Management of convergence in innovation: strategies and capabilities for value creation beyond blurring industry boundaries*. Springer Science & Business Media.

Hacklin, F., & Wallin, M. W. (2013). Convergence and interdisciplinarity in innovation management: A review, critique, and future directions. *Service Industries Journal*, *33*(7-8), 774–788. doi:10.1080/02642069.2013.740471

Hagedoorn, J., & Cloodt, M. (2003). Measuring innovative performance: Is there an advantage in using multiple indicators? *Research Policy*, *32*(8), 1365–1379. doi:10.1016/S0048-7333(02)00137-3

Hahn, G. J., & Packowski, J. (2015). A perspective on applications of in-memory analytics in supply chain management. *Decision Support Systems*, *76*, 45–52. doi:10.1016/j.dss.2015.01.003

Hair, J. F., Black, W. C., Babin, B. J., & Anderson, R. E. (2010). *Multivariate data analysis* (7th ed.). Prentice Hall.

Hair, N., Wetsch, L. R., Hull, C. E., Perotti, V., & Hung, Y.-T. C. (2012). Market Orientation in Digital Entrepreneurship: Advantages and Challenges in A Web 2.0 Networked World. *International Journal of Innovation and Technology Management*, *9*(6), 1250045. doi:10.1142/S0219877012500459

Hall, B. H., Jaffe, A. B., & Trajtenberg, M. (2001). *The NBER patent citation data file: Lessons, insights and methodological tools (No. w8498)*. National Bureau of Economic Research. doi:10.3386/w8498

Hamel, G., & Prahalad, C. K. (1989). Strategic intent. *Harvard Business Review*, *67*(3), 63–78. PMID:10303477

Hanafizadeh, P., Behboudi, M., Abedini Koshksaray, A., & Jalilvand Shirkhani Tabar, M. (2014). Mobilebanking adoption by Iranian bank clients. *Telematics and Informatics*, *31*(1), 62–78. doi:10.1016/j.tele.2012.11.001

Handoko, I. P., & Gaol, F. L. (2012). Performance evaluation of CRM system based on cloud computing. *Applied Mechanics and Materials*, *234*, 110–123. doi:10.4028/www.scientific.net/AMM.234.110

Han, E. J., & Sohn, S. Y. (2016). Technological convergence in standards for information and communication technologies. *Technological Forecasting and Social Change*, *106*, 1–10. doi:10.1016/j.techfore.2016.02.003

Hang, C. C., Chen, J., & Yu, D. (2011). An Assessment Framework for Disruptive Innovation. *Foresight*, *13*(5), 4–13. doi:10.1108/14636681111170185

Hanna, N. K. (Ed.). (2016). Mastering digital transformation: Towards a smarter society, economy, city and nation. In Hanna, N. K. (Ed.), Mastering Digital Transformation: Towards a Smarter Society, Economy, City and Nation (i-xxvi). Emerald Group Publishing Limited. doi:10.1108/978-1-78560-465-220151009

Hansen, R., & Pauleit, S. (2014). From Multifunctionality to Multiple Ecosystem Services? A Conceptual Framework for Multifunctionality in Green Infrastructure Planning for Urban Areas. *Ambio*, *43*(4), 516–529. doi:10.100713280-014-0510-2 PMID:24740622

Hansen, R., & Sia, S. K. (2015). Hummel's Digital Transformation Toward Omnichannel Retailing: Key Lessons Learned. *MIS Quarterly Executive*, *14*(2).

Harrigan, K. R. (1985). Vertical integration and corporate strategy. *Academy of Management Journal*, *28*(2), 397–425.

Harrigan, K. R., & Di Guardo, M. C. (2017). Sustainability of patent-based competitive advantage in the US communications services industry. *The Journal of Technology Transfer*, *42*(6), 1334–1361. doi:10.100710961-016-9515-2

Harrigan, K. R., Di Guardo, M. C., & Marku, E. (2018). Patent value and the Tobin's q ratio in media services. *The Journal of Technology Transfer*, *43*(1), 1–19. doi:10.100710961-017-9564-1

Harrigan, K. R., Di Guardo, M. C., Marku, E., & Velez, B. N. (2017). Using a distance measure to operationalise patent originality. *Technology Analysis and Strategic Management*, *29*(9), 988–1001. doi:10.1080/09537325.2016.1260106

Hartmann, P. M., Hartmann, P. M., Zaki, M., Zaki, M., Feldmann, N., Feldmann, N., ... Neely, A. (2016). Capturing value from big data–a taxonomy of data-driven business models used by start-up firms. *International Journal of Operations & Production Management*, *36*(10), 1382–1406. doi:10.1108/IJOPM-02-2014-0098

Hashem, I. A. T., Yaqoob, I., Anuar, N. B., Mokhtar, S., Gani, A., & Khan, S. U. (2015). The rise of "big data" on cloud computing: Review and open research issues. *Information Systems*, *47*, 98–115. doi:10.1016/j.is.2014.07.006

Hatch, M. (2013). *The maker movement manifesto: Rules for innovation in the new world of crafters, hackers, and tinkerers*. New York: McGraw-Hill.

Hazen, H. T., Boone, C. A., Ezell, J. D., & Jones-Farmer, L. A. (2014). Data quality for data science, predictive analytics, and big data in supply chain management: An introduction to the problem and suggestions for research and applications. *International Journal of Production Economics*, *154*, 72–80. doi:10.1016/j.ijpe.2014.04.018

Hedman, J., & Kalling, T. (2003). The Business Model concept: Theoretical underpinnings and empirical illustrations. *European Journal of Information Systems*, *12*(1), 49–59. doi:10.1057/palgrave.ejis.3000446

Helfat, C. E., & Raubitschek, R. S. (2018). Dynamic and integrative capabilities for profiting from innovation in digital platform-based ecosystems. *Research Policy*, *47*(8), 1391–1399. doi:10.1016/j.respol.2018.01.019

Henderson, J., & Venkatraman, N. (1993). Strategic alignment: Leveraging information technology for transforming organizations. *IBM Systems Journal*, *32*(1), 4–16. doi:10.1147j.382.0472

Hendricks, K. B., Singhal, V. R., & Stratman, J. K. (2007). The impact of enterprise systems on corporate performance: A study of ERP, SCM, and CRM system implementations. *Journal of Operations Management*, *25*(1), 65–82. doi:10.1016/j.jom.2006.02.002

Henningsson, S., & Carlsson, S. (2011). The DySIIM model for managing IS integration in mergers and acquisitions. *Information Systems Journal*, *21*(5), 441–476.

Henriette, E., Feki, M., & Boughzala, I. (2015). The shape of digital transformation: a systematic literature review. *MCIS 2015 Proceedings*, 431-443.

Henriette, E., Feki, M., & Boughzala, I. (2016). Digital transformation challenges. *MCIS 2016 Proceedings*, *33*. Retrieved from http://aisel.aisnet.org/mcis2016/33

Hernández, E., Fuentes, M. L., & Marrero, S. (2012). Una aproximación a la reputación en línea de los establecimientos hoteleros españoles. *Papers de Turisme*, *52*, 63–88.

Hernandez, M., Wang, Y., Sheng, H., Kalliny, M., & Minor, M. (2017). Escaping the corner of death? An eye-tracking study of reading direction influence on attention and memory. *Journal of Consumer Marketing*, *34*(1), 1–10. doi:10.1108/JCM-02-2016-1710

Herrera, F., Guerrero, M., & Urbano, D. (2018). Entrepreneurship and Innovation Ecosystem's Drivers: The Role of Higher Education Organizations. In *Entrepreneurial, Innovative and Sustainable Ecosystems* (pp. 109–128). Cham: Springer. doi:10.1007/978-3-319-71014-3_6

Hess, T., Matt, C., Benlian, A., & Wiesböck, F. (2016). Options for Formulating a Digital Transformation Strategy. *MIS Quarterly Executive*, *15*(2).

Hess, T., Matt, C., Wiesböck, F., & Benlian, A. (2016). Options for Formulating a Digital Transformation Strategy. *MIS Quarterly Executive*, *15*(2), 103–119.

Higgins-Desbiolles, F. (2018). Event tourism and event imposition: A critical case study from Kangaroo Island, South Australia. *Tourism Management*, *64*, 73–86. doi:10.1016/j.tourman.2017.08.002

Hileman, G., & Rauchs, M. (2017). *Global Cryptocurrency Benchmarking Study*. Cambridge Judge Business School.

Hinings, B., Gegenhuber, T., & Greenwood, R. (2018). Digital innovation and transformation: An institutional perspective. *Information and Organization*, *28*(1), 52–61. doi:10.1016/j.infoandorg.2018.02.004

Hoehle, H., Scornavacca, E., & Huff, S. (2012). Three decades of research on consumer adoption and utilization of electronic banking channels: A literature analysis. *Decision Support Systems*, *54*(1), 122–132. doi:10.1016/j.dss.2012.04.010

Hofmann, E. (2015). Big data and supply chain decisions: The impact of volume, variety and velocity properties on the bullwhip effect. *International Journal of Production Research*. doi:10.1080/00207543.2015.1061222

Hogarth, R.M., and Soyer, E. (2015). Using Simulated Experience to Make Sense of Big Data. *MIT Sloan Management Review,* (spring), 5-10.

Hogg, G., Laing, A., & Winkelman, D. (2003). The professional service encounter in the age of the Internet: An exploratory study. *Journal of Services Marketing*, *17*(5), 476–494. doi:10.1108/08876040310486276

Hollebeek, L., Glynn, M., & Brodie, R. (2014). Consumer brand engagement in social media: Conceptualization, scale development and validation. *Journal of Interactive Marketing*, *28*(2), 149–165. doi:10.1016/j.intmar.2013.12.002

Höller, N., Schrammel, J., Tscheligi, M., & Paletta, L. (2010). The perception of information and advertisement screens mounted in public transportation vehicles – Results from a mobile eye-tracking study. *Lecture Notes in Informatics*, 4007-4021. Retrieved from https://www.tobiipro.com/fields-of-use/marketing-consumer-research/

Hollister, S. J., Flanagan, C. L., Zopf, D. A., Morrison, R. J., Nasser, H., Patel, J. J., ... Green, G. E. (2015). Design control for clinical translation of 3D printed modular scaffolds. *Annals of Biomedical Engineering*, *43*(3), 774–786. doi:10.100710439-015-1270-2 PMID:25666115

Hoopes, D. G., Madsen, T. L., & Walker, G. (2003). Guest Editors' Introduction to the Special Issue: Why is there a Resource-Based View? Toward a Theory of Competitive Heterogeneity. *Strategic Management Journal*, *24*(10), 889–902. doi:10.1002mj.356

Hsieh, Y.-J., & Wu, Y. (2018). Entrepreneurship through the platform strategy in the digital era: Insights and research opportunities. *Computers in Human Behavior*, 1–9.

Huang, C. H. (2015). Continued evolution of automated manufacturing–cloud-enabled digital manufacturing. *International Journal of Automation and Smart Technology*, *5*(1), 2–5. doi:10.5875/ausmt.v5i1.861

Huang, Y., Leu, M. C., Mazumder, J., & Donmez, A. (2015). Additive manufacturing: Current state, future potential, gaps and needs, and recommendations. *Journal of Manufacturing Science and Engineering*, *137*(1), 014001. doi:10.1115/1.4028725

Hui, G. (2014). How the Internet of Things changes Business Models. *Harvard Business Review*, 29.

Hull, C. E., Hung, Y.-T. C., Hair, N., Perotti, V., & DeMartino, R. (2007). Taking advantage of digital opportunities: A typology of digital entrepreneurship. *International Journal of Networking and Virtual Organisations*, *4*(3), 290–303. doi:10.1504/IJNVO.2007.015166

Hunter, L. W., Bernhardt, A., Hughes, K. L., & Skuratowicz, E. (2001). It's not just the ATMs: Technology, firm strategies, jobs, and earnings in retail banking. *Industrial & Labor Relations Review*, *54*(2), 402–424. doi:10.1177/001979390105400222

Hu, P. J., Chau, P. Y., Sheng, O. R. L., & Tam, K. Y. (1999). Examining the technology acceptance model using physician acceptance of telemedicine technology. *Journal of Management Information Systems*, *16*(2), 91–112. doi:10.1080/07421222.1999.11518247

Iacocca Institute. (1991). *21st century manufacturing enterprise strategy*. Bethlehem, PA: Lehigh University.

Iansiti, M., & Lakhani, K. R. (2014). Digital Ubiquity: How Connections, Sensors, and Data Are Revolutionizing Business. *Harvard Business Review*, *92*(11), 91–99.

IDC. (2017). Worldwide Semiannual Big Data and Analytics Spending Guide. Retrieved from https://www.idc.com/getdoc.jsp?containerId=prUS42371417

Illia, A., Ngniatedema, T., & Huang, Z. (2015). A conceptual model for mobile banking adoption. *Journal of Management Information and Decision Sciences*, *18*(1), 111–122.

Interactive Advertising Bureau. (n.d.). *Standards, Guidelines, and Best Practice Documents in Public Comment*. Retrieved from https://www.iab.com/guidelines/iab-standards-guidelines-best-practice-documents-in-public-comment/

i-SCOOP. (2016). *Digitization, digitalization and digital transformation: the differences*. Retrieved from https://www.iscoop.eu/digitization-digitalization-digital-transformation-disruption/

Isenberg, D. (2011). The entrepreneurship ecosystem strategy as a new paradigm for economic policy: Principles for cultivating entrepreneurship. *Presentation at the Institute of International and European Affairs*. Retrieved from https://doc.uments.com/download/s-the-entrepreneurship-ecosystem-strategy-as-a-new-paradigm-for.pdf

Isenberg, D. J. (2010). How to start an entrepreneurial revolution. *Harvard Business Review*, *88*(6), 40–50.

Issacs, L. (2013). Rolling the Dice with Predictive Coding: Leveraging Analytics Technology for Information Governance. *Information & Management*, *47*(1), 22–26.

Ivens, B. S. (2004). How relevant are different forms of relational behavior? An empirical test based on Macneil's exchange framework. *Journal of Business and Industrial Marketing*, *19*(5), 300–309. doi:10.1108/08858620410549929

Iwai, K. (2000). A contribution to the evolutionary theory of innovation, imitation and growth. *Journal of Economic Behavior & Organization*, *43*(2), 167–198. doi:10.1016/S0167-2681(00)00115-3

Iyiola, O., & Akintunde, O. (2011). Perceptions as influencer of consumer choice behavior: The case of tourism in Nigeria. *Journal of Marketing Development and Competitiveness*, *5*(7), 27–36.

Jaffe, A. B. (1986). *Technological opportunity and spillovers of R&D: evidence from firms' patents, profits and market value*. Academic Press.

Javalgi, R. G., Todd, P. R., Johnston, W. J., & Granot, E. (2012). Entrepreneurship, muddling through, and Indian Internet-enabled SMEs. *Journal of Business Research*, *65*(6), 740–744. doi:10.1016/j.jbusres.2010.12.010

Jeong, S., Kim, J. C., & Choi, J. Y. (2015). Technology convergence: What developmental stage are we in? *Scientometrics*, *104*(3), 841–871. doi:10.100711192-015-1606-6

Jha, M., Jha, S., & O'Brien, L. (2016). Combining Big Data Analytics with Business Process using Reengineering.

John, A. J., Clark, C. E. F., Freeman, M. J., Kerrisk, K. L., Garcia S. C., & Halachmi I. (2016). Review: Milking robot utilization, a successful precision livestock farming evolution. *Animal*, *10*(9), 1484-1492.

Johnson, M. W., Christensen, C. M., & Kagermann, H. (2008). Reinventing your Business Model. *Harvard Business Review*, *86*, 50–59. PMID:18681297

Johnston, M. (2015). *What is (and isn't) marketing?* Retrieved from http://regions.cim.co.uk/south-east/home/news/2015/january/what-is-marketing

Joskow, P. L. (2003). Vertical integration. In *Handbook of New Institutional Economics*. Boston, MA: Kluwer.

Julien, P. A., & Raymond, L. (1994). Factors of New Technology Adoption in The Retail Sector. *Entrepreneurship Theory and Practice*, *18*(4), 79–90. doi:10.1177/104225879401800405

Kadamudimatha. (2016). Digital Wallet: The Next Way of Growth. *International Journal of Commerce and Management Research, 2*(12).

Kahn, K. B. (2014). Solving the problems of new product Forecasting. *Business Horizons*, *57*(5), 607–615. doi:10.1016/j.bushor.2014.05.003

Kalinic, Z., & Marinkovic, V. (2016). Determinants of users' intention to adopt m-commerce: An empirical analysis. *Information Systems and e-Business Management*, *14*(2), 367–387. doi:10.100710257-015-0287-2

Kane, G. C., Palmer, D., Phillips, A. N., Kiron, D., & Buckley, N. (2015). Strategy, not technology, drives digital transformation. MIT Sloan Management Review and Deloitte University Press.

Kanellopoulos, D. N., & Panagopoulos, A. A. (2008). Exploiting tourism destinations' knowledge in an RDF-based P2P network. *Journal of Network and Computer Applications*, *31*(2), 179–200. doi:10.1016/j.jnca.2006.03.003

Karvonen, M., & Kässi, T. (2013). Patent citations as a tool for analysing the early stages of convergence. *Technological Forecasting and Social Change*, *80*(6), 1094–1107. doi:10.1016/j.techfore.2012.05.006

Kearns, G. S., & Sabherwal, R. (2006). Strategic alignment between business and information technology: A knowledge-based view of behaviors, outcome, and consequences. *Journal of Management Information Systems*, *23*(3), 129–162. doi:10.2753/MIS0742-1222230306

Kelly, K. (2016). *The Inevitable: Understanding the 12 Technological Forces That Will Shape Our Future*. Viking.

Kemp, R. (2014). Legal Aspects of Managing Big Data. *Computer Law & Security Review*, *30*(5), 482–491. doi:10.1016/j.clsr.2014.07.006

Kenney, M., & Zysman, J. (2015, June). Choosing a future in the platform economy: the implications and consequences of digital platforms. *Kauffman Foundation New Entrepreneurial Growth Conference*.

Khare, A., Stewart, B., & Schatz, R. (2017). *Phantom Ex Machina*. Springer International Publishing. doi:10.1007/978-3-319-44468-0

Kidwell, B., & Jewell, R. D. (2003). An examination of perceived behavioral control: Internal and external influences on intention. *Psychology and Marketing*, *20*(7), 625–642. doi:10.1002/mar.10089

Kim, G., Shin, B., & Lee, H. G. (2009). Understanding dynamics between initial trust and usage intentions of mobile banking. *Information Systems Journal*, *19*(3), 283–311. doi:10.1111/j.1365-2575.2007.00269.x

Kim, J., & Lee, S. (2017). Forecasting and identifying multi-technology convergence based on patent data: The case of IT and BT industries in 2020. *Scientometrics*, *111*(1), 47–65. doi:10.100711192-017-2275-4

Kim, N., Lee, H., Kim, W., Lee, H., & Suh, J. H. (2015). Dynamic patterns of industry convergence: Evidence from a large amount of unstructured data. *Research Policy*, *44*(9), 1734–1748. doi:10.1016/j.respol.2015.02.001

Kiron, D., Prentice, P. K., & Ferguson, R. B. (2014). The analytics mandate. *MIT Sloan Management Review*, *55*(4), 1–25.

Kissinger, G., Brasser, A., & Gross, L. (2013). *Reducing Risk: Landscape Approaches to Sustainable Sourcing. In Eco-Agriculture Partners, on behalf of the Landscapes for People*. Washington, DC: Food and Nature Initiative.

Klerkx, L., & Aarts, N. (2013). The interaction of multiple champions in orchestrating innovation networks: Conflicts and complementarities. *Technovation*, *33*(6-7), 193–210. doi:10.1016/j.technovation.2013.03.002

Knickrehm, M., Berthon, B., & Daugherty, P. (2016). *Digital disruption: The growth multiplier.* Accenture Strategy, Tech. Rep.

Koenig-Lewis, N., Palmer, A., & Moll, A. (2010). Predicting young consumers' take up of mobile banking services. *International Journal of Bank Marketing*, *28*(5), 410–432. doi:10.1108/02652321011064917

Kohlborn, T., Mueller, O., Poeppelbuss, J., & Roeglinger, M. (2014). Interview with Michael Rosemann on ambidextrous business process management. *Business Process Management Journal*, *20*(4), 634–638. doi:10.1108/BPMJ-02-2014-0012

Kohli, R., & Grover, V. (2008). Business Value of IT: An Essay on Expanding Research Directions to Keep up with the Times. *Journal of the Association for Information Systems*, *9*(1), 23–39. doi:10.17705/1jais.00147

Kosnik, E. (2018). Production for consumption: Prosumer, citizen-consumer, and ethical consumption in a postgrowth context. *Economic Anthropology*, *5*(1), 124–134. doi:10.1002ea2.12107

Kotarba, M. (2017). Measuring Digitalization - Key Metrics. *Foundations of Management*, *9*(1), 123–138. doi:10.1515/fman-2017-0010

Kraus, S., Palmer, C., Kailer, N., Kallinger, F. L., & Spitzer, J. (2018). Digital entrepreneurship: A research agenda on new business models for the twenty-first century. *International Journal of Entrepreneurial Behavior & Research*. doi:10.1108/IJEBR-06-2018-0425

Krech, C., Rüther, F., & Gassmann, O. (2015). Profiting from invention: business models of patent aggregating companies. *International Journal of Innovation Management*, *19*(03), 1540005. doi:10.1142/S1363919615400058

Kritikos, M. (2017). *Precision agriculture in Europe. Legal, social and ethical considerations*. PE 603.207. European Parliamentary Research Service. Retrieved from http://www.lafermedigitale.fr/

Krumeich, J., Werth, D., & Loos, P. (2016). Prescriptive control of business processes. *Business & Information Systems Engineering*, *58*(4), 261–280. doi:10.100712599-015-0412-2

Kruskal, J. (1977). The relationship between multidimensional scaling and clustering. In Classification and clustering (pp. 17-44). Academic Press.

Kshetri, N. (2014). Big data's impact on privacy, security and consumer welfare. *Telecommunications Policy*, *38*(11), 1134–1145. doi:10.1016/j.telpol.2014.10.002

Kulkarni, S. S., Apte, U. M., & Evangelopoulos, N. E. (2014). The use of latent semantic analysis in operations management research. *Decision Sciences*, *45*(5), 971–994. doi:10.1111/deci.12095

Kumar, V., & Rajan, B. (2012). Customer lifetime value management: strategies to measure and maximize customer profitability. In V. Shankar & G.S. Carpenter (Eds.), Handbook of Marketing Strategy (pp. 107-134). Edward Elgar Publishing.

Kumar, V., Chattaraman, V., Neghina, C., Skiera, B., Aksoy, L., Buoye, A., & Henseler, J. (2013). Data-driven services marketing in a connected world. *Journal of Service Management*, *24*(3), 330–352. doi:10.1108/09564231311327021

Kumar, V., & Reinartz, W. J. (2006). *Customer Relationship Management: A Databased Approach*. Hoboken, NJ: Wiley.

Kwon, K., Kang, D., Yoon, Y., Sohn, J. S., & Chung, I. J. (2014). A real time process management system using RFID data mining. *Computers in Industry*, *65*(4), 721–732. doi:10.1016/j.compind.2014.02.007

Laing, A., Hogg, G., & Winkelman, D. (2005). The impact on professional relationships: The case of health care. *Service Industries Journal*, *25*(5), 675–687. doi:10.1080/02642060500101021

Lala, G. (2014). The emergence and development of the technology acceptance model (TAM). In *The Proceedings of the International Conference" Marketing-from Information to Decision"* (p. 149). Babes Bolyai University.

Lam, C. (2010). *Hadoop in Action*. Greenwich, CT, USA: Manning Publications Co.

Laney, D. (2001). *3D Data Management: Controlling Data Volume, Velocity and Variety, Application Delivery Strategy. Gartner*. Retrieved from http://blogs.gartner.com/doug-laney/files/2012/01/ad949-3D-Data-Management-Controlling-Data-Volume-Velocity-and-Variety.pdf

Lanzolla, G., & Anderson, J. (2008). Digital transformation. *London Business School Review*, *19*(2), 72–76.

Lanzolla, G., & Giudici, A. (2017). Pioneering strategies in the digital world. Insights from the Axel Springer case. *Business History*, *59*(5), 744–777. doi:10.1080/00076791.2016.1269752

Laudon, K.C. & Laudon, J.P. (2102). *Management Information Systems: Managing the digital firm* (12th ed.). New Jersey: Pearson Education.

Laukkanen, T. (2005). Comparing Consumer Value Creation in Internet and Mobile Banking. In *International Conference on Mobile Business*. Sydney, NSW, Australia: IEEE. 10.1109/ICMB.2005.28

Laukkanen, T. (2007). Internet vs mobile banking comparing customer value perceptions. *Business Process Management Journal*, *13*(6), 788–797. doi:10.1108/14637150710834550

Laukkanen, T., & Cruz, P. (2008). Barriers to mobile banking adoption: a cross-national study. *Proceedings of the International Conference on E-Business*, 1, 26-29.

Laukkanen, T., & Lauronen, J. (2005). Consumer value creation in mobile banking services. *International Journal of Mobile Communications*, *3*(4), 325–338. doi:10.1504/IJMC.2005.007021

Laurent, C., Maxime, F., Mazé, A., & Tichit, M. (2003). Multifonctionnalité de l'agriculture et modèles de l'exploitation agricole. *Economie Rurale*, *273/274*, 134-152. doi:10.3406/ecoru.2003.5395

Lavie, D., & Rosenkopf, L. (2006). Balancing exploration and exploitation in alliance formation. *Academy of Management Journal*, *49*(4), 797–818. doi:10.5465/amj.2006.22083085

Law, C. C. H., & Ngai, E. W. T. (2007). An investigation of the relationships between organisational factors, business process improvement, and ERP success. *Benchmarking: An International Journal*, *14*(3), 387–406. doi:10.1108/14635770710753158

Layne-Farrar, A., & Schmidt, K. M. (2010). Licensing complementary patents: Patent trolls, market structure, and excessive royalties. *Berkeley Technology Law Journal*, *25*, 1121–1143.

Le Dinh, T., Vu, M. C., & Ayayi, A. (2018). Towards a living lab for promoting the digital entrepreneurship process. *International Journal of Entrepreneurship*, *22*(1), 1–17.

Leadem, R. (2018). The History of Digital Content (Infographic). How did we get to where we are today? https://www.entrepreneur.com/article/309740

Lee, I. (2017). Big data: Dimensions, evolution, impacts, and challenges. *Business Horizons*. doi:10.1016/j.bushor.2017.01.004

Lee, C., Kang, B., & Shin, J. (2015). Novelty-focused patent mapping for technology opportunity analysis. *Technological Forecasting and Social Change*, *90*, 355–365. doi:10.1016/j.techfore.2014.05.010

Lee, C., Park, G., & Kang, J. (2018). The impact of convergence between science and technology on innovation. *The Journal of Technology Transfer*, *43*(2), 522–544. doi:10.100710961-016-9480-9

Leeflang, P. S. H., Verhoef, P. C., Dahlström, P., & Freundt, T. (2014). Challenges and solutions for marketing in a digital era. *European Management Journal*, *32*(1), 1–12. doi:10.1016/j.emj.2013.12.001

Lee, J., & Ahn, J.-H. (2014, December 8). Attention to Banner Ads and Their Effectiveness: An Eye-Tracking Approach. *International Journal of Electronic Commerce*, 119–137.

Lee, J., Siau, K., & Hong, S. (2003). Enterprise integration with ERP and EAI. *Communications of the ACM*, *46*(2), 54–60. doi:10.1145/606272.606273

Lee, K. S., Lee, H. S., & Kim, S. Y. (2007). Factors influencing the adoption behavior of mobile banking: A South Korean perspective. *Journal of Internet Banking and Commerce*, *12*(2), 1–9.

Lee, M. C. (2009). Factors influencing the adoption of internet banking: An integration of TAM and TPB with perceived risk and perceived benefit. *Electronic Commerce Research and Applications*, *8*(3), 130–141. doi:10.1016/j.elerap.2008.11.006

Lee, M. K., Cheung, C. M., & Chen, Z. (2007). Understanding user acceptance of multimedia messaging services: An empirical study. *Journal of the Association for Information Science and Technology*, *58*(13), 2066–2077.

Lee, S. M., Olson, D. L., & Trimi, S. (2010). The impact of convergence on organizational innovation. *Organizational Dynamics*, *39*(3), 218–225. doi:10.1016/j.orgdyn.2010.03.004

Lee, S., & Kim, W. (2017). The knowledge network dynamics in a mobile ecosystem: A patent citation analysis. *Scientometrics*, *111*(2), 717–742. doi:10.100711192-017-2270-9

Lee, S., Kim, W., Lee, H., & Jeon, J. (2016). Identifying the structure of knowledge networks in the US mobile ecosystems: Patent citation analysis. *Technology Analysis and Strategic Management*, *28*(4), 411–434. doi:10.1080/09537325.2015.1096336

Leila, T., & Beaudry, C. (2015). Does government funding have the same impact on academic publications and patents? The case of nanotechnology in Canada. *International Journal of Innovation Management*, *19*(03), 1540001. doi:10.1142/S1363919615400010

Leloup, L. (2017), DLT: alternative aux blockchains Bitcoin et Ethereum. Retrieved from http://www.finyear.com/DLT-alternative-aux-blockchains-Bitcoin-et-Ethereum_a36182.html

Lemieux, C. (2012). Peut-on ne pas être constructiviste? *Politix*, *100*(4), 169–187. doi:10.3917/pox.100.0169

Leong, C., Pan, S. L., & Liu, J. (2016). Digital Entrepreneurship of Born Digital and Grown Digital Firms: Comparing the Effectuation Process of Yihaodian and Suning Research-in-Progress. In *Thirty Seventh International Conference on Information Systems*, Dublin, Ireland.

Levin, H. M., & Rumberger, R. W. (1986). Education and Training Needs for Using Computers in Small Businesses. *Educational Evaluation and Policy Analysis*, *8*(4), 423–434. doi:10.3102/01623737008004423

Leydesdorff, L. (2008). Patent classifications as indicators of intellectual organization. *Journal of the Association for Information Science and Technology*, *59*(10), 1582–1597.

Leydesdorff, L., & Vaughan, L. (2006). Co-occurrence matrices and their applications in information science: Extending ACA to the Web environment. *Journal of the Association for Information Science and Technology*, *57*(12), 1616–1628.

Liao, S. H., Chu, P. H., & Hsiao, P. Y. (2012). Data mining techniques and applications – A decade review from 2000 to 2011. *Expert Systems with Applications*, *39*(12), 11303–11311. doi:10.1016/j.eswa.2012.02.063

Libert, B., Beck, M., & Wind, Y. (2016). 7 Questions to ask before your next digital transformation. *Harvard Business Review*. Retrieved from https://hbr.org/2016/07/7-questions-to-ask-before-your-next-digital-transformation

Liébana-Cabanillas, F., Ramos de Luna, I., & Montoro-Ríos, F. (2017). Intention to use new mobile payment systems: A comparative analysis of SMS and NFC payments. *Economic Research-Ekonomska Istraživanja*, *30*(1), 892–910. do i:10.1080/1331677X.2017.1305784

Lien, J., Hughes, L., Kina, J., & Villasenor, J. (2015). Mobile money solutions for a smartphone-dominated world. *Journal of Payments Strategy and Systems*, *9*(3), 341–350.

Liere-Netheler, K., Packmohr, S., & Vogelsang, K. (2018, January). Drivers of Digital Transformation in Manufacturing. *Proceedings of the 51st Hawaii International Conference on System Sciences*. 10.24251/HICSS.2018.493

Li, F. (2018). The digital transformation of business models in the creative industries: A holistic framework and emerging trends. *Technovation*. doi:10.1016/j.technovation.2017.12.004

Li, F., & Whalley, J. (2002). Deconstruction of the telecommunications industry: From value chains to value networks. *Telecommunications Policy*, *26*(9-10), 451–472. doi:10.1016/S0308-5961(02)00056-3

Lin, C., & Kim, T. (2016). Predicting user response to sponsored advertising on social media via the technology acceptance model. *Computers in Human Behavior, 64*, 710-718. doi:10.1016/j.chb.2016.07.027

Lindh, C., & Rovira, E. R. (2018). New service development and digitalization: Synergies of personal interaction and IT integration. *Services Marketing Quarterly*, *39*(2), 108–123. doi:10.1080/15332969.2018.1436777

Lin, F., & Rohm, C. E. T. (2009). Managers' and end-users' concerns on innovation implementation: A case of an ERP implementation in China. *Business Process Management Journal*, *15*(4), 527–547. doi:10.1108/14637150910975525

Lin, H. F. (2011). An empirical investigation of mobile banking adoption: The effect of innovation attributes and knowledge-based trust. *International Journal of Information Management*, *31*(3), 252–260. doi:10.1016/j.ijinfomgt.2010.07.006

Li, Q., Luo, H., Xie, P. X., Feng, X. Q., & Du, R. Y. (2015). Product whole life-cycle and omni-channels data convergence oriented enterprise networks integration in a sensing environment. *Computers in Industry*, *70*, 23–45. doi:10.1016/j.compind.2015.01.011

Li, W., Du, W., & Yin, J. (2017). Digital entrepreneurship ecosystem as a new form of organizing: The case of Zhongguancun. *Frontiers of Business Research in China*, *11*(1), 5. doi:10.118611782-017-0004-8

Loebbecke, C., & Picot, A. (2015). Reflections on Societal and Business Model Transformation arising from Digitization and Big Data Analytics: A Research Agenda. *The Journal of Strategic Information Systems*, *24*(3), 149–157. doi:10.1016/j.jsis.2015.08.002

Lohse, G. L. (1997). Consumer eye movement patterns on yellow pages advertising. *Journal of Advertising*, *26*(1), 61–73. doi:10.1080/00913367.1997.10673518

Loi, M., Castriotta, M., & Di Guardo, M. C. (2016). The theoretical foundations of entrepreneurship education: How co-citations are shaping the field. *International Small Business Journal*, *34*(7), 948–971. doi:10.1177/0266242615602322

Longchamp, J. Y., & Pagès, B. (2012). Charges de mécanisation et structure d'exploitation. *Document de travail - Commission des comptes de l'agriculture de la nation*, 7. Retrieved from http://agreste.agriculture.gouv.fr/IMG/pdf/doctravail70712-2.pdf

Loshin, D. (2013). *Market and Business Drivers for Big Data Analytics*.

Lotfizadeh, F., & Ghorbani, A. (2015). A Multi-dimensional Model of Acceptance of Mobile Banking, *International Journal of Management. Accounting and Economics*, 2(5), 414–427.

Loubet, N., & Epié, C. (2015). *Blockchain and Beyond*. Retrieved from .https://blockchainfrance.files.wordpress.com/2015/12/cellabz-blockchain-beyond.pdf

Luan, C., Hou, H., Wang, Y., & Wang, X. (2014). Are significant inventions more diversified? *Scientometrics*, *100*(2), 459–470. doi:10.100711192-014-1303-x

Luan, C., Liu, Z., & Wang, X. (2013). Divergence and convergence: Technology-relatedness evolution in solar energy industry. *Scientometrics*, *97*(2), 461–475. doi:10.100711192-013-1057-x

Luarn, P., & Lin, H. (2005). Toward an understanding of the behavioral intention to use mobile banking. *Computers in Human Behavior*, *21*(6), 873–891. doi:10.1016/j.chb.2004.03.003

Lucky, E. O.-I. (2013). Exploring the ineffectiveness of government policy on entrepreneurship in Nigeria. *International Journal of Entrepreneurship and Small Business*, *19*(4), 471–487. doi:10.1504/IJESB.2013.055487

Lucky, E. O.-I., Rahman, H. A., & Minai, M. S. (2013). *A conceptual framework for a successful co-operative entrepreneurship development. In Handbook of entrepreneurship and co-operative development. Co-operative and Entrepreneurship Development institute (CEDI), Universiti Utara Malaysia (UUM)*. Malaysia: Sintok.

Luftman, J. (2000). Assessing business-information technology alignment maturity. *Communications of the Association for Information Systems*, *4*(1), 1–49.

Luftman, J., Kempaiah, K., & Nash, E. (2006). Key issues for information technology executives 2005. *MIS Quarterly Executive*, *5*(2), 81–99.

Luftman, J., Lewis, P., & Oldach, S. (1993). Transforming the Enterprise: The Alignment of Business and Information Technology Strategies. *IBM Systems Journal*, *32*(1), 198–222. doi:10.1147j.321.0198

Lu, J. (2012). The myths and facts of patent troll and excessive payment: have nonpracticing entities (NPEs) been overcompensated? *Business Economics (Cleveland, Ohio)*, *47*(4), 234–249. doi:10.1057/be.2012.26

Luo, X., Lee, C. P., Mattila, M., & Liu, L. (2012). An exploratory study of mobile banking services resistance. *International Journal of Mobile Communications*, *10*(4), 366–385. doi:10.1504/IJMC.2012.048136

Luo, X., Li, H., Zhang, J., & Shim, J. P. (2010). Examining multi-dimensional trust and multi-faceted risk in initial acceptance of emerging technologies: An empirical study of mobile banking services. *Decision Support Systems*, *49*(2), 222–234. doi:10.1016/j.dss.2010.02.008

Lustig, I., Dietrich, B., Johnson, C., & Dziekan, C. (2010). The analytics journey. *Analytics Magazine*, November/December, 11–13.

Lu, X. J., Huang, L. H., & Heng, M. S. H. (2006). Critical success factors of inter-organisational information systems: A case study of Cisco and Xiao Tong in China. *Information & Management*, *43*(3), 395–408. doi:10.1016/j.im.2005.06.007

Lyman, K. B., Caswell, N., & Biem, A. (2009). Business value network concepts for the extended enterprise. In P. H. M. Vervest, D. W. Liere, & L. Zheng (Eds.), *Proc. of the Network Experience*. Berlin: Springer. doi:10.1007/978-3-540-85582-8_9

Lynch, R. (2003). *Corporate strategy (3rd ed.)*. Harlow: Prentice-Hall Financial Times.

Madden, S. (2012). From databases to Big Data. *IEEE Internet Computing*, *16*(3), 4–6. doi:10.1109/MIC.2012.50

Madu, C. N., & Kuei, C. (2004). *ERP and Supply Chain Management*. Fairfield, CT: Chi Publishers.

Maes, R., Rijsenbrij, D., Truijens, O., & Goedvolk, H. (2000). Redefining business: IT alignment through a unified framework. (PrimaVera working paper; No. 2000-19). Amsterdam: Universiteit van Amsterdam, Department of Information Management.

Mahoney, J. T. (1992). The choice of organisational form: Vertical financial ownership versus other methods of vertical integration. *Strategic Management Journal, 13*(8), 559–584. doi:10.1002mj.4250130802

Maingueneau, D. (1991). *L'Analyse du Discours: Introduction aux Lectures de l'Archive*. Paris: Hachette.

Maitland, C. F., Bauer, J. M., & Westerveld, R. (2002). The European market for mobile data: Evolving value chains and industry structures. *Telecommunications Policy, 26*(9-10), 485–504. doi:10.1016/S0308-5961(02)00028-9

Maldonado, S., Montoya, R., & Weber, R. (2015). Advanced conjoint analysis using feature selection via support vector machines. *European Journal of Operational Research, 241*(2), 564–574. doi:10.1016/j.ejor.2014.09.051

Malhotra, M. K., & Kher, H. V. (1996). Institutional research productivity in production and operations management. *Journal of Operations Management, 14*(1), 55–77. doi:10.1016/0272-6963(95)00037-2

Mallat & Tuunainen. (2008). Exploring Merchant Adoption of Mobile Payment Systems: An Empirical Study. *e-Service Journal, 6*(2), 24-57. Doi:10.2979/esj.2008.6.2.24

Manyika, J., Chui, M., Brown, B., Bughin, J., Dobbs, R., Roxburgh, C., & Byers, A. H. (2011). *Big data: The next frontier for innovation, competition and productivity*. New York: McKinsey Global Institute.

Marchand, D., Kettinger, W., & Rollins, J. (2001). *Information orientation: The link to business performance*. Oxford: Oxford University Press.

Marine-Roig, E. (2017). Measuring destination image through travel reviews in search engines. *Sustainability, 9*(8), 1–18. doi:10.3390u9081425

Marku, E., Castriotta, M., & Di Guardo, M. C. (2017). Disentangling the Intellectual Structure of Innovation and M&A Literature. *Technological Innovation Networks: Collaboration and Partnership*, 47.

Marku, E., & Zaitsava, M. (2018). Smart Grid Domain: Technology Structure and Innovation Trends. International Journal of Economics. *Business and Management Research, 2*(4), 390–403.

Markus, M. L., & Loebbecke, C. (2013). Commoditized digital processes and business community platforms: New opportunities and challenges for digital business strategies. *Management Information Systems Quarterly, 37*(2), 649–654.

Marraccini, E., Rapey, H., Galli, M., Lardon, S., & Bonari, E. (2013). Assessing the Potential of Farming Regions to Fulfill Agro-Environmental Functions: A Case Study in Tuscany (Italy). *Environmental Management, 51*(3), 759–776. doi:10.100700267-012-9997-0 PMID:23263567

Martinez, M. T., Fouletier, P., Park, K. H., & Faurel, J. (2001). Virtual enterprise: Organisation, evolution and control. *International Journal of Production Economics, 74*(1-3), 225–238. doi:10.1016/S0925-5273(01)00129-3

Mathew, M., Sulphey, M. M., & Prabhakaran, J. (2014). Perceptions and Intentions of Customers towards Mobile Banking Adoption, *Journal. Contemporary Management Research, 8*(1), 83–101.

Matt, C., Hess, T., & Benlian, A. (2015). Digital transformation strategies. *Business & Information Systems Engineering, 57*(5), 339–343. doi:10.100712599-015-0401-5

Maurizio, A., Girolami, L., & Jones, P. (2007). EAI and SOA: Factors and methods influencing the integration of multiple ERP systems (in an SAP environment) to comply with the Sarbanes-Oxley Act. *Journal of Enterprise Information Management, 20*(1), 14–31. doi:10.1108/17410390710717110

Mazzone, D. M. (2014). *Digital or death: digital transformation: the only choice for business to survive smash and conquer*. Smashbox Consulting Inc.

McAdam, M., Crowley, C., & Harrison, R. T. (2018). "To boldly go where no [man] has gone before"-Institutional voids and the development of women's digital entrepreneurship. *Technological Forecasting and Social Change*.

McAfee, A. (2002). The impact of enterprise information technology adoption on operational performance: An empirical investigation. *Production and Operations Management, 11*(1), 33–53. doi:10.1111/j.1937-5956.2002.tb00183.x

McAfee, A., & Brynjolfsson, E. (2013). Le Big Data, une revolution du management. *Harvard Business Review*, (Avril-Mai), 1–9.

McAfee, A., Brynjolfsson, E., Davenport, T. H., Patil, D. J., & Barton, D. (2012). Big data: The management revolution. *Harvard Business Review, 90*(10), 61–67. PMID:23074865

McCain, K. W. (1990). Mapping authors in intellectual space: A technical overview. *Journal of the American Society for Information Science, 41*(6), 433–443. doi:10.1002/(SICI)1097-4571(199009)41:6<433::AID-ASI11>3.0.CO;2-Q

McCoy, S., Everard, A., Polak, P., & Galletta, D. F. (2007). The effects of online advertising. *Communications of the ACM, 50*(3), 84–88. doi:10.1145/1226736.1226740

Mccutcheon, D. M., & Meredith, J. R. (1993). Conducting case study research in operations management. *Journal of Operations Management, 11*(3), 239–256. doi:10.1016/0272-6963(93)90002-7

McDonald, J., & Léveillé, V. (2014). Whither the retention schedule in the era of big data and open data? *Records Management Journal, 24*(2), 99–121. doi:10.1108/RMJ-01-2014-0010

McGrath, R. G. (2013). *The End of Competitive Advantage: How to Keep Your Strategy Moving as Fast as Your Business*. Boston, Mass: Harvard Business Review Press.

McKelvey, B., Tanriverdi, H., & Yoo, Y. (2016). Complexity and Information Systems. Research in the Emerging Digital World. *Management Information Systems Quarterly*.

McKelvie, A., & Picard, R. G. (2008). The growth and development of new and young media firms. *Journal of Media Business Studies, 5*(1), 1–8. doi:10.1080/16522354.2008.11073458

McKinsey & Company. (2013). *Disruptive Technologies: Advances That Will Transform Life*. Business, and the Global Economy.

McKinsey Global Institute. (2012). *Big data: The Next Frontier for Innovation*, Competition, and Productivity.

McQuail, D. (2000). *McQuail's Mass Communication Theory* (4th ed.). London: Sage.

McQuivey, J. (2013). *Digital disruption: Unleashing the next wave of innovation*. Academic Press.

Mêgnigbêto, E. (2017). Controversies arising from which similarity measures can be used in co-citation analysis. *Malaysian Journal of Library and Information Science, 18*(2), 25–31.

Meltzer, J. P., & Pérez, C. (2016). *Digital Colombia: Maximizing the global internet and data for sustainable and inclusive growth*. Washington, DC: Global Economy and Development at Brookings. Working Paper No. 96.

Metallo, C., Agrifoglio, R., Schiavone, F., & Mueller, J. (2018). (in press). Understanding Business Model in the Internet of Things industry. *Technological Forecasting and Social Change*. doi:10.1016/j.techfore.2018.01.020

Mezghani, K. (2014). Switching toward Cloud ERP: A research model to explain intentions. *International Journal of Enterprise Information Systems, 10*(3), 48–64. doi:10.4018/ijeis.2014070104

Mezghani, K., & Ayadi, F. (2016). Factors explaining IS managers' attitudes toward Cloud Computing adoption. *International Journal of Technology and Human Interaction*, *12*(1), 1–20. doi:10.4018/IJTHI.2016010101

Miller, H. G., & Mork, P. (2013). *From data to decisions: a value chain for big data. In IT Pro* (pp. 57–59). IEEE.

Ministère de l'Education Nationale, de l'Enseignement Supérieur et de la Recherche (2011), *Stratégie de recherche pour le numérique*. Retrieved from https://cache.media.enseignementsup-recherche.gouv.fr/file/Strategie_Recherche/28/1/Rapport_atelier_7_314281.pdf

Mintzberg, H. (1994). The fall and rise of strategic planning. *Harvard Business Review*, *72*(1), 107–114.

Mishra, D., Gunasekaran, A., Papadopoulos, T., & Childe, S. (2016). Big Data and Supply Chain Management: A Review and Bibliometric Analysis. *Annals of Operations Research*. doi:10.100710479-016-2236-y

Mitchel, D., & Coles, C. (2003). The ultimate competitive advantage of continuing Business Model innovation. *The Journal of Business Strategy*, *24*(5), 15–21. doi:10.1108/02756660310504924

Mithas, S., Tafti, A., & Mitchell, W. (2013). How a firm's competitive environment and digital strategic posture influence digital business strategy. *Management Information Systems Quarterly*, *37*(2), 511–536. doi:10.25300/MISQ/2013/37.2.09

Moller, C. (2005). ERPII: A conceptual framework for next-generation enterprise systems? *Journal of Enterprise Information Management*, *18*(4), 483–497. doi:10.1108/17410390510609626

Mortenson, M. J., Doherty, N. F., & Robinson, S. (2015). Operational research from Taylorism to terabytes: A research agenda for the analytics age. *European Journal of Operational Research*, *241*(3), 583–595. doi:10.1016/j.ejor.2014.08.029

Mou, J., & Shin, D. (2018). Effects of social popularity and time scarcity on online consumer behaviour regarding smart healthcare products: An eye-tracking approach. *Computers in Human Behavior*, *78*, 74–89. doi:10.1016/j.chb.2017.08.049

Munroe, D. K., McSweeney, K., Olson, J. L., & Mansfield, B. (2014). Using economic geography to reinvigorate land-change science. *Geoforum*, *52*, 12–21. doi:10.1016/j.geoforum.2013.12.005

Nagel, R. N. (1992). *21st Century Manufacturing Enterprise Strategy. An Industry-Led View. Prepared for the Office of Naval Research Arlington, VA*. Bethlehem, PA: Iacocca Institute. Lehigh University. doi:10.21236/ADA257032

Nah, F. F., Tan, X., & Beethe, M. (2005) End-users' acceptance of Enterprise Resource Planning (ERP) Systems: an investigation using grounded theory approach. *Paper presented at the Eleventh Americas Conference on Information Systems*, Omaha, NB.

Nambisan, S. (2016) Digital Entrepreneurship: Toward a Digital Technology Perspective of Entrepreneurship. SAGE Publications Inc. doi:. doi:10.1111/etap.12254

Nambisan, S., Lyytinen, K., Majchrzak, A., & Song, M. (2017). Digital innovation management: Reinventing innovation management research in a digital world. *Management Information Systems Quarterly*, *41*(1), 223–238. doi:10.25300/MISQ/2017/41:1.03

Nambisan, S., Siegel, D., & Kenney, M. (2018). On Open Innovation, Platforms, and Entrepreneurship. *Strategic Entrepreneurship Journal*, *12*(3), 354–368. doi:10.1002ej.1300

Negash, S. (2011). Mobile banking adoption by under-banked communities in the United States: adapting mobile banking features from low-income countries. *11th International Conference on Mobile Business (ICMB)*.

Nepelski, D., Bogdanowicz, M., Biagi, F., Desruelle, P., De Prato, G., Gabison, G., . . . Van Roy, V. (2017). *7 ways to boost digital innovation and entrepreneurship in Europe*. Available at: http://publications.jrc.ec.europa.eu/repository/bitstream/JRC104899/jrc104899_formatted_final_20170426.pdf

Ngai, E. W. T., & Wat, F. K. T. (2002). A literature review and classification of electronic commerce research. *Information & Management, 39*(5), 415–429. doi:10.1016/S0378-7206(01)00107-0

Ng, I., Scharf, K., Pogrebna, G., & Maull, R. (2015). Contextual variety, Internet-of-Things and the choice of tailoring over platform: Mass customisation strategy in supply chain management. *International Journal of Production Economics, 159*(0), 76–87. doi:10.1016/j.ijpe.2014.09.007

Ngoasong, M. Z. (2017). Digital entrepreneurship in a resource-scarce context: A focus on entrepreneurial digital competencies. *Journal of Small Business and Enterprise Development*. doi:10.1108/JSBED-01-2017-0014

Nguyen-Duc, A., Cruzes, D. S., & Conradi, R. (2015). The impact of Global Dispersion on Coordination, Team Performance and Software Quality: A Systematic Literature Review. *Information and Software Technology, 57*, 277–294. doi:10.1016/j.infsof.2014.06.002

Nguyen, N., & Gaston, L. (2002). Contact personnel, physical environment and the perceived corporate image of intangible services by new clients. *International Journal of Service Industry Management, 13*(3), 242–262. doi:10.1108/09564230210431965

Nicita, A., Ramello, G. B., & Scherer, F. M. (2005). Intellectual property rights and the organization of industries: New perspectives in law and economics. *International Journal of the Economics of Business, 12*(3), 289–296. doi:10.1080/13571510500299029

Nicosia, F. (1974). *Advertising, Management, and Society: A Business Point of View*. New York: McGraw-Hill.

Nielsen, J. (2005). *Scrolling and Scrollbars*. Retrieved from https://www.nngroup.com/articles/scrolling-and-scrollbars

Nielsen, J. (2006). *F-Shaped Pattern For Reading Web Content*. Retrieved from https://www.nngroup.com/articles/f-shaped-pattern-reading-web-content-discovered

Nielsen, J. (2010). *Scrolling and Attention (Original Research Study)*. Retrieved from https://www.nngroup.com/articles/scrolling-and-attention-original-research/

Nielsen. (2015). *Global Trust in Advertising – 2015*. Retrieved from http://www.nielsen.com/us/en/insights/reports/2015/global-trust-in- advertising-2015.html

Niemann, H., Moehrle, M. G., & Frischkorn, J. (2017). Use of a new patent text-mining and visualization method for identifying patenting patterns over time: Concept, method and test application. *Technological Forecasting and Social Change, 115*, 210–220. doi:10.1016/j.techfore.2016.10.004

Nieto, M., & Pérez, W. (2000). The development of theories from the analysis of the organisation: Case studies by the patterns of behaviour. *Management Decision, 38*(10), 723–734. doi:10.1108/00251740010360588

Nurmilaakso, J. M. (2008). Adoption of e-business functions and migration from EDI-based to XML-based e-business frameworks in supply chain integration. *International Journal of Production Economics, 113*(2), 721–733. doi:10.1016/j.ijpe.2007.11.001

Nurunnabi, M. (2017). Transformation from an Oil-based Economy to a Knowledge-based Economy in Saudi Arabia: The Direction of Saudi Vision 2030. *Journal of the Knowledge Economy, 8*(2), 536–564. doi:10.100713132-017-0479-8

Nwankpa, J. K., & Roumani, Y. (2016). IT Capability and Digital Transformation: A Firm Performance Perspective. In Proceedings of the *Thirty Seventh International Conference on Information Systems, Dublin*. Retrieved from https://pdfs.semanticscholar.org/e8c4/16395a5d6690550b4aa74d81950eaa28bd84.pdf

Nzembayie, K. F. (2017). Using insider action research the study of digital entrepreneurial processes: a pragmatic design choice? In *European Conference on Research Methodology for Business and Management Studies*. Academic Conferences International Limited.

OECD. (2016). Stimulating Digital Innovation for Growth and Inclusiveness: The Role of Policies for the Successful Diffusion of ICT. In *OECD Digital Economy Papers, No. 256*. Paris: OECD Publishing.

Oh, C., Cho, Y., & Kim, W. (2015). The effect of a firm's strategic innovation decisions on its market performance. *Technology Analysis and Strategic Management*, *27*(1), 39–53. doi:10.1080/09537325.2014.945413

Ojala, A., Evers, N., & Rialp, A. (2018). Extending the international new venture phenomenon to digital platform providers: A longitudinal case study. *Journal of World Business*, *53*(5), 725–739. doi:10.1016/j.jwb.2018.05.001

Olander, H., & And Hurmelinna-Laukkanen, P. (2015). Perceptions of Employee Knowledge Risks in Multinational, Multilevel Organisations: Managing Knowledge Leaking And Leaving. *International Journal of Innovation Management*, *19*(03), 1540006. doi:10.1142/S136391961540006X

Olsen, K. A., & Sætre, P. (2007). IT for niche companies: Is an ERP system the solution? *Information Systems Journal*, *17*(1), 37–58. doi:10.1111/j.1365-2575.2006.00229.x

Onetti, A., Zucchella, A., Jones, M., & McDougall-Covin, P. (2012). Internationalization, innovation and entrepreneurship: Business models for new technology-based firms. *The Journal of Management and Governance*, *16*(3), 337–368. doi:10.100710997-010-9154-1

Osterwalder, A. (2004). *The Business Model ontology: A proposition in a design science approach* (Dissertation). University of Lausanne, Switzerland.

Osterwalder, A., & Pigneur, Y. (2010). *Business Model generation: a handbook for visionaries, game changers, and challengers*. John Wiley & Sons.

O'Sullivan, A., & Sheffrin, S. M. (2003). *Economics: Principles in action*. Upper Saddle River, NJ: Pearson Prentice Hall.

Oura, J., & Kijima, K. (2002). Organisation design initiated by information system development: A methodology and its practice in Japan. *System Research and Business Science*, *19*(1), 77–86. doi:10.1002res.415

Owen, L., Goldwasser, C., Choate, K., & Blitz, A. (2008). Collaborative innovation throughout the extended enterprise. *Strategy and Leadership*, *36*(1), 39–45. doi:10.1108/10878570810840689

Owens, J. W., Chaparro, B. S., & Palmer, E. M. (2011). Text advertising blindness: The new banner blindness? *Journal of Usability Studies*, *6*(3), 172–197.

Özüdoğru, A. G., Ergün, E., Ammari, D., & Görener, A. (2018). How industry 4.0 changes business: A commercial perspective. *International Journal of Commerce and Finance*, *4*(1), 84–95.

Pagani, M. (2013). Digital Business Strategy and Value Creation: Framing the Dynamic Cycle of Control Points. *Management Information Systems Quarterly*, *37*(2), 617–632. doi:10.25300/MISQ/2013/37.2.13

Palaniswamy, R., & Frank, T. (2000). Enhancing manufacturing performance with ERP systems. *Information Systems Management*, *17*(3), 1–13. doi:10.1201/1078/43192.17.3.20000601/31240.7

Pan, B., MacLaurin, T., & Crotts, J. C. (2007). Travel blogs and the implications for destination marketing. *Journal of Travel Research*, *46*(1), 35–45. doi:10.1177/0047287507302378

Pantano, E., & Di Pietro, L. (2012). Understanding Consumer's Acceptance of Technology-Based Innovations in Retailing. *Journal of Technology Management & Innovation*, *7*(4), 1–19. doi:10.4067/S0718-27242012000400001

Parker, G., Alstyne, M., & Choudary, S. (2016). *Platform Revolution: How Networked Markets are Transforming the Economy, and How to Make Them Work for You*. W. W. Norton & Company.

Park, G., Chung, L., & Khan, L. (2017). A Modeling Framework for Business Process Reengineering Using Big Data Analytics and A Goal-Orientation.

Park, H., & Yoon, J. (2014). Assessing coreness and intermediarity of technology sectors using patent co-classification analysis: The case of Korean national R&D. *Scientometrics*, *98*(2), 853–850. doi:10.100711192-013-1109-2

Patel, K., & McCarthy, M. P. (2000). *Digital Transformation: The Essentials of E-Business Leadership*. New York: McGraw-Hill Professional.

Pauwels, K., Ambler, T., Clark, B. H., LaPointe, P., Reibstein, D., Skiera, B., ... Wiesel, T. (2009). Dashboards as a service: Why, how, and what research is needed? *Journal of Service Research*, *12*(2), 175–189. doi:10.1177/1094670509344213

Payne, A., & Frow, P. (2005). A strategic framework for customer relationship management. *Journal of Marketing*, *69*(4), 167–176. doi:10.1509/jmkg.2005.69.4.167

Pearce, J. M. (2015). Applications of Open Source 3-D Printing on Small Farms. *Organic Farming*, *1*(1), 19–35. doi:10.12924/of2015.01010019

Penrose, E. G. (1959). *The Theory of the Growth of the Firm*. New York: Wiley.

Peppard, J. (2016). A Tool for Balancing Your Company's Digital Investments, *Harvard Business Review Digital Articles*, October 18, 2016, 2-5.

Peppard, J., & Rylander, A. (2006). From value chain to value network: Insights for mobile operators. *European Management Journal*, *24*(2-3), 128–141. doi:10.1016/j.emj.2006.03.003

Peppard, J., & Ward, J. (2016). *The strategic management of information systems: Building a digital strategy*. John Wiley & Sons.

Pernice, K. (2017). *F-Shaped Pattern of Reading on the Web: Misunderstood, But Still Relevant (Even on Mobile)*. Retrieved from https://www.nngroup.com/articles/f-shaped-pattern-reading-web-content

Peteraf, M. A. (1993). The cornerstone of competitive advantage: A resource-based view. *Strategic Management Journal*, *14*(3), 179–191. doi:10.1002mj.4250140303

Pham, C., Ferrero, F., Diop, M., Lizzi, L., Dieng, O., & Thiaré, O. (2017). Low-cost Antenna Technology for LPWAN IoT in Rural Applications. In *Proceedings of the 7th IEEE International Workshop on Advances in Sensors and Interfaces (IWASI'17)*. IEEE. Retrieved from http://cpham.perso.univ-pau.fr/Paper/IWASI17.pdf

Pham, C., Rundle-Thiele, S., Parkinson, J., & Li, S. (2017). Alcohol Warning Label Awareness and Attention: A Multi-method Study. *Alcohol and Alcoholism (Oxford, Oxfordshire)*, 1–7. doi:10.1093/alcalc/agx087 PMID:29136096

Piccinini, E., Hanelt, A., Gregory, R., & Kolbe, L. (2015). Transforming industrial business: the impact of digital transformation on automotive organizations. *ICIS 2015 Conference Proceedings*.

Pieters, R., & Wedel, M. (2004). Attention capture and transfer in advertising: Brand, pictorial and text-size effects. *Journal of Marketing*, *68*(2), 36–50. doi:10.1509/jmkg.68.2.36.27794

Pignatti, E., Carli, G., & Canavari, M. (2015). What really matters? A qualitative analysis on the adoption of innovations in agriculture. *Agrárinformatika. Agrárinformatika Folyóirat*, *6*(4), 73–84. doi:10.17700/jai.2015.6.4.212

Pilkington, M. (2016). Blockchain technology principles and applications. In F. X. Olleros & M. Zhegu (Eds.), *Research Handbook on Digital Transformation*. Cheltenham, UK: Edward Elgar. doi:10.4337/9781784717766.00019

Piller, F. T., Weller, C., & Kleer, R. (2015). Business Models with Additive Manufacturing—Opportunities and Challenges from the Perspective of Economics and Management. In *Advances in Production Technology* (pp. 39–48). Springer International Publishing.

Pinto, I. R., Marques, J. M., Levine, J. M., & Abrams, D. (2016). Membership role and subjective group dynamics: Impact on evaluative intragroup differentiation and commitment to prescriptive norms. *Group Processes & Intergroup Relations*, *19*(5), 570–590.

Ployhart, R. E. (2006). Staffing in the 21st century: New challenges and strategic opportunities. *Journal of Management*, *32*(6), 868–897. doi:10.1177/0149206306293625

Pohlmann, T., & Opitz, M. (2013). Typology of the Patent Troll Business. *R & D Management*, *43*(2), 103–120. doi:10.1111/radm.12003

Pongnumkul, S., Chaovalit, P., & Surasvadi, N. (2015). Applications of Smartphone-Based Sensors in Agriculture: A Systematic Review of Research. *Journal of Sensors*. doi:10.1155/2015/195308

Porteous, D. (2007). *Just how transformational is m-banking?* Commissioned by Finmark. Retrieved September 1, 2018, from https://www.microfinancegateway.org/sites/default/files/mfg-en-paper-just-how transformational-is-m-banking-feb-2007.pdf

Porter, M. E. (1985). *Competitive Advantage: Creating and Sustaining Superior Performance*. New York: The Free Press.

Porter, M. E. (1996). What is Strategy? *Harvard Business Review*, *74*(6), 61–78. PMID:10158474

Porter, M. E., & Heppelmann, J. E. (2014). How Smart, Connected Products Are Transforming Competition. *Harvard Business Review*, *92*(11), 64–88.

Porter, M. E., & Heppelmann, J. E. (2015). How Smart, Connected Products Are Transforming Companies. *Harvard Business Review*, *93*(10), 98–114.

Pousttchi, K., & Schurig, M. (2004). Assessment of Today's Mobile Banking Applications from The View of Consumer Requirements. *37th Hawaii International Conference on System Sciences*.

Prahalad, C. K., & Ramaswamy, V. (2004). Cocreation experience: The next practice in value creation. *Journal of Interactive Marketing*, *18*(3), 5–14. doi:10.1002/dir.20015

Primdahl, J., Kristensen, L. S., & Busck, A. G. (2013). The Farmer and Landscape Management: Different Roles, Different Policy Approaches: The Farmer as a Landscape Manager. *Geography Compass*, *7*(4), 300–314. doi:10.1111/gec3.12040

Prinsloo, C., & James, I. (2015). *Digital Disruption: Changing the rules of business for a hyper-connected world*. Gordon Institute of Business Science, University of Pretoria.

Prodanov, H. (2018). Social Entrepreneurship and Digital Technologies. *Economic Alternatives*, (1), 123-138.

Purchase, V., Parry, G., Valerdi, R., Nightingale, D., & Mills, J. (2011). Enterprise transformation: Why are we interested, what is it, and what are the challenges? *Journal of Enterprise Transformation*, *1*(1), 14–33. doi:10.1080/19488289.2010.549289

Puschel, J., Mazzon, J. A., & Hernandez, J. M. C. (2010). Mobile banking: Proposition of an integrated adoption intention framework. *International Journal of Bank Marketing, Vol.*, *28*(5), 389–409. doi:10.1108/02652321011064908

Radach, R., Lemmer, S., Vorstius, C., Heller, D., & Radach, K. (2003). Eye movements in the processing of print advertisements. In J. Hyönä, R. Radach, & H. Deubel (Eds.), *The Mind's Eyes: Cognitive and Applied Aspects of Eye Movement Research* (pp. 609–623). Oxford, UK: Elsevier Science. doi:10.1016/B978-044451020-4/50032-3

Radcliffe, D., & Lam, A. (2018). Social Media in the Middle East: The Story of 2017. Retrieved from https://papers.ssrn.com/sol3/papers.cfm?abstract_id=3124077

Radu, M., & Redien-Collot, R. (2008). The Social Representation of Entrepreneurs in the French Press: Desirable and Feasible Models? *International Small Business Journal*, *26*(3), 259–298. doi:10.1177/0266242608088739

Raghavan, V., Wani, M., & Abraham, D. M. (2018). Exploring E-Business in Indian SMEs: Adoption, Trends and the Way Forward. In *Emerging Markets from a Multidisciplinary Perspective. Advances in Theory and Practice of Emerging Markets*. Cham: Springer. doi:10.1007/978-3-319-75013-2_9

Raguseo, E. (2018). Big data technologies: An empirical investigation on their adoption, benefits and risks for companies. *International Journal of Information Management*, *38*(1), 187–195. doi:10.1016/j.ijinfomgt.2017.07.008

Rai, A., Pavlou, P. A., Im, G., & Du, S. (2012). Inter firm IT Capability Profiles and Communications for Cocreating Relational Value: Evidence from the Logistics Industry. *Management Information Systems Quarterly*, *36*(1), 233–262.

Ranyard, J. C., Fildes, R., & Hu, T. I. (2015). Reassessing the scope of OR practice: The influences of problem structuring methods and the analytics movement. *European Journal of Operational Research*, *245*(1), 1–13. doi:10.1016/j.ejor.2015.01.058

Rappa, M. A. (2004). The utility business model and the future of computing services. *IBM Systems Journal*, *43*(1), 32–42. doi:10.1147j.431.0032

Rauniar, R., Rawski, G., Yang, J., & Johnson, B. (2014). Technology acceptance model (TAM) and social media usage: An empirical study on Facebook. *Journal of Enterprise Information Management*, *27*(1), 6–30. doi:10.1108/JEIM-04-2012-0011

Raymond, C. M., Bieling, C., Fagerholm, N., Martin-Lopez, B., & Plieninger, T. (2015). The farmer as a landscape steward: Comparing local understandings of landscape stewardship, landscape values, and land management actions. *Ambio*, 1–12. doi:10.100713280-015-0694-0 PMID:26346276

Rayner, K., Miller, B., & Rotello, C. M. (2008). Eye movements when looking at print advertisements: The goal of the viewer matters. *Applied Cognitive Psychology*, *22*(5), 697–707. doi:10.1002/acp.1389 PMID:19424446

Redien-Collot, R., & O'Shea, N. (2015). Battling with institutions: How novice female entrepreneurs contribute to shaping public policy discourse. *Revue de l'Entrepreneuriat*, *15*(2-3), 57–80.

Reichardt, M., Jürgens, C., Klöble, U., Hüter, J., & Moser, K. (2009). Dissemination of Precision Farming in Germany: Acceptance, Adoption, Obstacles, Knowledge Transfer and Training Activities. *Precision Agriculture*, *10*(6), 525–545. doi:10.100711119-009-9112-6

Reich, B., & Benbasat, I. (1996). Measuring the linkage between business and information technology objectives. *Management Information Systems Quarterly*, *20*(1), 55–81. doi:10.2307/249542

Reich, B., & Benbasat, I. (2000). Factors that influence the social dimension of alignment between business and information technology objectives. *Management Information Systems Quarterly*, *24*(1), 81–113. doi:10.2307/3250980

Reis, J., Amorim, M., Melão, N., & Matos, P. (2018, March). Digital Transformation: A Literature Review and Guidelines for Future Research. In *World Conference on Information Systems and Technologies* (pp. 411-421). Springer.

Resnick, M., & Albert, W. (2013). The Impact of Advertising Location and User Task on the Emergence of Banner Ad Blindness: An Eye-Tracking Study. *International Journal of Human-Computer Interaction*, *30*(3), 206–219. doi:10.1080/10447318.2013.847762

Reynolds, J. (September 26, 2017). Difference Between Developing Countries & Emerging Countries. Retrieved from https://bizfluent.com/info-10002682-difference-between-developing-countries-emerging-countries.html

Richardson, J. (1996). Vertical integration and rapid response in fashion apparel. *Organization Science*, *7*(4), 400–412. doi:10.1287/orsc.7.4.400

Richter. (2014). Prosumption: Evolution, revolution, or eternal return of the same? *Journal of Consumer Culture, 14*(1), 3-24.

Richter, C., Kraus, S., Brem, A., Durst, S., & Giselbrecht, C. (2017). Digital entrepreneurship: Innovative business models for the sharing economy. *Creativity and Innovation Management*, *26*(3), 300–310. doi:10.1111/caim.12227

Richter, C., Kraus, S., & Syrjä, P. (2015). The shareconomy as a precursor for digital entrepreneurship business models. *International Journal of Entrepreneurship and Small Business*, *25*(1), 18–35. doi:10.1504/IJESB.2015.068773

Riedl, R., Benlian, A., Hess, T., Stelzer, D., & Sikora, H. (2017). On the relationship between information management and digitalization. *Business & Information Systems Engineering*, *59*(6), 475–482. doi:10.100712599-017-0498-9

Riivari, J. (2005). Mobile banking: A powerful new marketing and CRM tool for financial services companies all over Europe? *Journal of Financial Services Marketing*, *10*(1), 11–20. doi:10.1057/palgrave.fsm.4770170

Rippa, P., & Secundo, G. (2018). Digital academic entrepreneurship: The potential of digital technologies on academic entrepreneurship. *Technological Forecasting and Social Change*. doi:10.1016/j.techfore.2018.07.013

Riquelme, H. E., & Rios, R. E. (2010). The moderating effect of gender in the adoption of mobile banking. *International Journal of Bank Marketing*, *28*(5), 328–341. doi:10.1108/02652321011064872

Rodon, J., Sese, F., & Christiaanse, E. (2011). Exploring users' appropriation and post-implementation managerial intervention in the context of industry IOIS. *Information Systems Journal*, *21*(3), 223–248. doi:10.1111/j.1365-2575.2009.00339.x

Rogers, D. L. (2016). *The digital transformation playbook: rethink your business for the digital age*. Columbia University Press. doi:10.7312/roge17544

Rogers, E. M. (1995). *Diffusion of Innovations*. New York: The Free Press.

Rohrer, C., & Boyd, J. (2004). The rise of intrusive online advertising and the response of user experience research at Yahoo! In CHI'04 Extended Abstracts on Human Factors in Computing Systems (pp. 1085-1086). ACM.

Rosenbaum, H., & Cronin, B. (1993). Digital entrepreneurship: Doing business on the information superhighway. *International Journal of Information Management*, *13*(461-463).

Rosenberg, N. (1963). Technological change in the machine tool industry, 1840–1910. *The Journal of Economic History*, *23*(04), 414–443. doi:10.1017/S0022050700109155

Ross, J. W., Beath, C. M., & Quaadgras, A. (2013). You may not need big data after all. *Harvard Business Review*, *91*(12), 90–98. PMID:23593770

Rothaermel, F. T., Hitt, M. A., & Jobe, L. A. (2006). Balancing vertical integration and strategic outsourcing: Effects on product portfolio, product success, and firm performance. *Strategic Management Journal*, *27*(11), 1033–1056. doi:10.1002mj.559

Royal Academy of Engineering. (2013). *Additive Manufacturing: Opportunities and Constraints*. London: Royal Academy of Engineering.

Rushkoff, D. (2016). *Reprogramming money – Bank vaults to blockchain. In Throwing Rocks at the Google Bus, How Growth Became the Enemy of Prosperity*. NY, NY: Penguin.

Ryan, D. (2016). *Understanding Digital Marketing: Marketing Strategies for Engaging the Digital Generation*. New York: Kogan Page Publishers.

Safeena, R., Date, H., Hundewale, N., & Kammani, A. (2013). Combination of TAM and TPB in internet banking adoption. *International Journal of Computer Theory and Engineering*, *5*(1), 146–150. doi:10.7763/IJCTE.2013.V5.665

Safeena, R., Date, H., Kammani, A., & Hundewale, N. (2012). Technology adoption and Indian consumers: Study on mobile banking. *International Journal of Computer Theory and Engineering*, *4*(6), 1020–1024. doi:10.7763/IJCTE.2012.V4.630

Samaranayake, P. (2009). Business process integration, automation, and optimization in ERP: Integrated approach using enhanced process models. *Business Process Management Journal*, *15*(4), 504–526. doi:10.1108/14637150910975516

Sambamurthy, V., Bharadwaj, A., & Grover, V. (2003). Shaping Agility Through Digital Options: Reconceptualizing the Role of Information Technology in Contemporary Firms. *Management Information Systems Quarterly*, *27*(2), 237–263. doi:10.2307/30036530

Sammis, K., Lincoln, C., & Pomponi, S. (2016). *Influencer marketing for dummies*. John Wiley & Sons.

Sampler, J. L., & Earl, M. J. (2014). What's your information footprint? *Sloan Management Review*, *55*(2), 95–96.

Santana, M. (2017). *Digital entrepreneurship: expanding the economic frontier in the Mediterranean*. European institute of the Mediterranean.

Santana, M. (2017). *Digital entrepreneurship: expanding the economic frontier in the mediterranean. Iemed*. European Institute of the Mediterranean.

Sarasvathy, S. D. (2008). *Effectuation: Elements of entrepreneurial expertise*. Cheltenham, U.K.: Edward Elgar Publishing. doi:10.4337/9781848440197

Schallmo, D. R., & Williams, C. A. (2018). *Digital Transformation Now! Guiding the Successful Digitalization of Your Business Model*. Springer International Publishing.

Scherer, A. G., & Palazzo, G. (2006). Corporate legitimacy as deliberation, A Communicative Framework. *Journal of Business Ethics*, *66*(1), 71–88. doi:10.100710551-006-9044-2

Scherer, A. G., Palazzo, G., & Seidl, D. (2012). Managing legitimacy in complex and heterogeneous environments: Sustainable development in a globalized world. *Journal of Management Studies*, *50*(2), 259–284. doi:10.1111/joms.12014

Schreckling, E., & Steiger, C. (2017). Digitalize or Drown. In G. Oswald & M. Kleinemeier (Eds.), *Shaping the Digital Enterprise. Trends and use cases in digital innovation and transformation* (pp. 3–27). Cham: Springer.

Schuchmann, D., & Seufert, S. (2015). Corporate Learning in Times of Digital Transformation: A Conceptual Framework and Service Portfolio for the Learning Function in Banking Organisations. *International Journal of Corporate Learning*, *8*(1), 31–39. doi:10.3991/ijac.v8i1.4440

Schumpeter, J. A. (1934). *The theory of economic development*. Cambridge, MA: Harvard University Press.

Schwab, K. (2017). *The fourth industrial revolution*. Crown Business.

Schweer, D., & Sahl, J. C. (2017). The Digital Transformation of Industry–The Benefit for Germany. In *The Drivers of Digital Transformation* (pp. 23–31). Cham: Springer. doi:10.1007/978-3-319-31824-0_3

Scott, J. E., & Vessey, I. (2000). Implementing enterprise resource planning systems: The role of learning from failure. *Information Systems Frontiers*, *2*(2), 213–232. doi:10.1023/A:1026504325010

Seidle, R. (2015). Organizational Learning Sequences In Technological Innovation: Evidence From The Biopharmaceutical And Medical Device Sectors. *International Journal of Innovation Management*, *19*(03), 1540007. doi:10.1142/S1363919615400071

Self, R. J., & Voorhis, D. (2015). Tools and technologies for the implementation of big data, In Application of Big Data for National Security (pp. 140-154).

Sghari, A., Chaabouni, J., & Schiopoiu Burlea, A. (2017). Effets de l'usage des technologies mobiles par les clients sur le métier des chargés de clientèle dans les banques: Etude de cas. In *Colloque international La transformation numérique des entreprises & les modèles prédictifs sur Big Data*. Université de Med BOUDIAF.

Shaikh, A., Karjaluoto, H., & Chinje, N. B. (2015). Continuous mobile banking usage and relationship commitment-A multi-country assessment. *Journal of Financial Services Marketing*, *20*(3), 208–219. doi:10.1057/fsm.2015.14

Shaikh, D., & Lenz, K. (2006). Where's the Search? Re-Examining User Expectations of Web Objects. *Usability News*, *8*(1), 14. Retrieved from http://usabilitynews.org/wheres-the-search-re-examining-user-expectations-of-web-objects/

Shane, S., & Venkataraman, S. (2000). The promise of entrepreneurship as a field of research. *Academy of Management Review*, *25*(1), 217–226.

Shang, G., Saladin, B., Fry, T., & Donohue, J. (2015). Twenty-six years of operations management research (1985–2010): Authorship patterns and research constituents in eleven top rated journals. *International Journal of Production Research*, *53*(20), 6161–6197. doi:10.1080/00207543.2015.1037935

Shareef, M. A., Baabdullah, A., Dutta, S., Kumar, V., & Dwivedi, Y. K. (2018). Consumer adoption of mobile banking services: An empirical examination of factors according to adoption stages. *Journal of Retailing and Consumer Services*, *43*, 54–67. doi:10.1016/j.jretconser.2018.03.003

Sharif, A. M. (2010). It's written in the cloud: The hype and promise of cloud computing. *Journal of Enterprise Information Management*, *23*(2), 131–134. doi:10.1108/17410391011019732

Sharif, A. M., Irani, Z., & Love, P. E. D. (2005). Integrating ERP with EAI: A model for post-hoc evaluation. *European Journal of Information Systems*, *14*(2), 162–174. doi:10.1057/palgrave.ejis.3000533

Sharma, M. K. (2009). Receptivity of India's small and medium-sized enterprises to information system adoption. *Enterprise Information Systems*, *3*(1), 95–115. doi:10.1080/17517570802317901

Sharma, S., Padhy, S., & Verma, V. (2014). Multi-functional social CRM in cloud with cross-platform mobile application. *International Journal of Computers and Applications*, *93*(13), 9–15.

Sharp, J. M., Irani, Z., & Desai, S. (1999). Working towards agile manufacturing in the UK industry. *International Journal of Production Economics*, *62*(1-2), 155–169. doi:10.1016/S0925-5273(98)00228-X

Sheehan, K. (2014). *Controversies in contemporary advertising (2nd ed.)*. Thousand Oaks, CA: SAGE Publications, Inc.

Shih, Y. Y., & Fang, K. (2004). The use of a decomposed theory of planned behavior to study Internet banking in Taiwan. *Internet Research*, *14*(3), 213–223. doi:10.1108/10662240410542643

Shojaeizadeh, M., Djamasbi, S., & Trapp, A. C. (2016). Density of Gaze Points Within a Fixation and Information Processing Behavior. *Int. Conf. Univers. Access Human-Computer Interact.*, 465–71.

Shrestha, S. K. (2010). Trolls or Market-Makers? An Empirical Analysis of Nonpracticing Entities. *Columbia Law Review*, *110*, 114–160.

Singh, A., & Hess, T. (2017). How Chief Digital Officers Promote the Digital Transformation of their Companies. *MIS Quarterly Executive*, *16*(1).

Singh, S., Srivastava, V., & Srivastava, R. K. (2010). Customer acceptance of mobile banking: A conceptual framework. *SIES Journal of Management*, *7*(1), 55–64.

Sivertsson, O., & Tell, J. (2015). Barriers to Business Model innovation in Swedish agriculture. *Sustainability*, *7*(2), 1957–1969. doi:10.3390u7021957

Skilton, M. (2016). *Building the digital enterprise: a guide to constructing monetization models using digital technologies*. Springer.

Smaczny, T. (2001). Is an alignment between business and information technology the appropriate paradigm to manage IT in today's organizations. *Management Decision*, *39*(10), 797–802. doi:10.1108/EUM0000000006521

Smaoui Hachicha, Z., & Mezghani, K. (2018). Understanding intentions to switch toward cloud computing at firms' level: A multiple case study in Tunisia. *Journal of Global Information Management*, *26*(1), 136–165. doi:10.4018/JGIM.2018010108

Smith, D. (2014). Finding the signal in the noise of patent citations: How to focus on relevance for strategic advantage. *Technology Innovation Management Review*, *4*(9).

Smith, D. (2015). Disrupting the Disrupter: Strategic Countermeasures to Attack the Business Model of a Coercive Patent-Holding Firm. *Technology Innovation Management Review*, *5*(5), 5–16. Retrieved from http://timreview.ca/article/894

Smith, P. (2013). *Business Development, Marketing and Sales*. Retrieved from https://bizzmaxx2012.wordpress.com/2013/12/24/types-of-advertising-in-e-commerce

Smith, H. A., & McKeen, J. D. (2003). Developments in practice VII: Developing and delivering the IT value proposition. *Communications of the Association for Information Systems*, *11*, 25.

Sodhi, M. S., & Tang, C. S. (2010). *A Long View of Research and Practice in Operations Research and Management Science*. US: Springer. doi:10.1007/978-1-4419-6810-4

Soete, L. G., & Wyatt, S. M. (1983). The use of foreign patenting as an internationally comparable science and technology output indicator. *Scientometrics*, *5*(1), 31–54. doi:10.1007/BF02097176

Songini, M. L. (2002). J.D. Edwards pushes CRM, ERP integration. *Computerworld*, *36*(25), 4.

Soule, D. L., Puram, A. D., Westerman, G. F., & Bonnet, D. (2016). *Becoming a Digital Organization: The Journey to Digital Dexterity*. Academic Press.

Souza, G. C. (2014). Supply chain analytics. *Business Horizons*, *57*(5), 595–605. doi:10.1016/j.bushor.2014.06.004

Spekman, R., & Davis, E. W. (2016). The extended enterprise: A decade later. *International Journal of Physical Distribution & Logistics Management*, *46*(1), 43–61. doi:10.1108/IJPDLM-07-2015-0164

Spigel, B. (2017). The relational organization of entrepreneurial ecosystems. *Entrepreneurship Theory and Practice*, *41*(1), 4. doi:10.1111/etap.12167

Stam, E., & Spigel, B. (2018). Entrepreneurial Ecosystems. In R. Blackburn (Ed.), *The SAGE Handbook of Small Business and Entrepreneurship*. London: SAGE Publications Ltd. doi:10.4135/9781473984080.n21

Stemler, A. (2017). The myth of the sharing economy and its implications for regulating innovation. *Emory Law Journal*, *67*, 197.

Stevens, C. P. (2003). Enterprise resource planning: A trio of resources. *Information Systems Management*, *20*(3), 61–71. doi:10.1201/1078/43205.20.3.20030601/43074.7

Stolterman, E., & Fors, A. C. (2004). Information technology and the good life. In *Information Systems Research* (pp. 687–692). Boston, MA: Springer. doi:10.1007/1-4020-8095-6_45

Stone, B. (2017). *The Upstarts: How Uber, Airbnb and the Killer Companies of the New Silicon Valley are Changing the World*. Random House.

Straub, D. W., & Watson, R. T. (2001). Transformational Issues in Researching IS and Net-Enabled Organizations. *Information Systems Research*, *12*(4), 337–345. doi:10.1287/isre.12.4.337.9706

Stroh, P. J. (2015). Business Strategy—Creation, Execution and Monetization. *Journal of Corporate Accounting & Finance*, *26*(4), 101–105. doi:10.1002/jcaf.22055

Stylidis, D., & Cherifi, B. (2018). Characteristics of destination image: Visitors and non-visitors' images of London. *Tourism Review*, *73*(1), 55–67. doi:10.1108/TR-05-2017-0090

Subramaniam, M., & Venkatraman, N. V. (2001). Determinants of Transnational New Product Development Capability: Testing the Influence of Transferring and Deploying Tacit Overseas Knowledge. *Strategic Management Journal*, *22*(4), 359–378. doi:10.1002mj.163

Suciati, P., Maulidiyanti, M., & Lusia, A. (2018). Cultivation effect of tourism Tv program and influencer's instagram account on the intention of traveling. In *The 1st International Conference on Social Sciences University of Muhammadiyah Jakarta, Indonesia, 1–2 November 2017* (pp. 267–278). Indonesia: International Conference on Social Sciences.

Sugahara, S., Daidj, N., & Ushio, S. (2017). *Value Creation in Management Accounting and Strategic Management*. London, UK: ISTE-Wiley. doi:10.1002/9781119419921

Suh, J., & Sohn, S. Y. (2015). Analyzing technological convergence trends in a business ecosystem. *Industrial Management & Data Systems*, *115*(4), 718–739. doi:10.1108/IMDS-10-2014-0310

Suh, Y., & Lee, H. (2017). Developing ecological index for identifying roles of ICT industries in mobile ecosystems: The inter-industry analysis approach. *Telematics and Informatics*, *34*(1), 425–437. doi:10.1016/j.tele.2016.06.007

Sung, K., Kim, T., & Kong, H. K. (2010). Microscopic approach to evaluating technological convergence using patent citation analysis. In T.-h. Kim, J. Ma, W. c. Fang, B. Park, B.-H. Kang, & D. Ślęzak (Eds.), *U-and E-Service, Science and Technology* (pp. 188–194). Berlin: Springer. doi:10.1007/978-3-642-17644-9_21

Suoranta, M., Mattil, M., & Munnukka, J. (2005). Technology-based services: A study on the drivers and inhibitors of mobile banking. *International Journal of Management and Decision Making*, *6*(1), 33–46. doi:10.1504/IJMDM.2005.005964

Supply Chain Council. (2008). SCOR Framework – Introducing all elements of the supply chain reference model: Standard processes, metrics and best practices. Available at supplychainresearch.com/images/SCOR_Framework_2.1.ppt

Susarla, A., Oh, J. H., & Tan, Y. (2012). Social Networks and the Diffusion of User-Generated Content: Evidence from YouTube. *Information Systems Research*, *23*(1), 123–141. doi:10.1287/isre.1100.0339

Sussan, F., Autio, E., & Kosturik, J. (2016). *Leveraging ICTs for Better Lives: The Introduction of an Index on Digital Life*. Academic Press.

Sussan, F., & Acs, Z. J. (2017). The digital entrepreneurial ecosystem. *Small Business Economics*, *49*(1), 55–73. doi:10.100711187-017-9867-5

Sutton, S. G. (2006). Extended-enterprise systems' impact on enterprise risk management. *Journal of Enterprise Information Management*, *19*(1), 97–114. doi:10.1108/17410390610636904

Tahawultech (2018). Saudi Arabia IT spend to reach $40 billion. https://www.tahawultech.com/news/saudi-arabia-spend-hit-40-billion/ (Retrieved: 14/05/2018).

Tambe, P. (2014). Big data investment, skills, and firm value. *Management Science*, *60*(6), 1452–1469. doi:10.1287/mnsc.2014.1899

Tan, K. H., Zhan, Y. Z., Ji, G., Ye, F., & Chang, C. (2015). Harvesting big data to enhance supply chain innovation capabilities: An analytic infrastructure based on deduction graph. *International Journal of Production Economics*, *165*, 223–233. doi:10.1016/j.ijpe.2014.12.034

Tan, K.H., Zhan, Y.Z., & Ji, G. Ye F. and Chang, Ch. (2015). Harvesting big data to enhance supply chain innovation capabilities: An analytic infrastructure based on deduction graph. *International Journal of Production Economics*, 1–11.

Tanriverdi, H., & Venkatraman, N. V. (2005). Knowledge Relatedness and the Performance of Multibusiness Firms. *Strategic Management Journal*, *26*(2), 97–11. doi:10.1002mj.435

Tarantilis, C. D., Kiranoudis, C. T., & Theodorakopoulos, N. D, (2008). A web-based ERP system for business service and supply chain management: Application to real-world process scheduling. *European Journal of Operational Research*, *187*(3), 1310–1326. doi:10.1016/j.ejor.2006.09.015

Teece, D. J. (2010). Business Models, business strategy and innovation. *Long Range Planning*, *43*(2), 172–194. doi:10.1016/j.lrp.2009.07.003

Teece, D. J. (2018). Profiting from innovation in the digital economy: Enabling technologies, standards, and licensing models in the wireless world. *Research Policy*, *47*(8), 1367–1387. doi:10.1016/j.respol.2017.01.015

Tekic, Z., & Kukolj, D. (2013). Threat of Litigation and Patent Value: What Technology Managers Should Know. *Research Technology Management*, *56*(2), 18–25. doi:10.5437/08956308X5602093

Tenenhaus, M., Amato, S., & Esposito Vinzi, V. (2004). A global goodness-of-fit index for PLS structural equation modelling. In *Proceedings of the XLII SIS scientific meeting* (Vol. 1, pp. 739-742). CLEUP Padova.

Tennant, G. (2001). *Six Sigma: SPC and TQM in Manufacturing and Services*. Gower Publishing, Ltd.

Themistocleous, M., Irani, Z., & O'Keefe, R. (2001). ERP and application integration: Exploratory survey. *Business Process Management Journal*, *7*(3), 195–204. doi:10.1108/14637150110392656

Thong, J. Y. L. (1999, Spring). An Integrated Model of Information Systems Adoption in Small Businesses. *Journal of Management Information Systems*, *15*(4), 187–214. doi:10.1080/07421222.1999.11518227

Thong, J. Y. L., & Yap, C. S. (1995). CEO Characteristics, Organizational Characteristics and Information Technology Adoption in Small Businesses. *Omega*, *23*(4), 429–442. doi:10.1016/0305-0483(95)00017-I

Thong, J. Y., Hong, W., & Tam, K. Y. (2002). Understanding user acceptance of digital libraries: What are the roles of interface characteristics, organizational context, and individual differences? *International Journal of Human-Computer Studies*, *57*(3), 215–242. doi:10.1016/S1071-5819(02)91024-4

Thun, J. H. (2010). Angles of integration: An empirical analysis of the alignment of internet-based information technology and global supply chain integration. *The Journal of Supply Chain Management*, *46*(2), 30–44. doi:10.1111/j.1745-493X.2010.03188.x

Tian, W., & Zhao, Y. (2015). *Big Data Technologies and Cloud Computing*. In *Optimized Cloud Resource Management and Scheduling* (pp. 17–49). doi:10.1016/B978-0-12-801476-9.00002-1

Tijssen, R. J. (1992). A quantitative assessment of interdisciplinary structures in science and technology: Co-classification analysis of energy research. *Research Policy*, *21*(1), 27–44. doi:10.1016/0048-7333(92)90025-Y

Timmers, P. (1998). Business Models of electronic markets. *Electronic Markets*, *8*(2), 3–8. doi:10.1080/10196789800000016

Tiwana, A., Konsynski, B., & Bush, A. A. (2010). Research commentary - Platform evolution: Coevolution of platform architecture, governance, and environmental dynamics? *Information Systems Research*, *21*(4), 675–687. doi:10.1287/isre.1100.0323

Tiwari, S., Wee, H. M., & Daryanto, Y. (2018). Big data analytics in supply chain management between 2010 and 2016: Insights to industries. *Computers & Industrial Engineering*, *115*, 319–330. doi:10.1016/j.cie.2017.11.017

Tobii Studio 2.X User Manual. Manual release 1.0, September 2010.

Toffler, A. (1980). *The third wave*. New York: Bantam Books.

Torbacki, W. (2008). SaaS – direction of technology development in ERP/MRP systems. *Archives of Materials Science and Engineering*, *31*(1), 57–60.

Trainor, K. J., Andzulis, J. M., Rapp, A., & Agnihotri, R. (2014). Social media technology usage and customer relationship performance: A capabilities-based examination of social CRM. *Journal of Business Research*, *67*(6), 1201–1208. doi:10.1016/j.jbusres.2013.05.002

Traitler, H., Dubois, M. J. F., Heikes, K., Petiard, V., & Zilberman, D. (2018). *Megatrends in Food and Agriculture: Technology, Water Use and Nutrition*. John Wiley & Sons Ltd.

Trajtenberg, M. (1990). A penny for your quotes: Patent citations and the value of innovations. *The Rand Journal of Economics*, *21*(1), 172–187. doi:10.2307/2555502

Trauth, E. M., Farwell, D. W., & Lee, D. (1993). The IS expectation Gap: Industry Expectations versus Academic Preparation. *Management Information Systems Quarterly*, *17*(September), 293–307. doi:10.2307/249773

Trevor, J. (2018, January 12). Is anyone in your company paying attention to strategic alignment. *Harvard Business Review Digital Articles*.

Triantafillakis, A., Kanellis, P., & Martakos, D. (2004). Data warehousing interoperability for the extended enterprise. *Journal of Database Management*, *15*(3), 73–82. doi:10.4018/jdm.2004070105

Troxler, P., & Wolf, P. (2017). Digital maker-entrepreneurs open design: What activities make up their business model? *Business Horizons*, *60*(6), 807–817. doi:10.1016/j.bushor.2017.07.006

Tseng, F. M., Hsieh, C. H., Peng, Y. N., & Chu, Y. W. (2011). Using patent data to analyze trends and the technological strategies of the amorphous silicon thin-film solar cell industry. *Technological Forecasting and Social Change*, *78*(2), 332–345. doi:10.1016/j.techfore.2010.10.010

Tse, T., & Soufani, K. (2003). Business strategies for small firms in the new economy. *Journal of Small Business and Enterprise Development*, *10*(3), 306–320. doi:10.1108/14626000310489781

Tumbas, S., Berente, N., Seidel, S., & Brocke, V. J. (2015). The 'digital façade' of rapidly growing entrepreneurial organizations. In *International Conference on Information Systems*, Fort Worth, TX.

Turban, E., King, D., Lee, J. K., Liang, T. P., & Turban, D. C. (2015). Business-to-business E-commerce. In *Electronic Commerce* (pp. 161–207). Cham: Springer. doi:10.1007/978-3-319-10091-3_4

Udoh, E. E. (2010). *The adoption of grid computing technology by organizations: A quantitative study using technology acceptance model* [Doctoral Dissertation]. Capella University.

UniLaSalle. (2017). *Agriculture connectée : un bootcamp sur l'internet des objets à UniLaSalle*. Retrieved from https://web.archive.org/web/20180415163809/https://www.unilasalle.fr/lasalleactus/agriculture-connectee-bootcamp-linternet-objets-lagriculture-a-unilasalle/

Urbach, N., Drews, P., & Ross, J. (2017). Digital business transformation and the changing role of the IT Function. *Comments on the special issue. MIS Quarterly Executive, 16*(2).

Valentini, G., & Di Guardo, M. C. (2012). M&A and the profile of inventive activity. *Strategic Organization, 10*(4), 384–405. doi:10.1177/1476127012457980

Vallespir, B., & Kleinhans, S. (2001). Positioning a company in enterprise collaborations: Vertical integration and make-or-buy decisions. *Production Planning and Control, 12*(5), 478–487. doi:10.1080/09537280110042701

Van de Ven, A. H., & Poole, M. S. (1990). Methods for studying innovation development in the Minnesota Innovation Research Program. *Organization Science, 1*(3), 313–335. doi:10.1287/orsc.1.3.313

Van der Panne, G., Van Beers, C., & Kleinknecht, A. (2015). Success and Failure of Innovation: A Literature Review. *International Journal of Innovation Management, 7*(3), 309–338. doi:10.1142/S1363919603000830

van Eck, N. J., & Waltman, L. (2007). Bibliometric mapping of the computational intelligence field. *International Journal of Uncertainty, Fuzziness and Knowledge-based Systems, 15*(05), 625–645. doi:10.1142/S0218488507004911

van Eck, N. J., Waltman, L., Dekker, R., & van den Berg, J. (2010). A comparison of two techniques for bibliometric mapping: Multidimensional scaling and VOS. *Journal of the American Society for Information Science and Technology, 61*(12), 2405–2416. doi:10.1002/asi.21421

van Eck, N. J., Waltman, L., van den Berg, J., & Kaymak, U. (2006). Visualizing the computational intelligence field [Application Notes]. *IEEE Computational Intelligence Magazine, 1*(4), 6–10.

van Welsum, D. (2016). *Enabling Digital Entrepreneurs. WDR 2016 Background Paper*. Washington, DC: World Bank. Retrieved at www.openknowledge.worldbank.org/handle/10986/23646

Vanderschuren, H. (2012). Can genomics boost productivity of orphan crops? *Nature Biotechnology, 30*(12), 1172–1175. doi:10.1038/nbt.2440 PMID:23222781

Vannoy, S. A., & Salam, A. F. (2010). Managerial interpretations of the role of information systems in competitive actions and firm performance: A grounded theory investigation. *Information Systems Research, 21*(3), 496–515. doi:10.1287/isre.1100.0301

Vargo, R. F., & Lusch, S. L. (2004). Evolving for a new dominant logic for marketing. *Journal of Marketing, 68*(1), 1–17. doi:10.1509/jmkg.68.1.1.24036

Värlander, S., & Julien, A. (2010). The Effect of the Internet on Front-line Employee Skills: Exploring Banking in Sweden and France. *Service Industries Journal, 30*(8), 1245–1261. doi:10.1080/02642060802350979

Vasilchenko, E., & Morrish, S. (2011). The role of entrepreneurial networks in the exploration and exploitation of internationalization opportunities by information and communication technology firms. *Journal of International Marketing, 19*(4), 88–105. doi:10.1509/jim.10.0134

Vathanophas, V. (2007). Business process approach towards an inter-organisational enterprise system. *Business Process Management Journal, 13*(3), 433–450. doi:10.1108/14637150710752335

Vendrell-Herrero, F., Parry, G., Bustinza, O. F., & Gomes, E. (2018). Digital business models: Taxonomy and future research avenues. *Strategic Change, 27*(2), 87–90. doi:10.1002/jsc.2183

Venkatesh, V., & Davis, F. D. (1996). A model of the antecedents of perceived ease of use: Development and test. *Decision Sciences, 27*(3), 451–481. doi:10.1111/j.1540-5915.1996.tb01822.x

Venkatesh, V., Morris, M. G., Davis, G. B., & Davis, F. D. (2003). User acceptance of information technology: Toward a unified view. *Management Information Systems Quarterly, 27*(3), 425–478. doi:10.2307/30036540

Venkatesh, V., Speier, C., & Morris, M. G. (2002). User acceptance enablers in individual decision making about technology: Toward an integrated model. *Decision Sciences, 33*(2), 297–316. doi:10.1111/j.1540-5915.2002.tb01646.x

Venkatraman, N., & Henderson, J. C. (1998). Real strategies for virtual organizing. *Sloan Management Review, 40*(1), 33–48.

Verdouw, C. N., Beulens, A. J. M., & van der Vorst, J. G. A. J. (2013). Virtualisation of floricultural supply chains: A review from an Internet of Things perspective. *Computers and Electronics in Agriculture, 99*, 160–175. doi:10.1016/j.compag.2013.09.006

Verley, P. (1985). La révolution industrielle. Paris: Gallimard.

Vinzi, V. E., Trinchera, L., & Amato, S. (2010). PLS Path Modeling: From Foundations to Recent Developments and Open Issues for Model Assessment and Improvement. In *V.E. Vinzi, W.W. Chin, J. Henseler et al. (Eds.), Handbook of Partial Least Squares: Concepts, Methods and Applications*. Berlin: Springer-Verlag. doi:10.1007/978-3-540-32827-8_3

Voetglin, C., & Scherer, A. G. (2017). Responsible innovation and the innovation of responsibility: Governing sustainable development in a globalized world. *Journal of Business Ethics, 143*(2), 227–243. doi:10.100710551-015-2769-z

Vogel, C., Schindler, K., & Roth, S. (2015). 3D Scene Flow Estimation with a Piecewise rigid Scene Model. *International Journal of Computer Vision, 115*(1), 1–28. doi:10.100711263-015-0806-0

von Briel, F., Davidsson, P., & Recker, J. (2018). Digital technologies as external enablers of new venture creation in the IT hardware sector. *Entrepreneurship Theory and Practice, 42*(1), 47–69. doi:10.1177/1042258717732779

Von Hippel, E. (2009). Democratizing innovation: The evolving phenomenon of user innovation. *International Journal of Innovation Science, 1*(1), 29–40. doi:10.1260/175722209787951224

Von Leipzig, T., Gampa, M., Manza, D., Schöttlea, K., Ohlhausena, P., Oosthuizenb, G., & Palma, D. (2017). Initialising customer-orientated digital transformation in enterprises. *Procedia Manufacturing, 8*, 517–524. doi:10.1016/j.promfg.2017.02.066

Waaijer, C. J., van Bochove, C. A., & van Eck, N. J. (2010). Journal Editorials give indication of driving science issues. *Nature, 463*(7278), 157–157. doi:10.1038/463157a PMID:20075899

Wachal, R. (1971). Humanities and Computers: A Personal View. *The North American Review, 256*(1), 30–33.

Waller, M. A., & Fawcett, S. E. (2013). Data science, predictive analytics, and big data: A revolution that will transform supply chain design and management. *Journal of Business Logistics, 34*(2), 77–84. doi:10.1111/jbl.12010

Wamba, F. S., Akter, S., Edwards, A., Chopin, G., & Gnanzou, D. (2015). How 'big data' can make big impact: Findings from a systematic review and a longitudinal case study. *International Journal of Production Economics*, 1–33. doi:10.1016/j.ijpe.2014.12.031

Wamba, F. S., & Mishra, D. (2017). Big data integration with business processes: A literature review. *Business Process Management Journal*, *23*(3). doi:10.1108/BPMJ-02-2017-0047

Wang, G., Gunasekaran, A., Ngai, E. W., & Papadopoulos, T. (2016). Big data analytics in logistics and supply chain management: Certain investigations for research and applications. *International Journal of Production Economics*, *176*, 98–110. doi:10.1016/j.ijpe.2016.03.014

Wang, J., & Zhang, J. (2016). Big data analytics for forecasting cycle time in semiconductor wafer fabrication system. *International Journal of Production Research*, *54*(23), 7231–7244. doi:10.1080/00207543.2016.1174789

Wang, Y., Kung, L., & Byrd, T. A. (2018). Big data analytics: Understanding its capabilities and potential benefits for healthcare organizations. *Technological Forecasting and Social Change*, *126*, 3–13. doi:10.1016/j.techfore.2015.12.019

Wang, Y., Yu, Q., & Fesenmaier, D. R. (2002). Defining the virtual tourist community: Implications for tourism marketing. *Tourism Management*, *23*(4), 407–417. doi:10.1016/S0261-5177(01)00093-0

Wästlund, E., Shams, P., & Otterbring, T. (2018). Unsold is unseen … or is it? Examining the role of peripheral vision in the consumer choice process using eye-tracking methodology. *Appetite*, *120*, 49–56. doi:10.1016/j.appet.2017.08.024 PMID:28851559

Wedel, M., & Pieters, R. (2008). A Review of Eye-Tracking Research in Marketing. Review of Marketing Research.

Weill, P., & Broadbent, M. (1998). *Leveraging the New Infrastructure*. Boston: Harvard Business School Press.

Wernerfelt, B. (1984). The Resource-Based View of the Firm. *Strategic Management Journal*, *5*(2), 171–180. doi:10.1002mj.4250050207

Wernerfelt, B. (1989). From critical resources to corporate strategy. *Journal of General Management*, *14*(3), 4–12. doi:10.1177/030630708901400301

Wessels, L., & Drennan, J. (2010). An investigation of consumer acceptance of M-banking. *International Journal of Bank Marketing*, *28*(7), 547–568. doi:10.1108/02652321011085194

Westerlund, M., Leminen, S., & Rajahonka, M. (2014). Designing Business Models for the Internet of Things. *Technology Innovation Management Review*, *4*(7), 5–14. doi:10.22215/timreview/807

Westerman, G., Calméjane, C., Bonnet, D., Ferraris, P., & McAfee, A. (2011). *Digital Transformation: A Roadmap for Billion-Dollar Organizations*. Retrieved from https://www.capgemini.com/resource-file-access/resource/pdf/Digital_Transformation__A_Road-Map_for_Billion-Dollar_Organizations.pdf

Westerman, G., Bonnet, D., & McAfee, A. (2014). *Leading Digital: Turning Technology into Business Transformation*. Boston: Harvard Business Press.

Westerman, G., Bonnet, D., & McAfee, A. (2014). *Leading digital: Turning technology into business transformation*. Harvard Business Press.

Westerman, G., Calméjane, C., Bonnet, D., Ferraris, P., & McAfee, A. (2011). *Digital Transformation: A Roadmap for Billion-Dollar Organizations*. MIT Center for Digital Business and Capgemini Consulting.

West, J., & Bogers, M. (2017). Open innovation: Current status and research opportunities. *Innovation*, *19*(1), 43–50. doi:10.1080/14479338.2016.1258995

Weston, F.C., Jr. (2002). *A vision for the future of extended enterprise systems* [Presentation]. In *J.D. Edwards FOCUS Users Conference*, Denver, CO, June 12.

Wetzels, M., Odekerken-Schröder, G., & Van Oppen, C. (2009). Using PLS path modeling for assessing hierarchical construct models: Guidelines and empirical illustration. *Management Information Systems Quarterly*, *33*(1), 177–195. doi:10.2307/20650284

Whalen, P. S., & Akaka, M. A. (2015). A dynamic market conceptualization for entrepreneurial marketing: The co-creation of opportunities. *Journal of Strategic Marketing*, *24*(1), 61–75.

Wheeler, B. C. (2002). NEBIC: A Dynamic Capabilities Theory for Assessing Net-Enablement. *Information Systems Research*, *13*(2), 125–146. doi:10.1287/isre.13.2.125.89

White, M. (2012). Digital workplaces: Vision and reality. *Business Information Review*, *29*(4), 205–214. doi:10.1177/0266382112470412

Wilkes, L., & Veryard, R. (2004) Service-oriented architecture: considerations for agile systems. *Microsoft Architect Journal,* (April). Retrieved from www.msdn2.microsoft.com

Wöber, K. (2006). Domain specific search engines. In D. R. Fesenmaier (Ed.), *Destination recommendation systems: Behavioral foundations and applications*. Wallingford, UK: CABI. doi:10.1079/9780851990231.0205

Wong, D. (2012). *Data is the Next Frontier, Analytics the New Tool: Five Trends in Big Data and Analytics, and Their Implications for Innovation and Organisations*. London: Big Innovation Centre.

Word of Mouth Marketing Association-WOMMA. (2013). *Influencer guidebook 2013*. Retrieved from https://es.slideshare.net/svenmulfinger/womma-influencer-guidebook-2013-pdf

World Population Prospect. (2012). Available from: https://esa.un.org/unpd/wpp/publications/files/key_findings_wpp_2015.pdf

Xiang, Z., Schwartz, Z., Gerdes, J. H. Jr, & Uysal, M. (2015). What can big data and text analytics tell us about hotel guest experience and satisfaction? *International Journal of Hospitality Management*, *44*(January), 120–130. doi:10.1016/j.ijhm.2014.10.013

Xiang, Z., Wöber, K., & Fesenmaier, D. R. (2008). Representation of the online tourism domain in search engines. *Journal of Travel Research*, *47*(2), 137–150. doi:10.1177/0047287508321193

Xie, C., Bagozzi, R. P., & Troye, S. V. (2008). Trying to prosume: Toward a theory of consumers as co-creators of value. *Journal of the Academy of Marketing Science*, *36*(1), 109–122. doi:10.100711747-007-0060-2

Xu, W., Wei, Y., & Fan, Y. (2002). Virtual enterprise and its intelligence management. *Computers & Industrial Engineering*, *42*(2-4), 199–205. doi:10.1016/S0360-8352(02)00053-0

Yang, A. S. (2009). Exploring Adoption Difficulties in Mobile Banking Services. *Canadian Journal of Administrative Sciences*, *26*(13), 136–149. doi:10.1002/cjas.102

Yin, R. K. (1994). *Case study research : Design and methods*. London, UK: SAGE Publications.

Yin, S., & Kaynak, O. (2015). Big Data for Modern Industry: Challenges and Trends. *Proceedings of the IEEE*, *103*(2), 143–146. doi:10.1109/JPROC.2015.2388958

Yoo, Y., Boland, R. J., Jr., Lyytinen, K., & Majchrzak, A. (2012). Organizing for Innovation in the Digitized World. *Organization Science*, *23*(5), 1398-1408.

Yoo, Y., Boland, R. J. Jr, Lyytinen, K., & Majchrzak, A. (2012). Organizing for innovation in the digitized world. *Organization Science*, *23*(5), 1398–1408. doi:10.1287/orsc.1120.0771

Yoo, Y., Henfridsson, O., & Lyytinen, K. (2010). Research commentary-The new organizing logic of digital innovation: An agenda for information systems research. *Information Systems Research*, *21*(4), 724–735. doi:10.1287/isre.1100.0322

Yu, Y., Yi, W., Feng, Y., & Liu, J. (2018). Understanding the Intention to Use Commercial Bike-sharing Systems: An Integration of TAM and TPB. In *Proceedings of the 51st Hawaii International Conference on System Sciences*. 10.24251/HICSS.2018.082

Zhang, Y., Ren, S., Liu, Y., & Si, S. (2017). A big data analytics architecture for cleaner manufacturing and maintenance processes of complex products. *Journal of Cleaner Production*, *142*, 626–641. doi:10.1016/j.jclepro.2016.07.123

Zhao, F., & Collier, A. (2016). Digital entrepreneurship: Research and practice. In *9th Annual conference of the EuroMed academy of business*, Warsaw, Poland, September 14–16.

Zhao, F., & Collier, A. (2016). Digital entrepreneurship: Research and practice. *9th Annual Conference of the EuroMed Academy of Business*.

Zhong, R. Y., Huang, G. Q., Lan, S., Dai, Q. Y., Xu, C., & Zhang, T. (2015). A big data approach for logistics trajectory discovery from RFID-enabled production data. *International Journal of Production Economics*, *165*, 260–272. doi:10.1016/j.ijpe.2015.02.014

Zhong, R. Y., Lan, S., Xu, C., Dai, Q., & Huang, G. Q. (2016). Visualization of RFID-enabled shopfloor logistics big data in cloud manufacturing. *International Journal of Advanced Manufacturing Technology*, *84*(1–4), 5–16. doi:10.100700170-015-7702-1

Zhou, T. (2012). Examining mobile banking user adoption from the perspectives of trust and flow experience. *Information Technology Management*, *13*(1), 27–37. doi:10.100710799-011-0111-8

Zhou, T., Lu, Y., & Wang, B. (2010). Integrating TTF and UTAUT to explain mobile banking user adoption. *Computers in Human Behavior*, *26*(4), 760–767. doi:10.1016/j.chb.2010.01.013

Zhu, F., & Furr, N. (2016). Products to Platforms: Making the Leap. *Harvard Business Review*, 1.

Zhu, K., Kraemer, K. L., & Xu, S. (2006). The process of innovation assimilation by firms in different countries: A technology diffusion perspective on e-business. *Management Science*, *52*(10), 1557–1576. doi:10.1287/mnsc.1050.0487

Zupic, I., & Čater, T. (2015). Bibliometric methods in management and organization. *Organizational Research Methods*, *18*(3), 429–472. doi:10.1177/1094428114562629

About the Contributors

Karim Mezghani is an associate professor of Business Administration at Al Imam Mohammad Ibn Saud Islamic University, Saudi Arabia (College of Economics & Administrative Sciences). He received his M.Sc. degree and his Ph.D. degree from Sfax University, Tunisia (specialty: Management Information Systems). In addition to his experience as an associate professor in Tunisia and Saudi Arabia, Dr. Mezghani is a regular contributor to scientific conferences and indexed Journals iincluding "Journal of Global Information Management", "International Journal of Enterprise Information Systems", "International Journal of Technology and Human Interaction" and "Electronic Journal of Information Systems Evaluation". His researches focus on Enterprise Resource Planning (ERP) implementation, cloud computing adoption and consultant-client relationship.

Wassim J. Aloulou is an assistant professor at the Department of Business Administration of the College of Economics and Administrative Sciences, Al Imam Mohammad Ibn Saud Islamic University (Saudi Arabia) and of the ISAAS, Sfax University (Tunisia). He received his B.Sc. degree in Higher Commercial Studies from IHEC Carthage (1995), his M.Sc. degree in HR management from ISG Tunis (1998) in Tunisia, and his Ph.D. degree in Management Sciences (Specialty: Entrepreneurship) from Pierre Mendès France University (UPMF) of Grenoble 2, in France (2008). His research interests include (social) entrepreneurship, (social) entrepreneurial intentions and orientations of individuals and organizations, and entrepreneurship and innovation in the digital era. He has published in indexed journals including "Journal of Small Business and Enterprise Development", "Journal of Enterprising culture", "Journal of Entrepreneurship in Emerging Economies", "Journal of International Business and Entrepreneurship Development", "Middle East Journal of Management", "Journal of Family Business Management", and "European Journal of Innovation Management" among others. He has also contributed with books' chapters in strategic management in SMEs, chaos and complexity theories, (social and academic) entrepreneurship, women's economic empowerment, entrepreneurial intentions and behaviors; and with communications in international conferences including "International Conference on Innovation and Entrepreneurship ICIE" in 2013; 2015 and 2017 among others.

* * *

Bilal A. Al-Khateeb is an assistant professor in the Department of Business Administration, Faculty of Economics and Administrative Sciences, Al-Imam Mohammad Ibn Saud Islamic University, Riyadh 11564, Saud Arabia. He received his B.A degree in Business Administration specialization (MIS) from Sindh University in 2007, his M.B.A degree in Business Administration specialization (MIS) from Sindh

University in 2007, and his PhD degree in Management Information System from Universiti Utara Malaysia, Kedah, Malaysia in 2015. His research interest includes information sources and information channels, personal and situational characteristics, information choice strategies, information sharing, Blockchain.

Mohammed AbdulAziz Almansour is an assistant professor in the business administration department at Al Imam Mohammad Ibn Saud Islamic University in Riyadh in Saudia Arabia. He holds a PhD in Management Information Systems and E-Marketing from University of Wollongong in Australia. He holds MBA from King Abdulaziz Univeristy in Jeddah in Saudi Arabia. In addition to that, he got his bachelor degree in Management Information Systems and Risk Management from Washington State Univeristy in USA. He is very passionate about innovation and entrepreneurship.

Indira Ananth holds a Ph.D in Economics (Institute for Social and Economic Change, Bangalore) and M.Tech in Management Studies (Indian Institute of Science, Bangalore). She has over 25 years experience in action research, consultancy, training and teaching. Her research interests include technology diffusion, local governance and understanding macroeconomy. She currently teaches in Loyola Institute of Business Administration where she is Professor (Economics).

Manuel Castriotta is Research Fellow in Organization Studies at the Department of Economic and Business Sciences - University of Cagliari (Italy). His research interests are related to the influence of organizational studies on innovation and entrepreneurship. Methodologically, he combines both patent technological positioning and science mapping approaches by exploiting bibliometric techniques. He has published articles in high standing journals including International Small Business Journal and Scientometrics.

Ben Clegg (PhD) is a Professor of Operations Management at Aston Business School. He teaches, researches and consults in the area of operations improvement. He has worked with many companies both large and small. He has over 200 publications and brought in millions of pounds (GBP) of funding. He has held positions in senior management at Aston University and served on the European Operations Management Association (EurOMA) board. He is a Chartered Engineer and a Fellow of the Institute of Engineering and Technology (IET).

Nabyla Daidj, Associate Professor (PhD, HDR) of Strategy and Information Systems at Institut Mines -Télécom Business School. Her research interests are corporate and international strategy, inter-organizational relationships (strategic alliances, clusters, business ecosystems and coopetitive practices), innovation (open and disruptive innovation) and corporate governance. She published several books about "cooperation, game theory and strategic management", the sources of value creation by firms operating in the media sector within the context of ICT convergence (Developing Strategic Business Models and Competitive Advantage in the Digital Sector in 2015) and the evolution of governance (Strategy, Structure and Corporate Governance. Expressing Inter-Firms Networks in 2016). Several of her papers have been published in international journals (such as Journal of Media Business Studies, Journal of Media Economics, Communications & Strategies, Leadership, Journal of Business Ethics, Journal of High Technology Management Research). She is currently researching the evolution of business models of IT companies over the past ten years in a context of digital transformation.

Jérôme Dantan has a PhD in computer science and ENSEA French "Grande Ecole" Engineer. After working as a software engineer for Dassault group, he is currently Associate professor in computer science in UniLaSalle, specialist in Information Technology, and more particularly in IoT and Big Data issues. Member of INTERACT Research Unit UP 2018.C102.

Carlos de Laguno Alarcón is a Social Media Marketing specialist who help companies to reach their objectives by understanding their target market and designing tailored digital strategies to satisfy the needs of the audience. Working or pursuing university studies in Spain, United Kingdom, The Netherlands, South Korea and the United States, added to his passion for travel, has led Carlos to acquire a global approach across multiple communication and marketing disciplines. Carlos is now managing and mentoring influencer marketing campaigns internationally allowing tourism industry businesses to communicate in an effective way under the challenges of the digital world.

Maria Chiara Di Guardo is Full Professor of Innovation Management at the University of Cagliari (Italy). Her research focuses on how to organize efficiently the innovation process as well as the relationship between innovation and entrepreneurship. She has published a number of research articles for high standing international journals including Research Policy, International Small Business Journal, Journal of Business Research, Journal of Technology Transfer, Small Business Economics, Long Range Planning, Strategic Organization, Science and Public Policy, The Service Industries Journal, Management Research, Scientometrics, and Journal of Infometrics. She is also the author or co-author of books related to innovation management and entrepreneurship.

Michel J.F. Dubois is an expert in Agriculture Sciences at UniLaSalle, PhD in Plant Molecular Biology and in Philosophy and Habilitation to supervise research (HDR). Member of INTERACT Research Unit UP 2018.C102 (Innovation Territoire Agriculture et Agroindustrie, Connaissance et Technologie). He teaches Agriculture History and its challenges of today, Epistemology, History of technology. His main domains are Sustainable Development and Human and Technology Relationships in Agriculture.

Evanthia Faliagka holds a Diploma in Computer Engineering and Informatics (2006, University of Patras, School of Engineering), an MSc in Computer Science (2008, University of Patras, School of Engineering) and a PhD in Computer Science (2012, thesis: Web Mining and its applications in recommender systems). She is a teaching and research assistant at Technological Educational Institute of Western Greece. She teaches both undergraduate and graduate courses on Programming, Databases, Operating Systems and Software Engineering. Her research interests include the design and development of recommendations systems based on personality analysis, web mining, machine learning, opinion mining and usability analysis of web and mobile applications. She has supervised undergraduate theses on web and social mining and has authored many well-cited articles on web and social mining.

Mondher Feki is an Associate Researcher at LEMNA Research Center. His research interests include big data and business analytics, digital transformation, information system evaluation and business value. He received an MSc in Strategy and Organization, from the University of Paris Dauphine, and a Ph.D. in Management Information Systems, from the University of Paris Nanterre in France.

Dionysia Filiopoulou received her Computer Engineering Diploma and her MSc degree from the Computer Engineering and Informatics Department (CEID) of the University of Patras, between 2011 and 2017. Her diploma thesis is titled "Eye Tracking Metrics: Case Study on Display Advertising" (Supervisor: John Garofalakis, Professor). Between 2015 - 2016 she served as a professor's assistant in a laboratory based course called "Basic Electronics" in the department of Computer Engineering. Since March of 2018 she has been working as a Web Developer in the private sector.

Fatma Fourati-Jamoussi is an Associate Professor in Marketing and Strategic Intelligence at UniLaSalle, PhD in Management Sciences from the University Paris Dauphine. Member of INTERACT Research Unit UP 2018.C102 (Innovation Territoire Agriculture et Agroindustrie, Connaissance et Technologie). She teaches mainly the Economic and Business Intelligence. Her research interests concern the evaluation of business intelligence tools, and technology intelligence process, Sustainable Development and innovation in Higher Engineering Education.

Mehdi Jaber is an AgriLab Manager at UniLaSalle. His field interests deal with New Technologies and Digital Innovation Management.

Michela Loi, Ph.D., is Assistant Professor at the University of Cagliari (Italy), Department of Economic and Business Sciences. Her main research interests are on entrepreneurship and technology transfer. Some of her works have been published in Science and Public Policy, International Small Business Journal, and International Journal of Entrepreneurship Behaviour and Research.

Rim Louati is an assistant professor in management information systems in the Higher Institute of Business Administration at University of Sfax (Tunisia). She holds a PhD in Management Sciences from the University of Toulouse (France). Her research interests concern the evaluation of information systems.

Elona Marku is a Postdoctoral Research Fellow in Innovation Management at the University of Cagliari (Italy). She was visiting scholar at Columbia Business School (Columbia University) and CASS Business School (City-University of London) where she conducted research on innovation and strategy. Her current research interests focus on a better understanding of the firm strategies in the era of digital transformation. Marku's studies have been published in peer-reviewed journals such as Journal of Technology Transfer and Journal of Technology Analysis and Strategic Management. She also co-authored several book chapters and conference proceedings.

Sonia Mekadmi is a lecturer in management information systems in the Institute of Higher Commercial Studies at University of Carthage (Tunisia). She holds a Ph.D. In Management Sciences from the University of Toulouse (France). Her research interests concern the influence of organizational culture on the success of ERP of industrial groups, and the management of information systems in SMEs.

Madhava Priya is an Assistant Professor at LIBA. She was awarded the Research scholarship by WFI- Ingolstadt and AUDI, Germany. She has over 5 years of corporate experience in capital markets in various roles and 7 years of experience in teaching. Her research interests include Capital Market, Behavioural Finance and Financial Modelling. She has also delivered guest lectures and has published papers in International and National peer reviewed journals and presented papers in International and National conferences. She teaches Finance courses at LIBA.

Renaud Redien-Collot, specialist in entrepreneurship, has both a French and American educational background. He has a Ph.D from the Graduate School of Arts and Sciences of Columbia University as well as a degree as a Habilitation Professor in Management from Université de Reims Champagne-Ardenne. His research interests include innovation, the incubation process, entrepreneurship education and the renewal of SMEs' business models in the EU. As a member of the Board of Administration of the E. Roosevelt Foundation (1994-2000), he was in charge of major projects that have stimulated the development of entrepreneurship education and the emergence of innovative start-ups in the USA and in Eastern Europe. He was involved in several innovative projects promoting entrepreneurship such as "Women Entrepreneurs in European Union", "Enspire EU", "Antreman", and "Sofa". Prior to joining Institut Friedland, Renaud Redien-Collot was Director of International Affairs at Novancia, a business school of the CCI Paris Ile-de-France in Paris. Renaud Redien-Collot is also president of the scientific committee of Women Equity for Growth (WEG), the first French investment program dedicated to high growth businesses led by women. Since 2008, he has been a member of the Board of Administration of Académie de l'Entrepreneuriat.

Maria Rigou holds a Diploma in Computer Engineering and Informatics (1997, University of Patras, School of Engineering), an MSc in Computer Science (2000, thesis: Interactive Systems Evaluation), a PhD in Computer Science (2005, dissertation: Effective Algorithms for Web Personalization based on Web Mining), as well as a Master in Arts "Graphic Arts-Multimedia" (2011, Hellenic Open University, School of Humanitarian Studies, thesis: Learning and Entertainment by Casual Gaming). Currently she is a lecturer (for laboratory courses) at the Computer Engineering and Informatics Department (University of Patras) and a tutor at the Hellenic Open University (Postgraduate Programme in Information Systems). The last 20 years she has participated in the design, development and technical management of national and international R&D IT projects and has published research articles in scientific journals, books and conference proceedings. Her research interests are in the fields of Web technologies, Personalization techniques, Web Mining, Modeling Web Apps, User modeling, Semantic Web technologies, Human Computer Interaction, Usability evaluation and Graphics design.

Davide Rizzo is associate professor in agronomy and data science for the Chair in Agricultural Machinery and New Technologies backed by UniLaSalle (France). He is also a member of the INTERACT Research Unit UP 2018.C102 (Innovation Territoire Agriculture et Agroindustrie, Connaissance et Technologie). He holds a PhD in landscape agronomy from the Scuola Superiore Sant'Anna (Pisa, Italy) and has experience in research and teaching about the description of farmers' practices at the regional level.

Maria-Mercedes Rojas-de Gracia is a Professor of the Department of Economics and Business Administration, University of Malaga (Spain). Her primary line of research consists of the study of family vacation decision making. Currently, she is working on an investigation about the image of tourist destinations through social networks.

Vincent Sabourin is Full Professor of Strategy and Social Responsibility at the University of Quebec in Montreal. UQAM, Correspondence: Department of Strategy and Social Responsibility ESG School of Management UQAM, R-3555, 315 east St-Catherine Montreal Qc. Canada H3C 4P2. All suggestions are welcome through sabourin.vincent@uquam.ca

Loïc Sauvée is Professor in Management Sciences at UniLaSalle, PhD and Habilitation to supervise research (HDR), Head of INTERACT Research Unit UP 2018.C102 (Innovation Territoire Agriculture et Agroindustrie, Connaissance et Technologie) and expert in applied social sciences, Scientific Direction. His teaching fields are business organization and maketing in agrifood sectors. His research fields are governance of innovation processes in agrifood competitiveness cluster, alignment of quality management systems in agrifood chains and networks, implementation of CSR in food SMEs, dynamics of network governance ; member of editorial board of the Journal on Chain and Network Science.

Amira Sghari is Assistant Professor in Management studies at the Faculty of Economics and Management, University of Sfax-Tunisia. She received a Ph.D. degree in management science, in 2013, from the Aix-Marseille University, France. Dr. Sghari is member of a Research Unit on Management Studies. Her research interests include strategy and organizational structure, organizational change and information management.

Plácido Sierra-Herrezuelo is a professional in marketing and tourism. Specialized in market studies, strategic planning and development of both physical and digital promotion activities. He combines his activity in Tourism and Planning of the Costa del Sol with teaching at the University of Malaga.

Yi Wan has been a researcher at Aston Business School. He has an MSc from Warwick Business School and an BSc from Birmingham City University. He is now an ERP and operations manager in China.

Index

E

F

I

K

M

N

O

P

R

S

T

U

V

Y

Printed in the United States
By Bookmasters